Carlo Gaetan · Xavier Guyon

Spatial Statistics and Modeling

Translated by Kevin Bleakley

Springer

Carlo Gaetan
Dipartimento di Statistica
Università Ca' Foscari Venezia
San Giobbe - Cannaregio, 873
30121 Venezia VE
Italy
gaetan@unive.it

Xavier Guyon
SAMOS
Université Paris I
90 rue de Tolbiac
75634 Paris CX 13
France
guyon@univ-paris1.fr

ISBN 978-1-4614-2499-4 e-ISBN 978-0-387-92257-7
DOI 10.1007/978-0-387-92257-7
Springer New York Dordrecht Heidelberg London

Softcover reprint of the hardcover 1st edition 2010

Printed on acid-free paper

Springer is part of Springer Science+Business Media (www.springer.com)

Springer Series in Statistics

For other titles published in this series, go to
www.springer.com/series/692

Preface

Spatial analysis methods have seen a rapid rise in popularity due to demand from a wide range of fields. These include, among others, biology, spatial economics, image processing, environmental and earth science, ecology, geography, epidemiology, agronomy, forestry and mineral prospection.

In spatial problems, observations come from a spatial process $X = \{X_s, s \in S\}$ indexed by a spatial set S, with X_s taking values in a state space E. The positions of observation sites $s \in S$ are either fixed in advance or random. Classically, S is a 2-dimensional subset, $S \subseteq \mathbb{R}^2$. However, it could also be 1-dimensional (chromatography, crop trials along rows) or a subset of $\mathbb{R}^3$ (mineral prospection, earth science, 3D imaging). Other fields such as Bayesian statistics and simulation may even require spaces S of dimension $d \geq 3$. The study of spatial dynamics adds a temporal dimension, for example $(s,t) \in \mathbb{R}^2 \times \mathbb{R}^+$ in the 2-dimensional case.

This multitude of situations and applications makes for a very rich subject. To illustrate, let us give a few examples of the three types of spatial data that will be studied in the book.

Geostatistical data

Here, S is a *continuous* subspace of $\mathbb{R}^d$ and the random field $\{X_s, s \in S\}$ observed at n fixed sites $\{s_1, \ldots, s_n\} \subset S$ takes values in a real-valued state space E. The rainfall data in Figure 0.1-a and soil porosity data in Fig. 0.1-b fall into this category. Observation sites may or may not be regularly spaced. Geostatistics tries to answer questions about modeling, identification and separation of small and large scale variations, prediction (or kriging) at unobserved sites and reconstruction of X across the whole space S.

Lattice data and data on fixed networks

Here, S is a *fixed discrete* non-random set, usually $S \subset \mathbb{R}^d$ and X is observed at points in S. Points s might be geographical regions represented as a network with

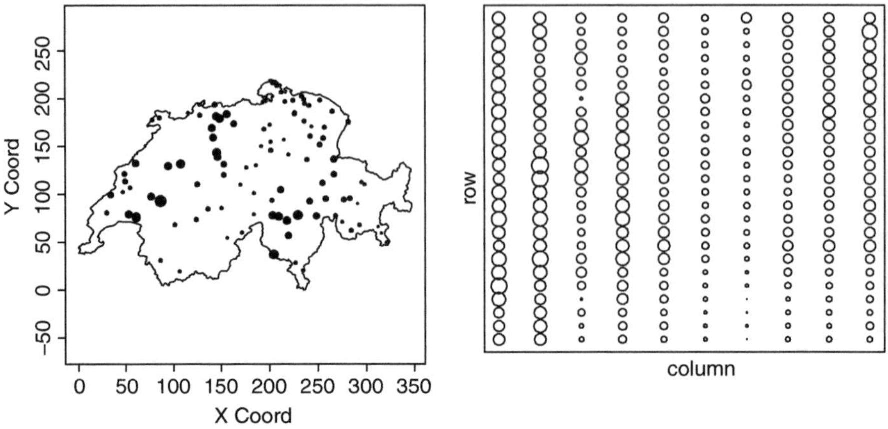

Fig. 0.1 (a) Rainfall over the Swiss meteorological network on May 8, 1986 (during the passage of Chernobyl's radioactive cloud. This is the `sic` dataset from the `geoR` package of R (178)); (b) Soil porosity (`soil` dataset from the `geoR` package). For both (a) and (b), the size of symbols are proportional to the value of X_s.

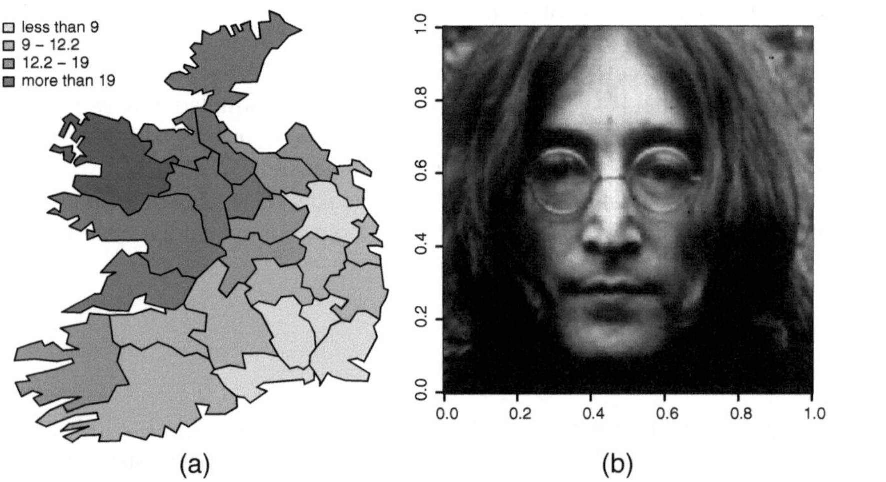

Fig. 0.2 (a) Percentage of people with blood group A in the 26 counties of Ireland (`eire` dataset from the `spdep` package); (b) Image of John Lennon (256×256 pixels in a 193-level grayscale, `lennon` dataset from the `fields` package).

given adjacency graph $\mathscr{G}$ (cf. the 26 counties of Ireland, Fig. 0.2-a) and X_s some value of interest measured at s. The state space E may or may not be real-valued. In image analysis, S is a regularly spaced set of pixels (cf. Fig. 0.2-b). Goals for these types of data include constructing and analyzing explicative models, quantifying spatial correlations, prediction and image restoration.

Point data

Figure 0.3-a shows the location of cell centers in a histological section seen under a microscope and Figure 0.3-b the location and size of pine trees in a forest. Here, the set of observation sites $x = \{x_1, x_2, \ldots, x_n\}$, $x_i \in S \subset \mathbb{R}^d$ is *random*, along with the number $n = n(x)$ of observation sites; x is the outcome of a spatial point process (PP) observed in window S. The process X is said to be *marked* if at each x_i we record a value, for example the diameter of the pine trees found at x_i. A central question in the statistical analysis of PPs is to know if the distribution of points is essentially regular (Figure 0.3-a), completely random (Poisson PP) or aggregated (Figure 0.3-b).

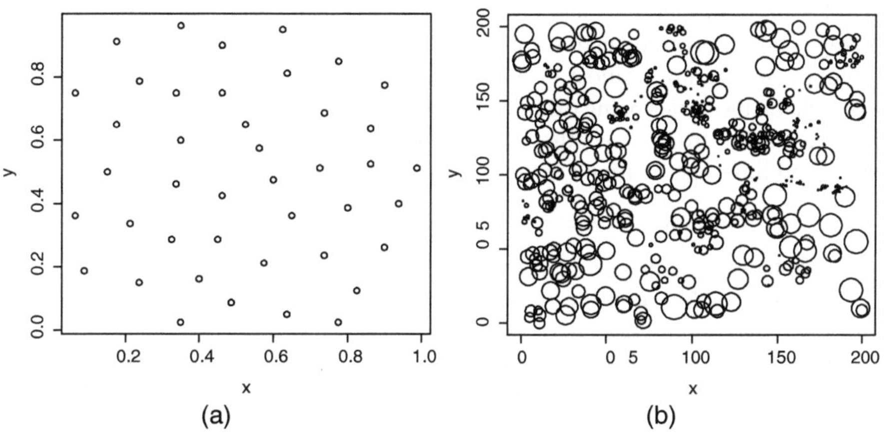

Fig. 0.3 (a) 42 cell centers of a histological section seen under a microscope (`cells` dataset from the `spatstat` package); (b) Position and diameter of 584 pine trees in a forest (`longleaf` dataset from the `spatstat` package).

As is the case for time series, spatial statistics differ from classical statistics due to non-independence of observations; throughout this book, we will generally call X a spatial process or *random field*.

This dependency structure means there is redundancy in available information that can be exploited when making predictions, though it also modifies statistical behavior. Unbiasedness, consistency, efficiency and convergence in distribution of estimators all have to be reexamined in this context. The originality of spatial statistics is to make use of *non-causal modeling*; in this sense, spatial statistics is radically different to time series statistics where causal models use the passage of time and a notion of the "past" (modeling river flows, stock prices, evolution of unemployment rates, etc.). Markov spatial modeling works with the idea of the spatial neighborhood of site s "in all directions." This includes dimension $d = 1$: for example, if $S \subseteq \mathbb{Z}^1$ and X_s is the quantity of corn harvested from each corn stalk along a row, a reasonable model would compare X_s with its *two neighbors*, the stalks to the "left" X_{s-1} and "right" X_{s+1}. We see that causal autoregressive modeling of X_s based on

X_{s-1} has no obvious meaning. If the crop is in a field, we could let the harvested quantity $X_{s,t}$ at site (s,t) depend on that of its 4 nearest neighbors $X_{s-1,t}$, $X_{s+1,t}$, $X_{s,t-1}$ and $X_{s,t+1}$, or even perhaps its 8 nearest neighbors.

These three types of spatial structure (cf. Cressie, (48)) provide the framework to this book. The first three chapters are devoted to modeling each in turn (Chapter 1: Second-order models, geostatistics, intrinsic models and autoregressive models; Chapter 2: Gibbs-Markov random fields over networks; Chapter 3: Spatial point processes). Due to the importance of simulation in spatial statistics, Chapter 4 presents *Monte Carlo Markov Chain* (MCMC) methods for spatial statistics. Chapter 5 then brings together the most important statistical methods for the various models and data types and investigates their properties. Four appendices round things off with a presentation of the most useful probabilistic and statistical tools in spatial statistics (simulation, limit theorems and minimum contrast estimation) as well as software packages for performing analyses presented in the book.

Numerous examples, most of them treated with the *R* software package (178), shed light on the topics being examined. When the data being studied are not directly available in *R* or from some other specified location, descriptions, relevant program scripts and links can be found at the website of the book:

`www.dst.unive.it/~gaetan/ModStatSpat.`

Each chapter ends with a set of exercises.

The bibliography gives the reader the chance to enrich their knowledge of notions only briefly presented here as well as several technical results whose proofs have been omitted. We also list reference books that fill gaps remaining after our intentionally reduced and non-exhaustive treatment of this multi-faceted subject undergoing great development (69).

Our thanks go to all our colleagues who have given us a taste for spatial analysis, for their ideas, remarks, contributions and those who have allowed us to use data collected from their own work. We would equally like to thank the *R Development Core Team* and authors of spatial packages for *R* (178) who have made their powerful and efficient software freely available to the public, indispensable when working with methods and tools described here. We thank reviewers for their careful rereading of the first draft; their remarks have helped to significantly improve the present version. Thanks to Bernard Ycart for encouraging us to expand an initially more modest project. Of course, we could never have undertaken this work without the patience and support of our families and the backing of our respective research teams, Dipartimento di Statistica - Università Ca' Foscari Venezia and Laboratoire SAMOS - Université Paris 1. Lastly, many thanks to Kevin Bleakley for the translation and English adaptation, done with much competence. Any remaining errors are ours.

Venice and Paris,
August 2009

Carlo Gaetan
Xavier Guyon

Contents

Abbreviations and notation

AR	autoregressive
ARMA	autoregressive moving average
a.s.	almost surely
CAR	conditional autoregressive
CFTP	coupling from the past
Ch.	chapter
CLS	conditional least squares
CLT	central limit theorem
c.n.d.	conditionally negative definite
CPL	conditional pseudo-likelihood
CSR	complete spatial randomness
ex.	example
Fig.	figure
GLM	generalized linear model
GLS	generalized least squares
GWN	Gaussian white noise
iff	if and only if
i.i.d.	independent and identically distributed
i.n.i.d.	independent non-identically distributed
LS	least squares
LSE	least squares estimation
MA	moving average
MAP	maximum a posteriori
MCMC	Monte Carlo Markov Chain
MCPL	maximum conditional pseudo-likelihood
MH	Metropolis-Hastings
ML	maximum likelihood
MPM	marginal posterior mode
MPP	marked point process
MSE	mean square error
MSNE	mean square normalized error

NN	nearest neighbor
OLS	ordinary least squares
p.d.	positive definite
PP	point process
PPP	Poisson point process
PRESS	prediction sum of squares
p.s.d.	positive-semidefinite
QGLS	quasi-generalized least squares
q.m.	quadratic mean
resp.	respectively
r.r.v.	real random variable
RSS	residual sum of squares
r.v.	random variable
SA	simulated annealing
SAR	simultaneous autoregressive
SARX	SAR with exogenous variables
SLLN	strong law of large numbers
s.t.	such that
STARMA	spatio-temporal ARMA
SWN	strong white noise
TV	total variation
WLS	weighted least squares
WN	white noise
w.r.t.	with respect to
WWN	weak white noise

$\mathscr{B}(S)$	Borel sets of S, $S \subseteq \mathbb{R}^d$
$\mathscr{B}_b(S)$	bounded Borel sets of S, $S \subseteq \mathbb{R}^d$
$\sharp(A)$	cardinality of (A)
$\|\Sigma\|$	determinant of Σ
$\delta(A)$	diameter of $A : \delta(A) = \sup_{x,y\in A} d(x,y)$
$\\|\cdot\\|$ or $\\|\cdot\\|_2$	Euclidean norm on $\mathbb{R}^p : \\|x\\| = \sqrt{\sum_1^p x_i^2}$
$\langle i, j\rangle$	i and j are neighbors
$[x]$	integer part of x
$d(A)$	interior diameter of A: $d(A) = \sup\{r : \exists x \text{ s.t. } B(x;r) \subseteq A\}$
$\cdots \doteq K(\theta,\alpha)$	$K(\theta,\alpha)$ equals, by definition, the left-hand side
$A \otimes B$	Kronecker product of matrices A and B
$\\|\cdot\\|_1$	l^1 norm: $\\|x\\|_1 = \sum_1^p \|x_i\|$
$\lambda_M(B)$	largest eigenvalue of B
ν	Lebesgue measure on $\mathbb{R}^d$
∂A (∂i)	neighborhood border of A (of site i)
$\pi \ll \mu$	π is absolutely continuous with respect to μ
${}^t uv$	scalar product on $\mathbb{R}^p$: ${}^t uv = \sum_{i=1}^p u_i v_i$

$\lambda_m(C)$	smallest eigenvalue of B
$\|\cdot\|_\infty$	sup norm: $\|x\|_\infty = \sup_i \lvert x_i \rvert$
$\|\cdot\|_{VT}$	total variation norm
${}^tC$	transpose of C
$X \sim \mathcal{N}(0,1)$	X has a $\mathcal{N}(0,1)$ distribution

Chapter 1
Second-order spatial models and geostatistics

Suppose $S \subseteq \mathbb{R}^d$ is a spatial set. A random field X on S taking values in a state space E means a collection $X = \{X_s,\, s \in S\}$ of random variables (r.v.) indexed by S taking values in E. This chapter is devoted to the study of *second-order random fields*, i.e., *real-valued* random fields where each X_s has finite variance. We also study the broader class of *intrinsic random fields*, that is, random fields with increments of finite variance. We consider two approaches.

In the *geostatistics* approach, S is a *continuous subset* of $\mathbb{R}^d$ and we model X in a "second-order" way with its *covariance* function or its *variogram*. For example, for $d = 2$, $s = (x,y) \in S$ is characterized by fixed geographic coordinates and if $d = 3$, we add altitude (or depth) z. Spatio-temporal evolution in space can also be modeled at space-time "sites" $(s,t) \in \mathbb{R}^3 \times \mathbb{R}^+$, where s represents space and t time. Initially developed for predicting mineral reserves in an exploration zone $S \subseteq \mathbb{R}^3$, geostatistics is today used in a variety of domains (cf. Chilès and Delfiner (43); Diggle and Ribeiro (63)). These include, among others, earth science and mining exploration (134; 152), epidemiology, agronomy and design of numerical experiments (193). A central goal of geostatistics is to predict X by *kriging* over all of S using only a finite number of observations.

The second approach involves *autoregressive* (AR) models, used when S is a *discrete network* of sites (we will also use the word "lattice"). S may have a regular form, for example $S \subset \mathbb{Z}^d$ (images, satellite data, radiography; (42), (224)) or it may not (econometrics, epidemiology; (45), (7), (105)). Here, the spatial correlation structure is induced by the AR model chosen. Such models are well adapted to situations where measurements have been aggregated over spatial zones: for example, in econometrics this might be the percentages of categories of a certain variable in an administrative unit, in epidemiology, the number of cases of an illness per district s and in agronomy, the total production in each parcel of land s.

C. Gaetan, X. Guyon, *Spatial Statistics and Modeling*, Springer Series in Statistics,
DOI 10.1007/978-0-387-92257-7_1, © Springer Science+Business Media, LLC 2010

1.1 Some background in stochastic processes

Let $(\Omega, \mathscr{F}, \mathbb{P})$ be a probability space, S a set of sites and $(E, \mathscr{E})$ a measurable state space.

Definition 1.1. Stochastic process

A stochastic process (or *process* or *random field*) taking values in E is a family $X = \{X_s, s \in S\}$ of random variables defined on $(\Omega, \mathscr{F}, \mathbb{P})$ and taking values in $(E, \mathscr{E})$. $(E, \mathscr{E})$ is called the state space of the process and S the (spatial) set of sites at which the process is defined.

For any integer $n \geq 1$ and n-tuple $(s_1, s_2, \ldots, s_n) \in S^n$, the distribution of $(X_{s_1}, X_{s_2}, \ldots, X_{s_n})$ is the image of $\mathbb{P}$ under the mapping $\omega \longmapsto (X_{s_1}(\omega), X_{s_2}(\omega), \ldots, X_{s_n}(\omega))$: that is, for $A_i \in \mathscr{E}$, $i = 1, \ldots, n$,

$$P_X(A_1, A_2, \ldots, A_n) = \mathbb{P}(X_{s_1} \in A_1, X_{s_2} \in A_2, \ldots, X_{s_n} \in A_n).$$

The event $(X_{s_1} \in A_1, X_{s_2} \in A_2, \ldots, X_{s_n} \in A_n)$ of $\mathscr{E}$ is a cylinder associated with the n-tuple $(s_1, s_2, \ldots, s_n)$ and events $A_i, i = 1, \ldots, n$ belonging to $\mathscr{F}$. The family of all finite-dimensional distributions of X is called the *spatial distribution* of the process; if $S \subseteq \mathbb{R}$, we say *time distribution*. More generally, the distribution of the process is uniquely defined as the extension of the spatial distribution to the sub-σ-algebra $\mathscr{A} \subseteq \mathscr{F}$ generated by the set of cylinders of $\mathscr{E}$ (32, Ch. 12), (180, Ch. 6).

For the rest of the chapter, we will be considering *real-valued* processes, $E \subseteq \mathbb{R}$ endowed with a Borel σ-field $\mathscr{E} = \mathscr{B}(E)$.

Definition 1.2. Second-order process

X is a *second-order process* (*random field*) if for all $s \in S$, $E(X_s^2) < \infty$. The mean of X (which necessarily exists) is the function $m : S \to \mathbb{R}$ defined by $m(s) = E(X_s)$. The covariance of X is the function $c : S \times S \to \mathbb{R}$ defined for all s, t by $c(s,t) = Cov(X_s, X_t)$.

With $L^2 = L^2(\Omega, \mathscr{F}, \mathbb{P})$ representing the set of real-valued and square integrable random variables on $(\Omega, \mathscr{F})$, $X \in L^2$ means that X is a second-order process. A process X is said to be *centered* if for all s, $m(s) = 0$.

Covariances are characterized by the *positive semidefinite* (p.s.d.) property:

$$\forall m \geq 1, \forall a \in \mathbb{R}^m \text{ and } \forall (s_1, s_2, \ldots, s_m) \in S^m : \sum_{i=1}^{m} \sum_{j=1}^{m} a_i a_j c(s_i, s_j) \geq 0.$$

This property is a consequence of non-negativity of the variance of linear combinations:

$$Var\left(\sum_{i=1}^{m} a_i X_{s_i}\right) = \sum_{i=1}^{m} \sum_{j=1}^{m} a_i a_j c(s_i, s_j) \geq 0.$$

We say that the covariance is *positive definite* (p.d.) if furthermore, for every m-tuple of distinct sites, $\sum_{i=1}^{m} \sum_{j=1}^{m} a_i a_j c(s_i, s_j) > 0$ whenever $a \neq 0$. *Gaussian processes* are an important class of L^2 processes.

Definition 1.3. Gaussian process

X is a *Gaussian process* on S if for every finite subset $\Lambda \subset S$ and real-valued sequence $a = (a_s, s \in \Lambda)$, $\sum_{s\in\Lambda} a_s X_s$ is a Gaussian random variable.

If $m_\Lambda = E(X_\Lambda)$ is the mean of $X_\Lambda = (X_s, s \in \Lambda)$ and Σ_Λ its covariance, then if Σ_Λ is invertible, the density (or likelihood) of X_Λ with respect to the Lebesgue measure on $\mathbb{R}^{\sharp\Lambda}$ is

$$f_\Lambda(x_\Lambda) = (2\pi)^{-\sharp\Lambda/2}(\det\Sigma_\Lambda)^{-1/2}\exp\left\{-\frac{1}{2}{}^t(x_\Lambda - m_\Lambda)\Sigma_\Lambda^{-1}(x_\Lambda - m_\Lambda)\right\},$$

where $\sharp U$ is the cardinality of U and x_Λ possible values of X_Λ. Such densities are well-defined and Kolmogorov's theorem ensures that for any mean function m and p.d. covariance c there exists a (Gaussian) random field with mean m and covariance c.

Example 1.1. Brownian motion on $\mathbb{R}^+$ and Brownian sheet on $(\mathbb{R}^+)^2$

X is a *Brownian motion* (180) on $S = \mathbb{R}^+$ if $X_0 = 0$, if for all $s > 0$, X_s follows a $\mathscr{N}(0,s)$ ($X_s \sim \mathscr{N}(0,s)$) and if increments $X(]s,t]) = X_t - X_s$, $t > s \geq 0$ are independent for disjoint intervals. The covariance of Brownian motion is $c(s,t) = \min\{s,t\}$ and the increment process $\Delta X_t = X_{t+\Delta} - X_t$, $t \geq 0$ is stationary (cf. Ch. 1.2) with marginal distribution $\mathscr{N}(0,\Delta)$.

This definition can be extended to the *Brownian sheet* (37) on the first quadrant $S = (\mathbb{R}^+)^2$ with: $X_{u,v} = 0$ if $u \times v = 0$, $X_{u,v} \sim \mathscr{N}(0, u \times v)$ for all $(u,v) \in S$ and independence of increments for disjoint rectangles; the increment on rectangle $]s,t]$, $s = (s_1, s_2)$, $t = (t_1, t_2)$, $s_1 < t_1$, $s_2 < t_2$ is given by

$$X(]s,t]) = X_{t_1,t_2} - X_{t_1 s_2} - X_{s_1 t_2} + X_{s_1 s_2}.$$

Brownian sheets are centered Gaussian processes with covariance $c(s,t) = \min\{s_1, s_2\} \times \min\{t_1, t_2\}$.

1.2 Stationary processes

In this section, we suppose that X is a second-order random field on $S = \mathbb{R}^d$ or $\mathbb{Z}^d$ with mean m and covariance c. The notion of stationarity of X can be more generally defined when S is an *additive subgroup* of $\mathbb{R}^d$: for example, S could be the triangular lattice of $\mathbb{R}^2$, $S = \{ne_1 + me_2,\ n \text{ and } m \in \mathbb{Z}\}$ with $e_1 = (1,0)$ and $e_2 = (1/2, \sqrt{3}/2)$; another example is the finite d-dimensional torus with p^d points, $S = (\mathbb{Z}/p\mathbb{Z})^d$.

1.2.1 Definitions and examples

Definition 1.4. Second-order stationary process

X is a second-order stationary process on S if it has constant mean and translation-invariant covariance c:

$$\forall s,t \in S\colon E(X_s) = m \text{ and } c(s,t) = Cov(X_s, X_t) = C(t-s).$$

$C : S \to \mathbb{R}$ is the stationary covariance function of X. Translation-invariance of c means:

$$\forall s,t,h \in S\colon c(s+h,t+h) = Cov(X_{s+h}, X_{t+h}) = C(s-t).$$

The correlation function of X is the function $h \mapsto \rho(h) = C(h)/C(0)$. The following properties hold:

Proposition 1.1. *Let X be a second-order stationary process with stationary covariance C. Then:*

1. $\forall h \in S$, $|C(h)| \leq C(0) = Var(X_s)$.
2. $\forall m \geq 1$, $a \in \mathbb{R}^m$ *and* $\{t_1, t_2, \ldots, t_m\} \subseteq S$: $\sum_{i=1}^m \sum_{j=1}^m a_i a_j C(t_i - t_j) \geq 0$.
3. *If* $A : \mathbb{R}^d \longrightarrow \mathbb{R}^d$ *is linear, the process* $X^A = \{X_{As}, s \in S\}$ *is stationary with covariance* $C^A(s) = C(As)$. C^A *is p.d. if C itself is and if A has full rank.*
4. *If C is continuous at the origin, then C is everywhere uniformly continuous.*
5. *If* C_1, C_2, $\ldots$ *are stationary covariances, the following functions are as well:*

   a. $C(h) = a_1 C_1(h) + a_2 C_2(h)$ *if* a_1 *and* $a_2 \geq 0$.
   b. *More generally, if* $C(\cdot; u)$, $u \in U \subseteq \mathbb{R}^k$ *is a stationary covariance for each u and if μ is a positive measure on* $\mathbb{R}^k$ *such that* $C_\mu(h) = \int_U C(h;u)\mu(du)$ *exists for all h, then C_μ is a stationary covariance.*
   c. $C(h) = C_1(h) C_2(h)$.
   d. $C(h) = \lim_{n\to\infty} C_n(h)$, *provided that the limit exists for all h.*

Proof. Without loss of generality, suppose that X is centered.
(1) is a consequence of the Cauchy-Schwarz inequality:

$$C(h)^2 = \{E(X_h X_0)\}^2 \leq \{E(X_0^2)E(X_h^2)\} = E(X_0^2)^2.$$

(2) follows from the fact that covariances are p.s.d. (3) can be shown directly. (4) can be inferred from the fact that $C(s+h) - C(s) = E[X_0(X_{s+h} - X_s)]$ and the Cauchy-Schwarz inequality,

$$|C(s+h) - C(s)| \leq \sqrt{C(0)}\sqrt{2[C(0) - C(h)]}.$$

(5) It is easy to show that the functions C defined by (a), (b) and (d) are p.s.d. Then, if X_1 and X_2 are stationary and independent with covariances C_1 and C_2, covariance C given in (5-a) (resp. (5-b)) is that of $X_t = \sqrt{a_1} X_{1,t} + \sqrt{a_2} X_{2,t}$ (resp. $X_t = X_{1t} X_{2t}$). □

The notion of stationarity can defined in two ways in L^2. The first, weaker, is that of stationary increment processes or *intrinsic processes* and is presented in Section 1.3. The second, stronger, is known as *strict stationarity*. We say that X is strictly stationary if for all $k \in \mathbb{N}$, all k-tuples $(t_1, t_2, \ldots, t_k) \in S^k$ and all $h \in S$, the distribution of $(X_{t_1+h}, X_{t_2+h}, \ldots, X_{t_k+h})$ is independent of h. In a sense, X is strictly stationary if the spatial distribution of the process is translation-invariant.

If X is strictly stationary and if $X \in L^2$, then X is stationary in L^2. The converse is generally not true but both notions represent the same thing if X is a Gaussian process.

Example 1.2. Strong White Noise (SWN) and Weak White Noise (WWN)

X is a *Strong White Noise* if the variables $\{X_s,\, s \in S\}$ are centered, independent and identically distributed (i.i.d.). X is a *Weak White Noise* if the variables $\{X_s,\, s \in S\}$ are centered and uncorrelated with finite constant variance: if $s \neq t, Cov(X_s, X_t) = 0$ and $Var(X_s) = \sigma^2 < \infty$. A SWN on S is strictly stationary; a WWN on S is a stationary process in L^2.

We denote $\|\cdot\|$ the Euclidean norm in $\mathbb{R}^d$: $\|x\| = \|x\|_2 = \sqrt{\sum_{i=1}^d x_i^2}$, $x = (x_1, x_2, \ldots, x_d)$.

Definition 1.5. Isotropic covariance

X has isotropic covariance if for each $s,t \in S$, $Cov(X_s, X_t)$ depends only on $\|s-t\|$:

$$\exists C_0 : \mathbb{R}^+ \rightarrow \mathbb{R} \text{ s.t.: } \forall t,s \in S,\ c(s,t) = C_0(\|s-t\|) = C(s-t).$$

Isotropic covariances are therefore stationary but isotropy imposes restrictions on the covariance. For example, if X is isotropic and centered in $\mathbb{R}^d$ and if we consider $d+1$ points mutually separated by distance $\|h\|$,

$$E\{\sum_{i=1}^{d+1} X_{s_i}\}^2 = (d+1)C_0(\|h\|)(1 + d\rho_0(\|h\|) \geq 0,$$

where $\rho_0 : \mathbb{R}^+ \rightarrow [-1, 1]$ is the isotropic correlation function. Therefore, for all h, this correlation satisfies

$$\rho_0(\|h\|) \geq -1/d. \tag{1.1}$$

1.2.2 Spectral representation of covariances

Fourier theory and Bochner's theorem (29; 43) together imply a bijection between stationary covariances C on S and their spectral measure F. It is thus equivalent to characterize a stationary model in L^2 by its stationary covariance C or its spectral measure F.

The $S = \mathbb{R}^d$ case

We associate with C a symmetric measure $F \geq 0$ bounded on the Borel sets $\mathscr{B}(\mathbb{R}^d)$ such that:

$$C(h) = \int_{\mathbb{R}^d} e^{i^t hu} F(du), \tag{1.2}$$

where ${}^t hu = \sum_{i=1}^d h_i u_i$. If C is integrable, F is absolutely continuous with density f (with respect to the Lebesgue measure ν on $\mathbb{R}^d$). f is called the *spectral density* of X. The inverse Fourier transform lets us express f in terms of C:

$$f(u) = (2\pi)^{-d} \int_{\mathbb{R}^d} e^{-i^t hu} C(h) dh.$$

If X has isotropic covariance C, its spectral density f does too and vice versa. Denote $r = \|h\|$, $h = (r, \theta)$ where $\theta = h\|h\|^{-1} \in S_d$ gives the direction of h in the unitary sphere S_d in $\mathbb{R}^d$ centered at 0, $\rho = \|u\|$ and $u = (\rho, \alpha)$, with $\alpha = u\|u\|^{-1} \in S_d$. For the polar coordinates $h = (r, \theta)$ and $u = (\rho, \alpha)$ of h and u, note $c_d(r) = C(h)$ and $f_d(\rho) = f(u)$ the covariance and isotropic spectral density. Integrating (1.2) over S_d with surface measure $d\sigma$, then over $\rho \in [0, \infty[$, we get:

$$\begin{aligned} C(h) = c_d(r) &= \int_{[0,\infty[} \left[\int_{S_d} \cos(r\rho\, {}^t\theta\alpha) d\sigma(\alpha) \right] \rho^{d-1} f_d(\rho) d\rho \\ &= \int_{[0,\infty[} \Lambda_d(r\rho) \rho^{d-1} f_d(\rho) d\rho. \end{aligned} \tag{1.3}$$

The Hankel transform $f_d \mapsto c_d$, analogous to a Fourier transform when dealing with isotropy shows that the variety of isotropic covariances is the same as that of the bounded positive measures on $[0, \infty[$. Furthermore (227), $\Lambda_d(v) = \Gamma(d/2)(v/2)^{-(d-2)/2} \mathscr{J}_{(d-2)/2}(v)$, where $\mathscr{J}_\kappa$ is the Bessel function of the first kind of order κ (2). For $n = 1$, 2 and 3, we have:

$$\begin{aligned} c_1(r) &= 2 \int_{[0,\infty[} \cos(\rho r) f_1(\rho) d\rho, \\ c_2(r) &= 2\pi \int_{[0,\infty[} \rho J_0(\rho r) f_2(\rho) d\rho, \\ c_3(r) &= \frac{2}{r} \int_{[0,\infty[} \rho \sin(\rho r) f_3(\rho) d\rho. \end{aligned}$$

Using (1.3), we obtain lower bounds:

$$C(h) \geq \inf_{v \geq 0} \Lambda_d(v) \int_{]0,\infty[} \rho^{d-1} f_d(\rho) d\rho = \inf_{v \geq 0} \Lambda_d(v) C(0).$$

In particular, we get the lower bounds (227; 184), tighter than those in (1.1): $\rho_0(\|h\|) \geq -0.403$ in $\mathbb{R}^2$, $\rho_0(\|h\|) \geq -0.218$ in $\mathbb{R}^3$, $\rho_0(\|h\|) \geq -0.113$ in $\mathbb{R}^4$ and $\rho_0(\|h\|) \geq 0$ in $\mathbb{R}^{\mathbb{N}}$.

Example 1.3. Exponential covariances in $\mathbb{R}^d$

For $t \in \mathbb{R}$, $\alpha, b > 0$, $C_0(t) = b\exp(-\alpha|t|)$ has the Fourier transform:

$$f(u)=\frac{1}{2\pi}\int_{]-\infty,\infty[} be^{-\alpha|t|-iut}dt=\frac{\alpha b}{\pi(\alpha^2+u^2)}.$$

As $f\geq 0$ is integrable over $\mathbb{R}$, it is a spectral density and C_0 therefore a covariance on $\mathbb{R}$. Also, as

$$\int_{]0,\infty[} e^{-\alpha x}\mathscr{J}_\kappa(ux)x^{\kappa+1}dx=\frac{2\alpha(2u)^\kappa\Gamma(\kappa+3/2)}{\pi^{1/2}(\alpha^2+u^2)^{\kappa+3/2}},$$

we see that

$$\phi(u)=\frac{\alpha b\Gamma[(d+1)/2]}{[\pi(\alpha^2+u^2)]^{(d+1)/2}}$$

is an isotropic spectral density of a process on $\mathbb{R}^d$ with covariance

$$C(h)=C_0(\|h\|)=b\exp(-\alpha\|h\|).$$

For any dimension d, C is therefore a covariance function, given the name *exponential*, with parameter b for the variance of X and $a=\alpha^{-1}$ the range.

The $S=\mathbb{Z}^d$ case

Note $\mathbb{T}^d=[0,2\pi[^d$ the d-dimensional torus. According to Bochner's theorem, any stationary covariance C on $\mathbb{Z}^d$ is associated with a measure $F\geq 0$ bounded on the Borel sets $\mathscr{B}(\mathbb{T}^d)$ such that:

$$C(h)=\int_{\mathbb{T}^d} e^{i^t uh}F(du).$$

If C is square summable ($\sum_{h\in\mathbb{Z}^d}C(h)^2<\infty$), the spectral measure F is absolutely continuous with density f (w.r.t. the Lebesgue measure) in $L^2(\mathbb{T}^d)$:

$$f(u)=(2\pi)^{-d}\sum_{h\in\mathbb{Z}^d}C(h)e^{-i^t uh}. \tag{1.4}$$

Furthermore, if $\sum_{h\in\mathbb{Z}^d}|C(h)|<\infty$, we have uniform convergence and f is continuous. Also, the greater the differentiability of f, the faster the convergence of C to 0 in the limit and vice versa: for example, if $f\in\mathscr{C}^k(\mathbb{T}^d)$ where $k=(k_1,\ldots,k_d)\in\mathbb{N}^d$,

$$\lim\sup_{h\longrightarrow\infty} h^k|C(h)|<\infty,$$

where $h=(h_1,h_2,\ldots,h_d)\longrightarrow\infty$ means at least one coordinate $h_i\rightarrow\infty$ and $h^k=h_1^{k_1}\times\ldots\times h_d^{k_d}$. In particular, if f is infinitely differentiable, C goes to zero faster than any power function. This is the case for ARMA models (cf. §1.7.1) which have rational spectral density f.

1.3 Intrinsic processes and variograms

1.3.1 Definitions, examples and properties

The stationarity property in L^2 may not be satisfied for various reasons: for example when $X_s = Y_s + Z$, where Y is stationary in L^2 but $Z \notin L^2$, or equally when X is in L^2 but not stationary, whether that be second-order (Brownian motion) or first-order ($X_s = a + bs + \varepsilon_s$ for a stationary centered residual process ε). A way to weaken the L^2 stationarity hypothesis is to consider the increment process $\{\Delta X_s^{(h)} = X_{s+h} - X_s,\, s \in S\}$ of X, which may be stationary in L^2 even when X is not stationary or not in L^2.

Definition 1.6. Intrinsic process

X is an intrinsically stationary process (or intrinsic process) if for each $h \in S$, the process $\Delta X^{(h)} = \{\Delta X_s^{(h)} = X_{s+h} - X_s : s \in S\}$ is second-order stationary. The semi-variogram of X is the function $\gamma : S \to \mathbb{R}$ defined by:

$$2\gamma(h) = Var(X_{s+h} - X_s).$$

Every stationary process in L^2 with covariance C is clearly an intrinsic process with variogram $2\gamma(h) = 2(C(0) - C(h))$. However, the converse is not true: Brownian motion in $\mathbb{R}$, with variogram $|h|$, is intrinsic but not stationary. Furthermore, processes with affine means and stationary residuals are intrinsic, differentiation having the effect (as for time series) of absorbing affine trends and rendering the process first-order stationary. If we differentiate k times, polynomial trends of degree k can be removed, the process X being called k-intrinsic if $\Delta^k X^{(h)}$ is stationary (cf. (43); in $\mathbb{Z}$, so-called ARIMA models are a generalization of ARMA). For instance, the Brownian sheet on $(\mathbb{R}^+)^2$ is not intrinsic as it can be easily verified that $Var(X_{(u,v)+(1,1)} - X_{(u,v)}) = u + v + 1$ depends on $h = (u,v)$.

If X is an intrinsic process and if the function $m(h) = E(X_{s+h} - X_s)$ is continuous at 0, then $m(\cdot)$ is linear: $\exists a \in \mathbb{R}^d$ s.t. $m(h) = \langle a, h \rangle$. In effect, m is additive, $m(h) + m(h') = E\{(X_{s+h+h'} - X_{s+h'}) + (X_{s+h'} - X_s)\} = m(h+h')$ and continuity of m at 0 implies linearity.

From now on, we will concentrate on intrinsic processes with centered increments: $\forall h,\, m(h) = 0$.

Proposition 1.2. *Properties of variograms*

1. $\gamma(h) = \gamma(-h)$, $\gamma(h) \geq 0$ *and* $\gamma(0) = 0$.
2. *Variograms are* conditionally negative definite *(c.n.d.):* $\forall a \in \mathbb{R}^n$ *s.t.* $\sum_{i=1}^n a_i = 0$, $\forall \{s_1, \ldots, s_n\} \subseteq S$, *we have:*

$$\sum_{i=1}^n \sum_{j=1}^n a_i a_j \gamma(s_i - s_j) \leq 0.$$

3. *If* A *is a linear transformation in* $\mathbb{R}^d$ *and* γ *a variogram, then* $h \mapsto \gamma(Ah)$ *is too.*

4. *Properties 5-(a,b,d) of covariances (cf. Prop. 1.1) remain true for variograms.*
5. *If γ is continuous at 0, then γ is continuous at every site s where γ is locally bounded.*
6. *If γ is bounded in a neighborhood of 0, $\exists a$ and $b \geq 0$ such that for any x, $\gamma(x) \leq a\|x\|^2 + b$.*

Proof. (1) is obvious. To prove (2), set $Y_s = (X_s - X_0)$. Y is stationary in L^2 with covariance $C_Y(s,t) = \gamma(s) + \gamma(t) - \gamma(s-t)$. Then, if $\sum_{i=1}^n a_i = 0$, we get $\sum_{i=1}^n a_i X_{s_i} = \sum_{i=1}^n a_i Y_{s_i}$ and

$$Var\left(\sum_{i=1}^n a_i X_{s_i}\right) = \sum_{i=1}^n \sum_{j=1}^n a_i a_j C_Y(s_i, s_j) = -\sum_{i=1}^n \sum_{j=1}^n a_i a_j \gamma(s_i - s_j) \geq 0.$$

(3) If X is an intrinsic process with variogram 2γ, then $Y = \{Y_s = X_{As}\}$ is intrinsic with variogram:

$$2\gamma_Y(h) = Var(X_{A(s+h)} - X_{As}) = 2\gamma(Ah).$$

(4) The proof is similar to that of Prop. 1.1. (5) $2\{\gamma(s+h) - \gamma(s)\} = E(A)$ where $A = (X_{s+h} - X_0)^2 - (X_s - X_0)^2$. It is easy to show that $A = B + C$ where $B = (X_{s+h} - X_s)(X_{s+h} - X_0)$ and $C = (X_{s+h} - X_s)(X_s - X_0)$. Applying the Cauchy-Schwarz inequality to each of the products B and C, the result follows from the upper bound:

$$|\gamma(s+h) - \gamma(s)| \leq \sqrt{\gamma(h)}[\sqrt{\gamma(s)} + \sqrt{\gamma(s+h)}].$$

Also, γ is uniformly continuous on any set over which γ is bounded. (6) We prove by induction that for each $n \in \mathbb{N}$ and $h \in \mathbb{R}^d$, $\gamma(nh) \leq n^2\gamma(h)$. This is true for $n = 1$; then, since

$$2\gamma((n+1)h) = E\{(X_{s+(n+1)h} - X_{s+h}) + (X_{s+h} - X_s)\}^2,$$

the Cauchy-Schwarz inequality gives

$$\gamma((n+1)h) \leq \gamma(nh) + \gamma(h) + 2\sqrt{\gamma(nh)\gamma(h)} \leq \gamma(h)\{n^2 + 1 + 2n\} = (n+1)^2\gamma(h).$$

Suppose next that $\delta > 0$ satisfies $\sup_{\|u\| \leq \delta} \gamma(u) = C < \infty$ and $x \in \mathbb{R}^d$ satisfies $n\delta \leq \|x\| \leq (n+1)\delta$, $n \geq 1$. Setting $\widetilde{x} = \delta\ \|x\|^{-1}$, the decomposition $x = n\widetilde{x} + \tau$ defines some τ satisfying $\|\tau\| \leq \delta$. We conclude by remarking that

$$\begin{aligned}\gamma(x) &= \gamma(n\widetilde{x} + \tau) \leq \gamma(n\widetilde{x}) + \gamma(\tau) + 2\sqrt{\gamma(n\widetilde{x})\gamma(\tau)} \\ &\leq Cn^2 + C + 2Cn = C(n+1)^2 \leq C\left(\frac{\|x\|}{\delta} + 1\right)^2.\end{aligned}$$

□

Unlike covariances, variograms are not necessarily bounded (for example, the variogram $\gamma(h) = |h|$ for Brownian motion). However, the previous proposition shows that variograms tend to infinity at a rate of at most $\|h\|^2$. One such example of quadratic growth $\gamma(t) = \sigma_1^2 t^2$ is that of the variogram of $X_t = Z_0 + tZ_1$, $t \in \mathbb{R}$, where Z_0 and Z_1 are centered and independent and $Var(Z_1) = \sigma_1^2 > 0$.

Characterizations exist to ensure a function γ is a variogram, one of them being the following (43): if γ is continuous and if $\gamma(0) = 0$, then γ is a variogram if and only if, for every $u > 0$, $t \mapsto \exp\{-u\gamma(t)\}$ is a covariance. For example, as $t \mapsto \exp\{-u \|t\|^2\}$ is a covariance on $\mathbb{R}^d$ for each $u > 0$ and dimension d, $\gamma(t) = \|t\|^2$ is a variogram on $\mathbb{R}^d$ that goes to infinity at a quadratic rate.

1.3.2 Variograms for stationary processes

If X is stationary with covariance C, then X is intrinsic with variogram

$$2\gamma(h) = 2(C(0) - C(h)). \tag{1.5}$$

In particular, variograms of stationary processes are bounded. Matheron (153) partially proved the converse, that is, if the variogram of intrinsic process X is bounded, then $X_t = Z_t + Y$ where Z is a stationary process of L^2 and Y some general real random variable.

If $C(h) \to 0$ as $\|h\| \to \infty$, then $\gamma(h) \to C(0)$ as $\|h\| \to \infty$. The variogram therefore has a *sill* at height $C(0) = Var(X)$ as $\|h\| \to \infty$. The *range* (resp. the *practical range*) is the distance at which the variogram reaches its sill (resp. 95% the value of the sill), cf. Fig. 1.1.

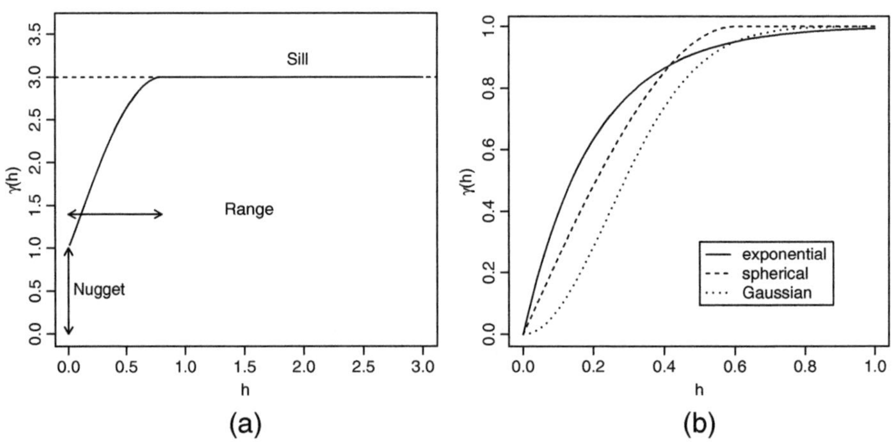

Fig. 1.1 (a) Semivariogram of a stationary model with a nugget effect component; (b) variogram models that have the same range.

Statistical methods for second-order stationary processes can be considered in terms of covariances or in terms of variograms. Statisticians prefer the first way, geostatisticians the second. We note that the advantage of working with variograms is that, unlike covariances, the mean does not have to be pre-estimated (cf. §5.1.4).

1.3.3 Examples of covariances and variograms

Isotropic variograms

The following examples are isotropic variograms on $\mathbb{R}^d$ traditionally used in geostatistics. Other models are presented in Yaglom (227), Chilès and Delfiner (43), Wackernagel (221) and the review article (195). The first five variograms, associated with stationary covariances $C(h) = C(0) - \gamma(h)$ are bounded with range parameter $a > 0$ and sill σ^2. Remember that $\|\cdot\|$ is the Euclidean norm on $\mathbb{R}^d$.

-*Nugget effect*: $\gamma(h;\sigma^2) = \sigma^2$ when $h > 0$, $\gamma(0) = 0$, associated with WWNs.
-*Exponential*: $\gamma(h;a,\sigma^2) = \sigma^2\{1-\exp(-\|h\|/a)\}$.
-*Spherical* ($d \le 3$):

$$\gamma(h;a,\sigma^2) = \begin{cases} \sigma^2\{1.5\|h\|/a - 0.5(\|h\|/a)^3\} & \text{if } \|h\| \le a \\ \sigma^2 & \text{if } \|h\| > a \end{cases}.$$

-*Generalized exponential, Gaussian* : $\gamma(h;a,\sigma^2,\alpha) = \sigma^2(1-\exp(-(\|h\|/a)^\alpha)$ if $0 < \alpha \le 2$; $\alpha = 2$ represents the Gaussian model.
-*Matérn*:

$$\gamma(h;a,\sigma^2,\nu) = \sigma^2\{1 - \frac{2^{1-\nu}}{\Gamma(\nu)}(\|h\|/a)^\nu \mathcal{K}_\nu(\|h\|/a)\},$$

where $\mathcal{K}_\nu(\cdot)$ is the modified Bessel function of the second kind with parameter $\nu > -1$ (2; 227; 200).
-*Power*: $\gamma(h;b,c) = b\|h\|^c$, $0 < c \le 2$.

The variogram shown in Figure 1.1-(a) can be interpreted as being from a process $Y_s = X_s + \varepsilon_s$ where ε is a white noise in L^2 (nugget effect at the origin) uncorrelated with X whose variogram is continuous and with sill

$$2\gamma_Y(h) = 2\sigma_\varepsilon^2(1-\delta_0(h)) + 2\gamma_X(h).$$

Comments

1. Spherical covariance can be interpreted in the following way: the volume $V(a,r)$ of the intersection of two spheres in $\mathbb{R}^3$ having the same diameter a and centers at a distance r apart is:

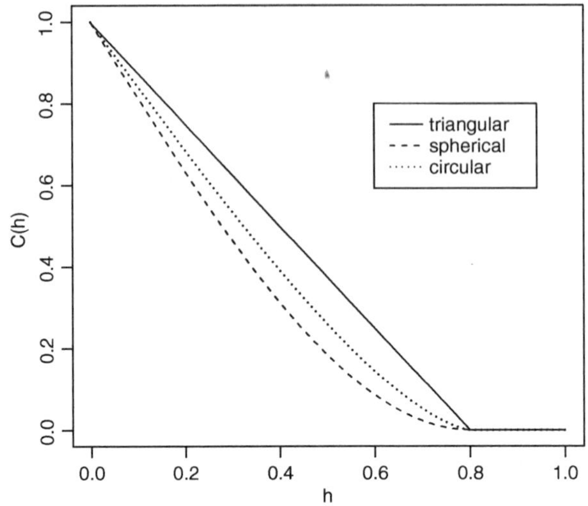

Fig. 1.2 Graph showing triangular, spherical and circular covariances with $\sigma^2 = 1$ and $a = 0.8$.

$$V(a,r) = \begin{cases} \nu(S_a)\left\{1 - 1.5(r/a) + 0.5(r/a)^3\right\} & \text{if } r \le a \\ 0 & \text{if } r > a \end{cases},$$

where $\nu(S_a)$ is the volume of a sphere of radius a. An example of a process leading to a spherical covariance is the process $X_s = N(S_a(s))$ counting the number of points of a homogeneous Poisson point process with intensity $\sigma^2/\nu(S_a)$ in the sphere $S_a(s)$ of diameter a centered at $s \in \mathbb{R}^3$ (cf. Ch. 3, §3.2).

2. The circular covariance C_{circ} on $\mathbb{R}^2$ is obtained in the same way by replacing spheres in $\mathbb{R}^3$ by disks in $\mathbb{R}^2$:

$$C_{circ}(h;a,\sigma^2) = \begin{cases} \frac{2\sigma^2}{\pi}\left(\arccos\frac{\|h\|}{a} - \frac{\|h\|}{a}\sqrt{1-\left(\frac{\|h\|}{a}\right)^2}\right) & \text{if } \|h\| \le a \\ 0 & \text{otherwise} \end{cases}. \quad (1.6)$$

Similarly, the triangular covariance C_{tri} on $\mathbb{R}^1$ can be obtained by simply replacing spheres in $\mathbb{R}^3$ by intervals $[-a,+a]$ in $\mathbb{R}^1$:

$$C_{tri}(h;a,\sigma^2) = \begin{cases} \sigma^2\left(1 - \frac{|h|}{a}\right) & \text{if } |h| \le a \\ 0 & \text{otherwise} \end{cases}.$$

Triangular, spherical and circular covariances are shown in Fig. 1.2.

3. As covariances on $\mathbb{R}^d$ remain positive semidefinite on any vectorial subspace, the restriction of a covariance to any subspace is still a covariance. In particular, the restriction of a spherical covariance to $\mathbb{R}^{d'}$, $d' \le 3$, is still a covariance. However, extending an isotropic covariance from $\mathbb{R}^d$ to $\mathbb{R}^{d'}$ for $d' > d$ does not generally give a covariance. Exercise 1.5 gives an example of this with respect to the triangular covariance.

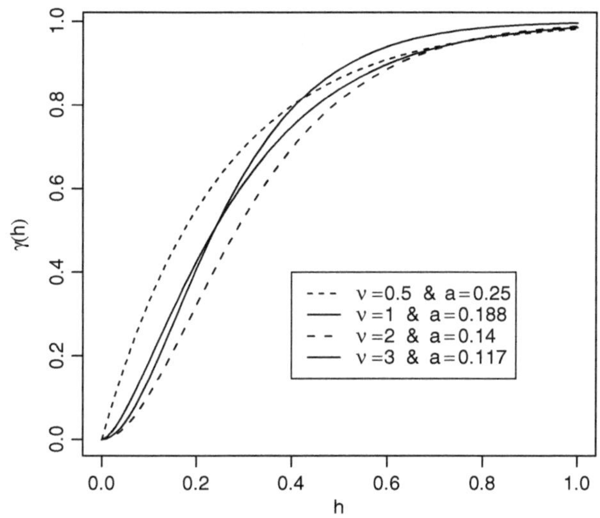

Fig. 1.3 Matérn semivariograms with the same range but different ν.

4. Our interest in Matérn covariance is due to its parameter ν which controls the variogram's regularity at 0 (cf. Fig. 1.3), which in turn controls the quadratic mean (q.m.) regularity of the process X (cf. §1.4) and its prediction $\widehat{X}$ using kriging (cf. §1.9): increasing ν increases regularity of γ at 0 and regularity of the process X (the kriging surface $\widehat{X}$). Taking $\nu = 1/2$ gives an exponential variogram which is continuous but not differentiable at 0, the associated process X being continuous but not differentiable in q.m.; $\nu = \infty$ corresponds to the infinitely differentiable Gaussian variogram associated with an infinitely differentiable process X. For integer $m \geq 1$ and taking $\nu > m$, the covariance is differentiable $2m$ times at 0 and X is differentiable m times in q.m. For example, if $\nu = 3/2$ and $r = \|h\|$, $C(h) = C(r) = \sigma^2(1 + (r/a)) \exp -(r/a)$ is twice differentiable at $r = 0$ and the associated random field differentiable in q.m.
5. The power model is self-similar, i.e., scale invariant: $\forall s > 0$, $\gamma(sh) = s^{\alpha}\gamma(h)$. It is therefore naturally associated with scale-free spatial phenomena and is the only model among those presented that has this property.
6. The generalized exponential model is identical to the exponential model when $\alpha = 1$ and the Gaussian model when $\alpha = 2$. Regularity of this type of variogram increases with α but the associated random field is only differentiable in quadratic mean when $\alpha = 2$.
7. Each of the previous models can be extended by taking positive linear combinations (or by integrating with respect to positive measures), in particular by adding a nugget effect variogram to any other variogram.

If X is a sum of K uncorrelated intrinsic processes (resp. stationary processes in L^2), it has the *nested* variogram (resp. covariance):

$$2\gamma(h) = \sum_{j=1}^{K} 2\gamma_j(h) \qquad (\text{resp. } C(h) = \sum_{j=1}^{K} C_j(h)).$$

This model can be interpreted as having independent spatial components acting on different scales with different sills. Statistically speaking, small-scale components can only be identified if the sampling grid is fairly dense and large-scale components only if the diameter of the sampling domain in S is relatively large.

1.3.4 Anisotropy

For a direction $\overrightarrow{e}$ in $\mathbb{R}^d$ such that $\|\overrightarrow{e}\| = 1$, the directional variogram of an intrinsic random field in direction $\overrightarrow{e}$ is defined as

$$2\gamma(h) = Var(X_{s+h\overrightarrow{e}} - X_s) \qquad \text{for } h \in \mathbb{R}.$$

We say that a variogram is *anisotropic* if at least two directional variograms differ.

We distinguish essentially two types of anisotropy: the first, geometric anisotropy is associated with a linear deformation of an isotropic model; the second corresponds to a nested variogram model over many subspaces of $\mathbb{R}^d$ (43; 77; 194).

Geometric anisotropy

The variogram 2γ on $\mathbb{R}^d$ exhibits geometric anisotropy if it results from an A-linear deformation of an isotropic variogram $2\gamma_0$:

$$\gamma(h) = \gamma_0(\|Ah\|),$$

i.e., if $\gamma(h) = \gamma_0(\sqrt{{}^t hQh})$, where $Q = {}^tAA$. Such variograms have the same sill in all directions (cf. Fig. 1.4-a) but with ranges that vary depending on the direction. In the orthonormal basis of eigenvectors of Q associated with eigenvalues $(\lambda_k, k = 1,\ldots,d)$, $\gamma(\widetilde{h}) = \gamma_0(\sum_{k=1}^d \lambda_k \widetilde{h}_k)$ in these new coordinates $\widetilde{h}$.

For example, if A is a rotation of angle ϕ around the origin in $\mathbb{R}^2$ followed by dilation by factor $0 \le e \le 1$ with respect to the new y axis, the set of ranges forms an ellipse with eccentricity e in this new basis. Figure 1.4-a gives an example of geometric anisotropy in $\mathbb{R}^2$ when γ_0 is an exponential model with parameters $a = 0.5$ and $\sigma^2 = 1$, with deformation A the parameters $\phi = 45^0$ and $e = 0.7$.

We note that Sampson and Guttorp (192) propose a non-stationary model

$$Var(X_s - X_{s'}) = 2\gamma_0(g(s) - g(s')),$$

where g is a bijective (or anamorphic) deformation of the space S (cf. (170; 171) for examples of such deformations).

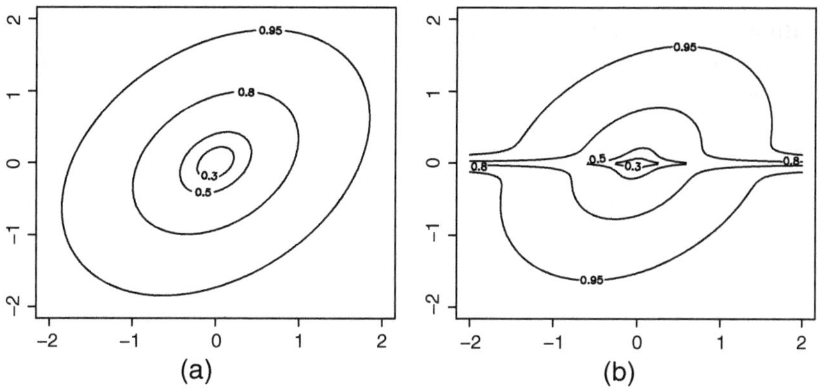

Fig. 1.4 (a) Geometric anisotropy and (b) zonal (or stratified) anisotropy.

Stratified anisotropy

We talk of *support anisotropy* if variogram $h \to 2\gamma(h)$, possibly after a change of coordinates, depends only on certain coordinates of h: for example, if $\mathbb{R}^d = E_1 \oplus E_2$, where $\dim(E_1) = d_1$ and if $2\gamma_0$ is an isotropic variogram on $\mathbb{R}^{d_1}$, $\gamma(h) = \gamma_0(h_1)$ for $h = h_1 + h_2$, $h_1 \in E_1$, $h_2 \in E_2$. The sill (and possibly the range) of γ will thus be direction-dependent (cf. Fig. 1.4-b). We say we have *zonal anisotropy* or *stratified anisotropy* if γ is the sum of several components, each with support anisotropy. For example,

$$\gamma(h) = \gamma_1\left(\sqrt{h_1^2 + h_2^2}\right) + \gamma_2(|h_2|)$$

has a sill of height $\sigma_1^2 + \sigma_2^2$ in the $(0,1)$ direction and σ_1^2 in the $(1,0)$ direction, where σ_i^2 are the sills of γ_i, $i = 1,2$.

Chilès and Delfiner (43) suggest to avoid using separable models like $\gamma(h) = \gamma_1(h_1) + \gamma_1(h_2)$ in $\mathbb{R}^2$ or $\gamma(h) = \gamma_1(h_1, h_2) + \gamma_2(h_3)$ in $\mathbb{R}^3$ as certain linear combinations of X can end up with zero variance: for example, if $X_s = X_x^1 + X_y^2$, with $Cov(X_x^1, X_y^2) = 0$ and $s = {}^t(x,y)$, then $\gamma(h) = \gamma_1(h_1) + \gamma_1(h_2)$ and for $h_x = {}^t(d_x, 0)$, $h_y = {}^t(0, d_y)$, $X_s - X_{s+h_x} - X_{s+h_y} + X_{s+h_x+h_y} \equiv 0$.

More generally, anisotropy can be obtained by combining other anisotropies. Figure 1.4-b gives an example where γ_1 is the exponential model with geometric anisotropy and parameters $a_1 = 0.5$, $\sigma_1^2 = 0.7$, $\phi = 45^0$, $e = 0.7$ and γ_2 a different exponential model with parameters $a_2 = 0.05$, $\sigma_2^2 = 0.3$.

1.4 Geometric properties: continuity, differentiability

Let us now associate the set of L^2 processes with the following notion of mean square convergence:

Definition 1.7. Quadratic mean (q.m.) continuity

We say that a second-order process $X = \{X_s, s \in S\}$ on $S \subseteq \mathbb{R}^d$ is quadratic mean continuous at $s \in S$ if for any converging sequence $s_n \longrightarrow s$ in S, $E(X_{s_n} - X_s)^2 \to 0$.

The following proposition characterizes q.m. continuity.

Proposition 1.3. *Let X be a centered L^2 process with covariance $C(s,t) = Cov(X_s, X_t)$. Then X is everywhere q.m. continuous iff its covariance is continuous on the diagonal of $S \times S$.*

Proof. If $C(s,t)$ is continuous at $s = t = s_0$, then $E(X_{s_0+h} - X_{s_0})^2 \to 0$ as $h \to 0$. In effect:

$$E(X_{s_0+h} - X_{s_0})^2 = C(s_0 + h, s_0 + h) - 2C(s_0 + h, s_0) + C(s_0, s_0).$$

To show the converse, we write:

$$\Delta = C(s_0 + h, s_0 + k) - C(s_0, s_0) = e_1 + e_2 + e_3,$$

with $e_1 = E[(X_{s_0+h} - X_{s_0})(X_{s_0+k} - X_{s_0})]$, $e_2 = E[(X_{s_0+h} - X_{s_0})X_{s_0}]$ and $e_3 = E[X_{s_0} (X_{s_0+k} - X_{s_0})]$. If X is q.m. continuous, then e_1, e_2 and $e_3 \to 0$ if h and $k \to 0$ and C is continuous on the diagonal. □

Almost sure (*a.s.*) continuity of trajectories is a result of a different nature and much harder to obtain. We have for example the following result (3): if X is a centered *Gaussian* process with continuous covariance, *a.s.* continuity of trajectories on $S \subseteq \mathbb{R}^d$ is assured if

$$\exists c < \infty \text{ and } \varepsilon > 0 \text{ s.t. } \forall s,t \in S\text{: } E(X_s - X_t)^2 \leq c\, |\log \|s - t\||^{-(1+\varepsilon)}.$$

When X is an intrinsic Gaussian process, this continuity holds if $\gamma(h) \leq c\,|\log \|h\||^{-(1+\varepsilon)}$ in a neighborhood of the origin. Apart from the nugget effect model, all variograms presented in §1.3.3 satisfy this property and the associated (Gaussian) models therefore have *a.s.* continuous trajectories.

We now examine differentiability in L^2 in given directions, or, equivalently, differentiability of processes in $\mathbb{R}^1$.

Definition 1.8. Quadratic mean differentiability

We say the process X on $S \subset \mathbb{R}^1$ is q.m. differentiable at s if there exists a real random variable (r.r.v.) $\dot{X}_s$ such that

$$\lim_{h \to 0} \frac{X_{s+h} - X_s}{h} = \dot{X}_s \text{ in } L^2.$$

We note that all trajectories of a process X might be extremely regular without X being q.m. differentiable (cf. Ex. 1.11).

Proposition 1.4. *Let X be a centered L^2 process with (not necessarily stationary) covariance $C(s,t) = Cov(X_s, X_t)$. If $\frac{\partial^2}{\partial s \partial t} C(s,t)$ exists and is finite on the diagonal* *of* $S \times S$*, then* X *is everywhere q.m. differentiable, the second-order mixed partial derivative* $\frac{\partial^2}{\partial s \partial t} C(s,t)$ *exists everywhere and the covariance of the derived process is* $Cov(\dot{X}_s, \dot{X}_t) = \frac{\partial^2}{\partial s \partial t} C(s,t)$.

Proof. Let $Y_s(h) = (X_{s+h} - X_s)/h$. To show existence of $\dot{X}_s$, we use Loève's criterion ((145), p. 135) which says that $Z_h \to Z$ in L^2 iff $E(Z_h Z_k) \to c < \infty$ whenever h and $k \to 0$ independently. Next let $\Delta_{s,t}(h,k) = E\,(Y_s(h) Y_t(k))$. It is easy to show that:

$$\Delta_{s,t}(h,k) = h^{-1}k^{-1}\{C(s+h,t+k) - C(s+h,t) - C(s,t+k) + C(s,t)\}. \tag{1.7}$$

Therefore, if $\frac{\partial^2}{\partial s \partial t} C(s,t)$ exists and is continuous at (s,s),

$$\lim_{h \to 0} \lim_{k \to 0} E\,(Y_s(h) Y_s(k)) = \frac{\partial^2}{\partial s \partial t} C(s,s).$$

Loève's criterion thus ensures convergence of $(Y_s(h))$ towards a limit denoted $\dot{X}_s$ whenever $h \to 0$ and the process $\dot{X} = \{\dot{X}_s,\, s \in S\}$ is in L^2. Let C^* be the covariance of $\dot{X}$. Using (1.7), $C^*(s,t)$ is the limit of $\Delta_{s,t}(h,k)$ when $h,k \longrightarrow 0$ and therefore $\frac{\partial^2}{\partial s \partial t} C(s,t) = C^*(s,t)$ exists everywhere. □

1.4.1 Continuity and differentiability: the stationary case

Continuity

We can deduce easily from the previous results that an intrinsic (resp. stationary in L^2) process is q.m. continuous if its variogram 2γ (resp. covariance C) is continuous at $h = 0$; in such cases the variogram 2γ (resp. covariance C) is continuous on any set where γ is bounded (resp. everywhere continuous; cf. Prop. 1.2). Matérn (154) has shown more precisely that if a random field allows a variogram that is everywhere continuous except at the origin, then this random field is the sum of two uncorrelated random fields, one associated with a pure nugget effect and the other with an everywhere continuous variogram.

Differentiability

$t \longmapsto X_t$ is q.m. differentiable in $\mathbb{R}$ if the second derivative $\gamma''(0)$ of the variogram exists. In this case, the second derivative γ'' exists everywhere, $\dot{X}$ is stationary with covariance γ'' and the bivariate process $(X, \dot{X}) \in L^2$ satisfies (227):

$$E(\dot{X}_{s+\tau}X_s) = \gamma'(\tau) \text{ and } E(X_{s+\tau}\dot{X}_s) = -\gamma'(\tau).$$

In particular, as $\gamma'(0) = 0$, X_s and $\dot{X}_s$ are uncorrelated for all s and independent if X is a Gaussian process. We remark that if X is stationary and supposing $C(s,t) = c(s-t)$, $C''_{s,t}(s,t) = -c''(s-t)$ and $c''(0)$ exists, then $\dot{X}$ is stationary with covariance $-c''$.

For integer $m \geq 1$, we say that X is m^{th}-order q.m. differentiable if $X^{(m-1)}$ exists in q.m. and if $X^{(m-1)}$ is q.m. differentiable. If we suppose X is stationary with covariance C, then X is m^{th}-order differentiable if $C^{(2m)}(0)$ exists and is finite. In this case, $X^{(m)}$ is stationary with covariance $t \mapsto (-1)^m C^{(2m)}(t)$. For example, a Matérn process is m^{th}-order q.m. differentiable whenever $\nu > m$ (200).

If γ is infinitely differentiable at the origin, X is infinitely q.m. differentiable. In this case, $X_t = \lim_{L^2} \sum_{k=0}^{n} t^k X_0^{(k)}/k!$ (200). X is "purely deterministic" as it suffices to know X is a (small) neighborhood of 0 to know it everywhere. This may lead us to put aside an infinitely differentiable variogram model (i.e., the Gaussian variogram) if we are not sure about the deterministic nature and/or hyperregularity of X.

Example 1.4. Quadratic mean regularity for processes on $\mathbb{R}^2$

Figure 1.5 gives an idea of process regularity for three different variograms. Simulations were performed using the `RandomFields` package (cf. §4.7).

(a) X is a GWN (Gaussian WN) with a nugget effect variogram γ that is not continuous at 0. X is not q.m. continuous, trajectories are extremely irregular.
(b) γ is exponential, isotropic and linear at the origin: $\gamma(h) = a + b\|h\| + o(\|h\|)$, continuous but not differentiable at 0. X is q.m. continuous but not differentiable.
(c) γ is a class $\mathscr{C}^2$ (in fact, $\mathscr{C}^\infty$) isotropic Gaussian variogram at the origin. Trajectories are q.m. continuous and differentiable. We would have the same regularity for any variogram of the form $a + b\|h\|^\alpha$ at the origin, for $\alpha \geq 2$.

Example 1.5. Separable covariances with cubic components

Separable covariances $C(h) = \prod_{k=1}^{d} C_k(h_k)$, where $h = (h_1, h_2, \ldots, h_d) \in \mathbb{R}^d$ are used for kriging due to ease of manipulation (cf. §1.9), particularly when performing simulations. They also help us to easily verify directional differentiability of the associated process. Separable covariances with cubic components (132) are associated with correlations in $[0,1]$ of the following type: for ρ, γ and $h \in [0,1]$,

$$C(h) = 1 - \frac{3(1-\rho)}{2+\gamma}h^2 + \frac{(1-\rho)(1-\gamma)}{2+\gamma}h^3. \tag{1.8}$$

C is p.d. if $\rho \geq (5\gamma^2 + 8\gamma - 1)(\gamma^2 + 4\gamma + 7)^{-1}$ (158). In this case, a process X with covariance C is q.m. differentiable and its derivative $\dot{X}$ has affine covariance in $[0,1]$:

$$C_{\dot{X}}(h) = -C''(h) = \tau^2\{1 - (1-\gamma)h\},$$

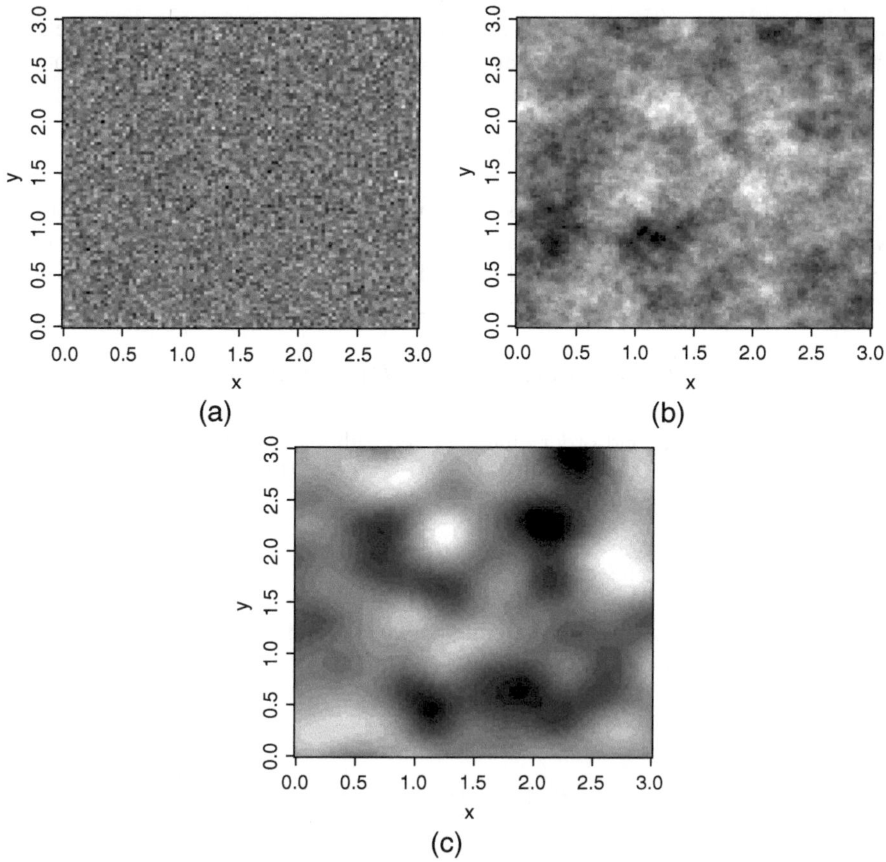

Fig. 1.5 Three Gaussian simulations for different variograms: (a) nugget effect, (b) isotropic exponential and (c) isotropic Gaussian.

where $\tau^2 = 6(1-\rho)/(2+\gamma)$. Parameters $\rho = \mathrm{Cor}(X_0, X_1) = C(1)$ and $\gamma = \mathrm{Cor}(\dot{X}_0, \dot{X}_1) = C_{\dot{X}}(1)/C_{\dot{X}}(0)$ can be respectively interpreted as correlations between the final observations and between their derivatives.

1.5 Spatial modeling using convolutions

1.5.1 Continuous model

A natural way to construct (Gaussian) models $X = (X_s, s \in S)$ over subsets S of $\mathbb{R}^d$ is to consider the convolution

$$X_s = \int_{\mathbb{R}^d} k(u,s) W(du), \tag{1.9}$$

where $\mathscr{K} = \{u \to k(u,s),\, s \in S\}$ is a family of non-random real-valued kernels on $\mathbb{R}^d$ and W a centered latent (Gaussian) random field with orthogonal increments on $\mathbb{R}^d$, that is: for δ the Dirac delta function,

$$E(W(du)W(dv)) = \delta(u,v)du \times dv.$$

A classic choice for W when $d = 1$ is Brownian motion (in this case, the convolution is a Wiener integral) or the Brownian sheet when $d = 2$ (cf. Example 1.1, (37) and Ex. 1.14). Convolution (1.9) is well-defined in L^2 if for any $s \in S$, $k(\cdot,s)$ is square integrable (227, p. 67-69). X_s is therefore a centered process with covariance:

$$C(s,t) = Cov(X_s, X_t) = \int_S k(u,s)k(u,t)du.$$

This model can be second-order characterized either by kernel family k or by its covariance C as X is a Gaussian process if W is. If $S = \mathbb{R}^d$ and if kernel family k is translation-invariant, $k(u,s) = k(u-s)$ and $\int k^2(u)du < \infty$, then X is stationary with covariance

$$C(h) = Cov(X_s, X_{s+h}) = \int_S k(u)k(u-h)du.$$

If k is isotropic, X is too and the mapping between C and k is bijective. Examples of such mappings $C \leftrightarrow k$ can be found in (219; 43, p. 646):

Gaussian covariance C, $d \geq 1$, $a > 0$:

$$k(u) = \sigma \exp\{-a\,\|u\|^2\} \leftrightarrow C(h) = \sigma^2 \left(\frac{\pi}{2a}\right)^{d/2} \exp\left\{-\frac{a}{2}\,\|h\|^2\right\};$$

Exponential covariance C, $d = 3$, $a > 0$:

$$k(u) = 2\sigma a^{-1/2}\left(1 - \frac{\|u\|}{a}\right)\exp\left(-\frac{\|h\|}{a}\right) \leftrightarrow C(h) = \sigma^2 \exp\left(-\frac{\|h\|}{a}\right);$$

Spherical covariance C, $d = 3$, $a > 0$:

$$k(u) = c\mathbf{1}\left\{\|u\| \leq \frac{a}{2}\right\} \leftrightarrow C(h) = V_d\left(\frac{a}{2}\right)\left(1 - \frac{3}{2}\left\|\frac{h}{a}\right\| + \frac{1}{2}\left\|\frac{h}{a}\right\|^3\right)\mathbf{1}\{\|h\| < a\}.$$

Such mappings are no longer bijective if X is stationary and non-isotropic as several different kernels k can give the same covariance C.

We now describe several advantages of representing X using convolutions (112):

1. Formula (1.9) allows us to deal with all second-order models without having to satisfy the positive definiteness condition for covariances (219).
2. With (1.9) we can generate *non-Gaussian models* whenever the convolution is well-defined. For example, if W is a Poisson process (cf. §3.2) (resp. Gamma process (225)) with independent increments, convolution allows us to model a process X with values in $\mathbb{N}$ (resp. values in $\mathbb{R}^+$).

3. For non-stationary but slowly-varying kernel families k we can propose parametrized types of *non-stationarity* for X (cf. (113) for modeling environmental data).
4. If we choose the latent process W to be a function of time t, convolution allows us to construct *spatio-temporal models* where the kernel k may or may not be a function of time. For example, a time-dependent model with a kernel that is constant for given t is $X_s(t) = \int_S k(u,s)W(du,t)$.
5. When observing a multivariate phenomenon $X \in \mathbb{R}^p$, *multivariate convolution* allows construction of spatially correlated components by choosing in (1.9) a kernel $k \in \mathbb{R}^p$. For example, if $S_0 \cup S_1 \cup S_2$ is a partition of S (112),

$$X_{1,s} = \int_{S_0 \cup S_1} k_1(u-s)W(du) \quad \text{and} \quad X_{2,s} = \int_{S_0 \cup S_2} k_2(u-s)W(du).$$

1.5.2 Discrete convolution

In practice, we have to use discrete convolutions of W at m sites $\mathcal{U} = \{u_1,u_2,\ldots,u_m\}$ of S: $\mathcal{U}$ is a convolution support that allows us to get reasonably close to spatial integral (1.9). Denoting $w = {}^t(w_1,w_2,\ldots,w_n)$ where $w_i = w(u_i)$, $i = 1,\ldots,m$, this model is written

$$X_s^w = (K * w)_s = \sum_{i=1}^{m} k(u_i,s)w_i, \qquad s \in S, \tag{1.10}$$

where w is a WWN with variance σ_w^2. Such models thus depend on the choice of support $\mathcal{U}$, though the spatial index s remains continuous. Though this formulation can be interpreted as a moving average (MA, cf. §1.7.1), the difference here is that there is no notion of closeness between s and the sites of $\mathcal{U}$, (1.10) being interpreted as an approximation to the continuous model (1.9).

Given n observations $X = {}^t(X_{s_1},X_{s_2},\ldots,X_{s_n})$ of X at $\mathcal{O} = \{s_1,s_2,\ldots,s_n\}$, a model that incorporates exogenous variables $z \in \mathbb{R}^p$ and a WWN measurement error ε can be written, for each site,

$$X_s = {}^t z_s \beta + X_s^w + \varepsilon_s, \qquad s \in \mathcal{O}, \beta \in \mathbb{R}^p. \tag{1.11}$$

This can be put in matrix form as:

$$X = Z\beta + Kw + \varepsilon,$$

where $K = (K_{l,i})$, $K_{l,i} = k(u_i,s_l)$, $l = 1,\ldots,n$ and $i = 1,\ldots,m$. The model's parameters are $\mathcal{U}$, $k(\cdot)$ and $(\beta,\sigma_w^2,\sigma_\varepsilon^2)$. Using statistics vocabulary, we say we are dealing with a random effects linear model where w is the cause of the random effect Kw and the deterministic trend is modeled using covariates z.

A possible choice for $\mathcal{U}$ is the regular triangular network with spacing δ; δ is a compromise between giving a good data fit (small spacing) and simple calculations (larger spacing). A compromise is to use a multiresolution model with two or more spacings. For example, the random component of a two-resolution model with triangular spacings δ and $\delta/2$ is written,

$$X^w = X^{1w} + X^{2w},$$

where X^{1w} (resp. X^{2w}) is component (1.10) associated with this δ-spacing and kernel k_1 (resp. $\delta/2$-spacing and kernel k_2).

In this context, a Bayesian formulation (cf. for example (143)) might be considered as it can incorporate uncertainty in the parameters determining the convolution.

Discrete convolutions equally allow us to construct non-stationary, non-Gaussian and multivariate models as well as spatio-temporal ones (208; 112). For example, (112) models the random component of the temporal evolution of ozone concentration over $T = 30$ consecutive days in a region of the United States by

$$X_s^w(t) = \sum k(u_i - s)w_i(t), \qquad s \in S, t = 1, \ldots, T,$$

with $\{w_i(t), t = 1, \ldots, T\}$ as T independent Gaussian random walks on a spatial support $\mathcal{U}$ of 27 sites.

1.6 Spatio-temporal models

We now present several spatio-temporal geostatistics models. This subject is undergoing significant expansion, in particular for applications in climatology and environmental sciences (136; 133; 142; 91). (126; 36) give models derived from stochastic differential equations, (148; 223; 208; 112) develop discrete-time approaches and (202) compare discrete and continuous-time approaches.

Suppose $X = \{X_{s,t},\ s \in S \subseteq \mathbb{R}^d \text{ and } t \in \mathbb{R}^+\}$ is a real-valued process with s representing space and t time. X is second-order stationary (resp. isotropic) if:

$$Cov(X_{s_1,t_1}, X_{s_2,t_2}) = C(s_1 - s_2, t_1 - t_2) \qquad (\text{resp. } = C(\|s_1 - s_2\|, |t_1 - t_2|)).$$

As $(s,t) \in \mathbb{R}^d \times \mathbb{R} = \mathbb{R}^{d+1}$, one possible approach is to consider time as an additional dimension and to reuse definitions and model properties studied earlier, this time in dimension $d+1$. However, this strategy does not take into account the fact that spatial and temporal variables work on different scales and have different meanings. For example, the isotropic exponential model $C(s,t) = \sigma^2 \exp\{-\|(s,t)\|/a\}$, where $s \in \mathbb{R}^d$ and $t \in \mathbb{R}$ is far from ideal; it is more natural to consider a geometric anisotropy model of the type $C(s,t) = \sigma^2 \exp\{-(\|s\|/b + |t|/c)\}$, $b, c > 0$. Proposition 1.1 then provides necessary tools to define more flexible stationary models in which spatial and temporal variables are treated separately. Furthermore, it may be pertinent to suggest semi-causal spatio-temporal models in which the concept of time has a well-defined meaning.

Covariances can be separable, as in these two examples:

(i) additive: $C(s,t) = C_S(s) + C_T(t)$;
(ii) factorizing: $C(s,t) = C_S(s)C_T(t)$,

where $C_S(\cdot)$ is a spatial covariance and $C_T(\cdot)$ a temporal one. Type (i) includes zonal anisotropy cases; these spatial and temporal anisotropies can be uncovered by performing variographic analysis (cf. §5.1.1) separately for space (considering pairs of sites (s_1, s_2) at the same time t) and time (considering pairs of times (t_1, t_2) at the same site s).

Separable space×time covariance

Case (ii) covers what are known as covariances that are *separable* in space and time.

The advantage of separable models is that they simplify the calculation of the covariance matrix, its inverse and its spectrum when X is observed on the rectangle $S \times T = \{s_1, s_2, \ldots, s_n\} \times \{t_1, t_2, \ldots, t_m\}$. More precisely, if $X = {}^t(X_{s_1,t_1}, \ldots, X_{s_n,t_1}, \ldots, X_{s_1,t_m}, \ldots, X_{s_n,t_m})$ is the vector of the $n \times m$ observations, $\Sigma = Cov(X)$ is the Kronecker product of Σ_T, an $m \times m$ temporal covariance matrix with Σ_S, the $n \times n$ spatial covariance matrix:

$$\Sigma = \Sigma_T \otimes \Sigma_S.$$

The product Σ is thus an $mn \times mn$ matrix made up of $m \times m$ blocks $\Sigma_{k,l}$ of size $n \times n$ equal to $C_T(k-l)\Sigma_S$. The inverse and determinant of Σ are then easily calculated:

$$(\Sigma)^{-1} = (\Sigma_T)^{-1} \otimes (\Sigma_S)^{-1}, \quad |\Sigma| = |\Sigma_T \otimes \Sigma_S| = |\Sigma_T|^n |\Sigma_S|^m$$

and the spectrum of Σ is the termwise product of the spectra of Σ_T and Σ_S. These properties simplify prediction, simulation and estimation of such models, especially when the spatial (n) or temporal (m) domain of observation is large.

The downside of separable models is that they do not allow spatio-temporal interactions $C_S(s_1 - s_2; u)$ between future instants of time u since $C(s_1 - s_2, t_1 - t_2) = C_S(s_1 - s_2)C_T(u)$. Also, separability implies reflective symmetry $C(s,t) = C(-s,t) = C(s,-t)$ of the covariance, a condition that is not generally needed.

Non-separable models

Cressie and Huang (50) propose constructing a non-separable model using the spectral density g:

$$C(h,u) = \int_{\omega \in \mathbb{R}^d} \int_{\tau \in \mathbb{R}} e^{i({}^t h\omega + u\tau)} g(\omega, \tau) d\omega d\tau. \tag{1.12}$$

If we express $g(\omega, \cdot)$ as the Fourier transform on $\mathbb{R}$ of some function $h(\omega, \cdot)$,

$$g(\omega, \tau) = \frac{1}{2\pi} \int_{\mathbb{R}} e^{-iu\tau} h(\omega, u) du,$$

where $h(\omega,u) = \int_{\mathbb{R}} e^{iu\tau} g(\omega,\tau)d\tau$, the spatio-temporal covariance can be written

$$C(h,u) = \int_{\mathbb{R}^d} e^{i^t h\omega} h(\omega,u)d\omega.$$

However, we can always write:

$$h(\omega,u) = k(\omega)\rho(\omega,u), \tag{1.13}$$

where $k(\cdot)$ is a spectral density on $\mathbb{R}^d$ and where for each ω, $\rho(\omega,\cdot)$ is an autocorrelation function on $\mathbb{R}$. Thus, under the following conditions:

1. For each ω, $\rho(\omega,\cdot)$ is a continuous autocorrelation function on $\mathbb{R}$ satisfying $\int_{\mathbb{R}} \rho(\omega,u)du < \infty$ and $k(\omega) > 0$;
2. $\int_{\mathbb{R}^d} k(\omega)d\omega < \infty$,

the function C defined as:

$$C(h,u) = \int_{\mathbb{R}^d} e^{i^t h\omega} k(\omega)\rho(\omega,u)d\omega \tag{1.14}$$

is a spatio-temporal covariance. If $\rho(\omega,u)$ is independent of ω, this model is separable.

Example 1.6. The Cressie-Huang model: if we take,

$$\rho(\omega,u) = \exp\left(-\frac{\|\omega\|^2 u^2}{4}\right)\exp\left(-\delta u^2\right), \ \delta > 0 \text{ and}$$
$$k(\omega) = \exp\left(-\frac{c_0\|\omega\|^2}{4}\right), \quad c_0 > 0,$$

then:

$$C(h,u) \propto \frac{1}{(u^2+c_0)^{d/2}} \exp\left(-\frac{\|h\|^2}{u^2+c_0}\right)\exp\left(-\delta u^2\right). \tag{1.15}$$

The condition $\delta > 0$ is needed to ensure that $\int \rho(0,u)du < \infty$, but the limit as $\delta \to 0$ of (1.15) remains a spatio-temporal covariance function.

The weakness of this approach is that it requires calculation of Fourier transforms on $\mathbb{R}^d$. Gneiting (90) proposes a different approach: let $\psi(t)$, $t \geq 0$ be a strictly monotone function and $t \mapsto \phi(t) > 0$, $t \geq 0$ a function for which $\phi'(t)$ is strictly monotone. Then the following function is a spatio-temporal covariance:

$$C(h,u) = \frac{\sigma^2}{\phi(|u|^2)^{d/2}} \psi\left(\frac{\|h\|^2}{\phi(|u|^2)}\right). \tag{1.16}$$

Example 1.7. Gneiting's spatio-temporal covariance

If $\psi(t) = \exp(-ct^{\gamma})$, $\phi(t) = (at^{\alpha}+1)^{\beta}$, with $a \geq 0$, $c \geq 0$, $\alpha, \gamma \in]0,1]$, $\beta \in [0,1]$ and $\sigma^2 > 0$, the following function is a spatio-temporal covariance on $\mathbb{R}^d \times \mathbb{R}$ (separable if $\beta = 0$):

$$C(h,u) = \frac{\sigma^2}{(a|u|^{2\alpha}+1)^{\beta d/2}} \exp\left(-\frac{c\|h\|^{2\gamma}}{(a|u|^{2\alpha}+1)^{\beta\gamma}}\right). \tag{1.17}$$

We can then infer non-separable covariances using covariance mixtures (cf. Prop. 1.1): if μ is a non-negative measure on some space W and for all $w \in W$, $C_S(\cdot,w)$ and $C_T(\cdot,w)$ are stationary covariances for which:

$$\int_W C_S(0,w)C_T(0,w)\mu(dw) < \infty,$$

then (58; 147):

$$C(h,u) = \int_W C_S(h,w)C_T(u,w)\mu(dw) < \infty$$

is a stationary covariance function that is in general non-separable. For example,

$$C(h,t) = \frac{\gamma^{n+1}}{\left(\frac{\|h\|^{\alpha}}{a} + \frac{|t|^{\beta}}{b} + \gamma\right)^{n+1}}, \qquad 0 < \alpha, \beta \leq 2 \tag{1.18}$$

is a mixture of this type for a Gamma distribution with mean $(n+1)/\gamma$ and spatial and temporal covariances respectively proportional to $\exp(-\|h\|^{\alpha}/a)$ and $\exp(-|t|^{\beta}/b)$.

1.7 Spatial autoregressive models

Spatial autoregressive models (AR) are useful for analyzing, characterizing and interpreting real-valued spatial phenomena $X = \{X_s, s \in S\}$ defined on *discrete spatial networks* S that have *neighborhood geometry*.

In domains like econometrics, geography, environmental studies and epidemiology, S does not have to be regular, because sites $s \in S$ represent centers of geographical units dispersed in space and observations X_s denote the value of the variable of interest at the site s. The irregularity of such networks S is an initial difference between spatial and temporal autoregressions, the latter usually having for S some interval in $\mathbb{Z}^1$.

In other domains such as imaging, radiography and remote sensing, S is indeed regular, typically being a subset of $\mathbb{Z}^d$. This property allows us to define stationary models and to bring the study of AR random fields on $\mathbb{Z}^d$ closer to that of time series on $\mathbb{Z}^1$. There is nevertheless a fundamental difference: spatial models are inherently non-causal in the sense that, unlike time series, they are not defined with respect to some order relation on S. While the use of temporal causality is entirely justified when modeling variables X_t such as inflation rate, stock price and rate of river flow, this is not so in the spatial case where autoregressive dependency exists in all directions in space. For example, presence/absence of a plant in some parcel of land can depend on the presence/absence of the plant in neighboring parcels, in all directions.

We begin by rapidly presently the stationary models known as MA, ARMA and AR on $\mathbb{Z}^d$ (cf. (96) for a more in-depth treatment). After this, we consider ARs on *finite general networks* and in particular two important classes of AR: SAR (S for *Simultaneous* AR) and CAR (C for *Conditional* AR) models.

1.7.1 Stationary MA and ARMA models

MA models

Let $(c_s, s \in \mathbb{Z}^d)$ be a sequence in $l^2(\mathbb{Z}^d)$ (i.e., satisfying: $\sum_{\mathbb{Z}^d} c_s^2 < \infty$) and η a SWN on $\mathbb{Z}^d$ with variance σ_η^2. An $MA(\infty)$ model on $\mathbb{Z}^d$ (MA for *Moving Average*) is a linear process defined in L^2 by:

$$X_t = \sum_{s \in \mathbb{Z}^d} c_s \eta_{t-s}. \tag{1.19}$$

X is the *infinite moving average* of the noise η with respect to weights c.

Proposition 1.5. *The covariance and spectral density of the MA process (1.19) on $\mathbb{Z}^d$ are respectively:*

$$C(h) = \sigma_\eta^2 \sum_{t \in \mathbb{Z}^d} c_t c_{t+h} \text{ and } f(u) = \frac{\sigma_\eta^2}{(2\pi)^d} \left| \sum_{t \in \mathbb{Z}^d} c_t e^{i\,{}^t ut} \right|^2.$$

Proof. C can be calculated using bilinearity of covariances and the fact that η is a SWN. As for the spectral density, it can be found using the Fourier inversion formula (1.4):

$$f(u) = \frac{\sigma_\eta^2}{(2\pi)^d} \sum_{h \in \mathbb{Z}^d} \sum_{t \in \mathbb{Z}^d} c_t c_{t+h} e^{i\,{}^t ut} = \frac{\sigma_\eta^2}{(2\pi)^d} \left| \sum_{t \in \mathbb{Z}^d} c_t e^{i\,{}^t ut} \right|^2.$$

□

We say we have an MA *model* if the support $M = \{s \in \mathbb{Z}^d : c_s \neq 0\}$ of the sequence of weights is finite. The covariance C is zero outside of its support $S(C) = M - M = \{h : h = t - s \text{ for } s,t \in M\}$. Even though when $d = 1$ any covariance process with finite support has an MA representation, this is not the case when $d \geq 2$ (cf. Prop. 1.8).

ARMA models

These models are an extension of temporal ($d = 1$) ARMA models: let P and Q be the following two polynomials of the d-dimensional complex variable $z \in \mathbb{C}^d$,

$$P(z) = 1 - \sum_{s \in R} a_s z^s \quad \text{and} \quad Q(z) = 1 + \sum_{s \in M} c_s z^s,$$

with R (resp. M) the support of the AR (resp. MA) being finite subsets of $\mathbb{Z}^d$ not containing the origin and for $s = (s_1, s_2, \dots, s_d)$, $z^s = z_1^{s_1} \dots z_d^{s_d}$. Let $B^s X_t = X_{t-s}$ denote the s-translation operator in L^2. In formal terms, an ARMA is associated with polynomials P and Q and an SWN η in L^2 by:

$$\forall t \in \mathbb{Z}^d : \ P(B)X_t = Q(B)\eta_t, \tag{1.20}$$

or alternatively,

$$\forall t \in \mathbb{Z}^d : \ X_t = \sum_{s \in R} a_s X_{t-s} + \eta_t + \sum_{s \in M} c_s \eta_{t-s}.$$

Let $\mathbb{T} = \{\xi \in \mathbb{C}, |\xi| = 1\}$ be the 1-dimensional torus. We have the following existence result:

Proposition 1.6. *Suppose that P has no zeros on the torus $\mathbb{T}^d$. Then equation (1.20) has a stationary solution X in L^2. Denoting $e^{iu} = (e^{iu_1}, \dots, e^{iu_d})$, the spectral density of X is:*

$$f(u) = \frac{\sigma^2}{(2\pi)^d} \left| \frac{Q}{P}(e^{iu}) \right|^2,$$

and its covariance is given by the Fourier coefficients of f.

Proof. As P has no zeros on the torus $\mathbb{T}^d$, $P^{-1}Q$ has Laurent series development,

$$P^{-1}(z)Q(z) = \sum_{s \in \mathbb{Z}^d} c_s z^s,$$

which converges in a neighborhood of the torus $\mathbb{T}^d$ and whose coefficients (c_s) decrease exponentially fast to 0. This ensures that $X_t = \sum_{s \in \mathbb{Z}^d} c_s \eta_{t-s}$ exists in L^2, that it satisfies (1.20) and has spectral density

$$f(u) = \frac{\sigma^2}{(2\pi)^d} \left| \sum_{s \in \mathbb{Z}^d} c_s e^{i^t su \rangle} \right|^2 = \frac{\sigma^2}{(2\pi)^d} \left| \frac{Q}{P}(e^{iu}) \right|^2.$$

□

MA models correspond to the choice $P \equiv 1$ and AR models to $Q \equiv 1$. As with time series, interest in ARMA models is due to the fact that they can get "close" to random fields that have continuous spectral densities: in effect, for any dimension d, the set of rational fractions is dense (e.g., for the sup norm) in the space of continuous functions on the torus $\mathbb{T}^d$.

As the spectral density of ARMA processes is rational, its covariance decreases exponentially fast to zero in the limit. Here again, as with time series, after a certain rank the covariances satisfy the linear recurrence relationships known as the Yule-Walker equations. In $\mathbb{Z}$, these equations can be solved analytically and provide a tool to identify the ranges R and M of the AR and MA parts and estimate parameters a and c. However, in dimension $d \geq 2$, the Yule-Walker equations cannot be solved analytically. Furthermore, unlike time series ARMA models generally do not have a finite unilateral (or causal) representation with respect to the lexicographic order when $d \geq 2$ (cf. (1.8)).

Even though there is no theoretical reason limiting their use (cf. for example (119)), the preceding remarks explain why, unlike for time series ARMA models are rarely used in spatial statistics.

Nevertheless we note that *semi-causal* spatio-temporal models (non-causal in space but causal in time) can turn out to be well adapted to studying spatial dynamics: STARMA (Spatio-Temporal ARMA) models, introduced by Pfeifer and Deutsch (76; 174) fall into this category (cf. also (48, §6.8)).

Two types of autoregressive model, SAR and CAR models are frequently used in spatial analyses. First, let us take a look at stationary models.

1.7.2 Stationary simultaneous autoregression

To simplify, suppose that X is centered. Let R be a finite subset of $\mathbb{Z}^d$ not containing the origin. A stationary SAR (*Simultaneous* AR) model relative to the SWN η and with parameters $a = \{a_s, s \in R\}$ is:

$$X_t = \sum_{s \in R} a_s X_{t-s} + \eta_t. \tag{1.21}$$

X_t is the weighted sum of the values X_u at the *R-neighbors* of t along with added noise η_t. X exists if the characteristic polynomial P of the autoregression has no zero on the torus $\mathbb{T}^d$, where

$$P(e^{i\lambda}) = 1 - \sum_{s \in R} a_s e^{i\,^t\lambda s}.$$

Equations (1.21) can be interpreted as a system of *simultaneous* AR *equations* with the usual econometrical meaning: $\{X_{t-s},\ s \in R\}$ are *"spatially lagged" endogenous variables* influencing the response X_t at t with site u having influence on t when $t - u \in R$. This *relation* defines the *oriented graph* $\mathscr{R}$ of the SAR model. We now present some examples.

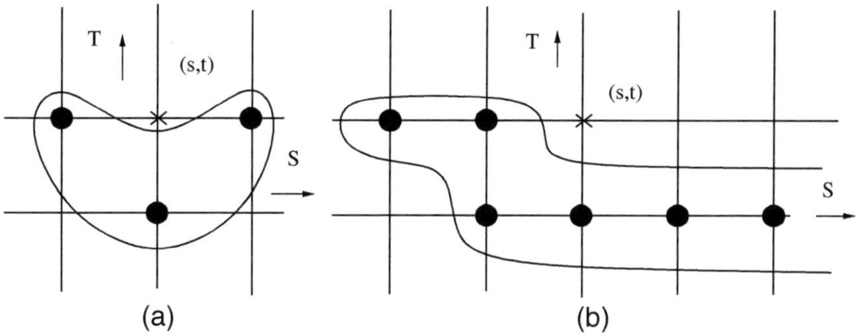

Fig. 1.6 (a) Semi-causal model; (b) Semi-causal model with lexicographic order.

Example 1.8. Several SAR models

Semi-causal Space×Time models

$s \in \mathbb{Z}$ gives the spatial coordinate and $t \in \mathbb{N}$ the temporal one; an example of Markov dynamics at time t and location s is:

$$\forall t \in \mathbb{N} \text{ and } s \in \mathbb{Z}\text{: } X_{s,t} = \alpha X_{s,t-1} + \beta(X_{s-1,t} + X_{s+1,t}) + \varepsilon_{s,t}.$$

The temporal connection $(s,t-1) \rightarrow (s,t)$ has a direction whereas instantaneous spatial links $(s,t) \longleftrightarrow (s \pm 1,t)$ do not. The lexicographic causal representation of this SAR is infinite (cf. Fig. 1.6). More precisely, for $\alpha = \beta = \delta/(1+\delta^2)$, this semi-causal model has the following infinite causal representation for the lexicographic order (defined by $(u,v) \preceq (s,t)$ if $v < t$ or if $v = t$ and $u \leq s$; (24)):

$$X_{s,t} = 2\delta X_{s-1,t} + \delta^2 X_{s-2,t} - \delta X_{s-1,t-1} + \delta(1-\delta^2) \sum_{j \geq 0} \delta^j X_{s+j,t-1} + \varepsilon_{s,t}.$$

Isotropic four nearest neighbor SAR models on $\mathbb{Z}^2$

$$X_{s,t} = \alpha(X_{s-1,t} + X_{s+1,t} + X_{s,t-1} + X_{s,t+1}) + \varepsilon_{s,t}.$$

Here, $\mathcal{R}$ is a symmetric graph; X exists if and only if

$$\forall \lambda, \mu \in [0, 2\pi[,\ P(\lambda,\mu) = 1 - 2\alpha(\cos\lambda + \cos\mu) \neq 0,$$

ensuring that spectral density $f(\lambda,\mu) = \sigma_\varepsilon^2\, P(\lambda,\mu)^{-2} \in L^2(\mathbb{T}^2)$; this condition is satisfied iff $|\alpha| < 1/4$.

Factorizing $SAR(1)$ *models*

An example of a factorizing SAR on $\mathbb{Z}^2$ is

$$X_{s,t} = \alpha X_{s-1,t} + \beta X_{s,t-1} - \alpha\beta X_{s-1,t-1} + \varepsilon_{s,t},\ |\alpha| \text{ and } |\beta| < 1. \tag{1.22}$$

Noting B_1 and B_2 the lag operators relative to coordinates s and t, equation (1.22) can be written:

$$(1-\alpha B_1)(1-\beta B_2)X_{s,t} = \varepsilon_{s,t}.$$

We deduce that the covariance c of X is separable:

$$c(s-s',t-t') = \sigma^2 \alpha^{|s-s'|} \beta^{|t-t'|},$$

where $\sigma^2 = \sigma_\varepsilon^2(1-\alpha^2)^{-1}(1-\beta^2)^{-1}$ is the product of a 1-dimensional $AR(1)$ covariance with parameter α and a covariance of the same type with parameter β.

It is easy to generalize these models to autoregressions of any order $p=(p_1,p_2)$ and any dimension $d \geq 2$. Being able to factorize the AR polynomial and the covariance makes these models easy to work with (cf. §1.6).

SAR models are frequently used for their simplicity and because they involve few parameters. However, the following two problems must be dealt with:

1. Without imposing constraints, SAR models are not (in general) identifiable. Recall that we say model $\mathscr{M}(\theta)$ is *identifiable* if the distributions it defines for two different θ are different; for example, in $\mathbb{Z}$, it is simple to show that the following three SAR models:

   (i) $X_t = aX_{t-1} + bX_{t+1} + \eta_t,\ t \in \mathbb{Z}^1,\ a \neq b,\ |a|,\ |b| < 1/2,$
   (ii) $X_t = bX_{t-1} + aX_{t+1} + \eta_t^*$ and
   (iii) $X_t = a_1X_{t-1} + a_2X_{t-2} + \varepsilon_t$

   are identical for appropriate choices of a_1, a_2 and variances of the WN errors η, η^* and ε (it suffices to identify each of their spectral densities and to realize that we can impose constraints allowing us to make equal the three of them). This said, if we impose the constraint $a < b$, the model becomes identifiable.
2. As with simultaneous equations in econometrics, estimation of SAR models by ordinary least squares (OLS) on the residuals is not consistent (cf. Prop. 5.6).

1.7.3 Stationary conditional autoregression

Suppose X is a centered stationary second-order process on $\mathbb{Z}^d$ with spectral density f. If $f^{-1} \in L^1(\mathbb{T}^d)$, X has the infinite non-causal linear representation (96, Th. 1.2.2):

$$X_t = \sum_{s \in \mathbb{Z}^d \setminus \{0\}} c_s X_{t-s} + e_t.$$

In this form, $c_s = c_{-s}$ for all s and e_t is a *conditional residual*, i.e., for any $s \neq t$, e_t and X_s are uncorrelated.

This leads us to the following definition of an L-Markov random field or $CAR(L)$ model: let L be a *finite symmetric* subset of $\mathbb{Z}^d$ not containing the origin 0 and L^+ the positive half-space of L with respect to the lexicographic order on $\mathbb{Z}^d$.

Definition 1.9. A stationary $CAR(L)$ model in L^2 is given by, for any $t \in \mathbb{Z}^d$,

$$X_t = \sum_{s \in L} c_s X_{t-s} + e_t \text{ with, if } s \in L^+ : c_s = c_{-s}; \tag{1.23}$$
$$\forall s \neq t : Cov(e_t, X_s) = 0 \text{ and } E(e_t) = 0.$$

The absence of correlation between X_s and e_t when $s \neq t$ translates the fact that $\sum_{s \in L} c_s X_{t-s} = \sum_{s \in L^+} c_s (X_{t-s} + X_{t+s})$ is the *best linear prediction* in L^2 of X_t using all other variables $\{X_s, s \neq t\}$; here, X is an L-Markov random field in the linear prediction sense. If X is a Gaussian random field, it is the best possible prediction and we say that X is an L-Markov Gaussian process. CAR and SAR models differ in several respects:

1. CAR models require parametric constraints: L must be symmetric and for all s, $c_s = c_{-s}$.
2. The conditional residuals e_t do not represent a white noise. We say that $\{e_t\}$ is a *colored noise*.
3. Residuals e_t are uncorrelated with X_s when $s \neq t$.

Proposition 1.7. *The model X defined in (1.23) exists in L^2 if the characteristic polynomial of the CAR models,*

$$P^*(\lambda) = (1 - 2 \sum_{s \in L^+} c_s \cos({}^t su))$$

is non-zero on the torus $\mathbb{T}^d$. In this case, the spectral density is

$$f_X(u) = \sigma_e^2 (2\pi)^{-d} P^*(\lambda)^{-1}$$

and the conditional residuals form a correlated noise with covariance:

$$Cov(e_t, e_{t+s}) = \begin{cases} \sigma_e^2 \text{ if } s = 0, \\ -\sigma_e^2 c_s \text{ if } s \in L \end{cases} \text{ and } Cov(e_t, e_{t+s}) = 0 \text{ otherwise.}$$

Proof. We first remark that $E(e_0 X_u) = 0$ if $u \neq 0$ and $E(e_0 X_0) = \sigma_e^2$ when $u = 0$. Since $e_0 = X_0 - \sum_{s \in L} c_s X_{-s}$, this orthogonality becomes, in the frequency domain:

$$\forall u \neq 0 : \int_{T^d} e^{-i\langle \lambda, u \rangle} [1 - \sum_{s \in L} c_s e^{-i\langle \lambda, s \rangle}] f_X(\lambda) d\lambda = 0.$$

Plancherel's theorem implies that $f_X(u) = \sigma_e^2 (2\pi)^{-d} P^*(u)^{-1}$. As residual $e_t = X_t - \sum_{s \in L} c_s X_{t-s}$ is a linear filter spectral density of X, it has spectral density

$$f_e(u) = \sigma_e^2 (2\pi)^{-d} P(u)^{-1} |P(u)|^2 = \sigma_e^2 (2\pi)^{-d} P(u).$$

The result is thus proved. □

Note that whereas the spectral density of a CAR model is proportional to $P^*(u)^{-1}$, that of a SAR model is proportional to $|P(u)|^{-2}$. In dimension $d \geq 3$, the condition "P^* has no zeros on the torus" is not necessary (cf. Ex. 1.12).

As for SAR models, the Yule-Walker equations for covariance of CAR models can be obtained by multiplying the equation defining X_t by X_s and then taking the expectation: for example, for an isotropic four nearest neighbor (4-NN) CAR model in $\mathbb{Z}^2$, these equations are:

$$\forall s\colon r(s) = \sigma_e^2\, \delta_0(s) + a \sum_{t:\|t-s\|_1=1} r(t).$$

There are three reasons justifying modeling using CAR models:

1. CAR representations are intrinsic: they give the best linear prediction of X_t from its other values $\{X_s, s \neq t\}$.
2. Estimating CAR models using OLS is consistent (cf. Prop. 5.6).
3. The family of stationary CAR models contains that of the SAR models, strictly so when $d \geq 2$.

Proposition 1.8. *Stationary SAR and CAR models on $\mathbb{Z}^d$.*

1. *Every SAR model is a CAR model. In $\mathbb{Z}$, both classes are identical.*
2. *When $d \geq 2$, the family of CAR models is larger than that of the SAR models.*

Proof. 1. To get the CAR representation of a SAR model: $P(B)X_t = \varepsilon_t$, we write the spectral density f of X and see that it is the spectral density of a CAR model by expanding $\left|P(e^{i\lambda})\right|^2 = \left|1 - \sum_{s\in R} a_s e^{i\langle\lambda,s\rangle}\right|^2$. We thus obtain the support L of the CAR model and its coefficients after imposing the normalization $c_0 = 0$:

$$f(u) = \frac{\sigma_\varepsilon^2}{(2\pi)^d\,|P(e^{iu})|^2} = \frac{\sigma_e^2}{(2\pi)^d C(e^{iu})}, \text{ with } c_0 = 1.$$

For $A - B = \{i - j : i \in A \text{ and } j \in B\}$, we get:

$$L = \{R^* - R^*\} \setminus \{0\}, \text{ where } R^* = R \cup \{0\} \text{ and}$$
$$\text{if } s \in L,\ c_s = (\sigma_e^2/\sigma_\varepsilon^2) \sum_{v,v+s\in R} a_v a_{v+s} \text{ if } s \neq 0 \text{ and } 0 \text{ otherwise.}$$

When $d = 1$ the SAR and CAR classes are identical due to Fejer's theorem which states that any trigonometric polynomial $P^*(e^{i\lambda})$ of one complex variable for which $P^* \geq 0$ is the square modulus of a trigonometric polynomial: if $P^*(e^{i\lambda}) \geq 0$, $\exists\, P$ such that $P^*(e^{i\lambda}) = \left|P(e^{i\lambda})\right|^2$. Thus the CAR-$P^*$ model can be equated with the SAR-P one.

2. We show that over $\mathbb{Z}^2$ the CAR model $X_t = c\sum_{s:\|s-t\|_1=1} X_s + e_t$, $c \neq 0$ has no SAR representation. The spectral density of X satisfies:

$$f_X^{-1}(\lambda_1, \lambda_2) = c(1 - 2c(\cos\lambda_1 + \cos\lambda_2)). \qquad (1.24)$$

If some SAR had spectral density f_X, its support R would satisfy $R \subseteq L$. Noting (a_s) the coefficients of the SAR, we must either have $a_{(1,0)} \neq 0$ or $a_{(-1,0)} \neq 0$, say $a_{(1,0)} \neq 0$; similarly, either $a_{(0,1)}$ or $a_{(0,-1)} \neq 0$, say $a_{(0,1)} \neq 0$. This implies that a non-zero term that depends on $\cos(\lambda_1 - \lambda_2)$ has to appear in f_X^{-1}, which is not the case. Thus, CAR model (1.24) has no SAR representation.

□

MA processes with finite support have covariances with bounded range. When $d = 1$, Fejer's theorem ensures that the converse is true: processes on $\mathbb{Z}$ having covariances with bounded range are MAs. This is no longer true for $d \geq 2$: for example, the random field with correlation ρ at distance 1 and 0 at distances > 1 has no MA representation; this can be proved using similar arguments to those in part (2) of the previous proposition.

Let us present several examples of CAR representations of SAR models on $\mathbb{Z}^2$.

Example 1.9. SAR → CAR mappings

1. The causal AR (cf. Fig. 1.7-a) with support $R = \{(1,0),(0,1)\}$:

$$X_{s,t} = \alpha X_{s-1,t} + \beta X_{s,t-1} + \varepsilon_{s,t}$$

is a $CAR(L)$ model with half-support $L^+ = \{(1,0),\ (0,1),\ (-1,1)\}$ and coefficients $c_{1,0} = \alpha\kappa^2$, $c_{0,1} = \beta\kappa^2$, $c_{-1,1} = -\alpha\beta\kappa^2$ and $\sigma_e^2 = \kappa^2\sigma_\varepsilon^2$, where $\kappa^2 = (1+\alpha^2+\beta^2)^{-1}$.

2. The non-causal SAR model:

$$X_{s,t} = a(X_{s-1,t} + X_{s+1,t}) + b(X_{s,t-1} + X_{s,t+1}) + \varepsilon_{s,t}$$

is a $CAR(L)$ model (cf. Fig. 1.7-b) with half-support $L^+ = \{(1,0),(2,0),(-1,1),(0,1),(0,2),(1,1),\ (0,2)\}$ and coefficients:

$$c_{1,0} = 2a\kappa^2,\ c_{0,1} = 2b\kappa^2,\ c_{2,0} = 2a^2\kappa^2,\ c_{0,2} = 2b^2\kappa^2$$
$$c_{-1,1} = -2ab\kappa^2,\ \sigma_e^2 = \sigma_\varepsilon^2\kappa^2 \text{ where } \kappa^2 = (1+2a^2+2b^2)^{-1}.$$

3. The factorizing SAR model:

$$X_{s,t} = \alpha X_{s-1,t} + \beta X_{s,t-1} - \alpha\beta X_{s-1,t-1} + \varepsilon_{s,t},\ |\alpha| \text{ and } |\beta| < 1$$

is an 8-NN CAR model with coefficients:

$$c_{1,0} = \alpha(1+\alpha^2)^{-1},\ c_{0,1} = \beta(1+\beta^2)^{-1},\ c_{1,1} = c_{-1,1} = -c_{1,0}\times c_{0,1}$$
$$\sigma_e^2 = \sigma_\varepsilon^2\kappa^2, \quad \text{where } \kappa^2 = (1+\alpha^2)^{-1}(1+\beta^2)^{-1}.$$

In these three examples, $\kappa^2 < 1$ is the gain in variance of the CAR prediction of X with respect to the SAR prediction.

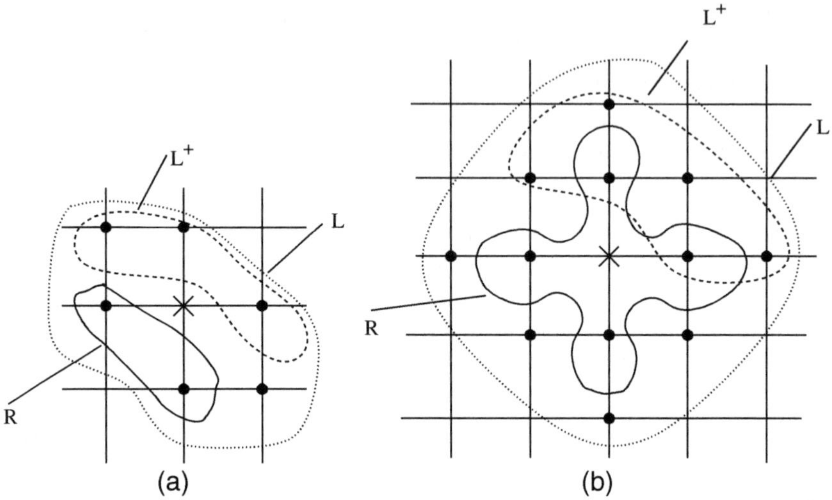

Fig. 1.7 (a) Support $R = \{(1,0),(0,1)\}$ of the causal SAR model and support L of the associated CAR model; (b) Support $R = \{(1,0),(0,1),(-1,0),(0,-1)\}$ of the non-causal SAR model and support L of the associated CAR model.

1.7.4 Non-stationary autoregressive models on finite networks S

A real-valued process on $S = \{1,2,\dots,n\}$ is a vectorial r.v. $X^* \in \mathbb{R}^n$. Non-stationarity of X^* can influence the vector of expectations $\mu = E(X^*)$, the network S and the covariance matrix $\Sigma = Cov(X^*)$. We only deal with second-order non-stationarity here, working with the centered process $X = X^* - \mu$ for which $\Sigma = Cov(X^*) = Cov(X)$.

Let $\varepsilon = (\varepsilon_t, t \in S)$ be a centered noise in L^2. MA, AR and ARMA representations of X, either site by site or in matrix notation in the basis ε are defined by the equations:

$$MA:\ X_t = \sum_{s\in S} b_{t,s}\varepsilon_s\ ,\ \text{or}\ X = B\varepsilon,$$

$$AR:\ X_t = \sum_{s\in S:s\neq t} a_{t,s}X_s + \varepsilon_t\ ,\ \text{or}\ AX = \varepsilon,$$

$$ARMA:\ X_t = \sum_{s\in S:s\neq t} a_{t,s}X_s + \sum_{s\in S} b_{t,s}\varepsilon_s,\ \text{or}\ AX = B\varepsilon,$$

where, for $s,t \in S$, we have $B_{t,s} = b_{t,s}$, $A_{s,s} = 1$, $A_{t,s} = -a_{t,s}$ when $t \neq s$. The MA representation always exists; the AR and ARMA ones do too as long as A is invertible. Noting $\Gamma = Cov(\varepsilon)$, these models are second-order characterized by their covariances Σ:

$$\begin{aligned} MA &: \Sigma = B\Gamma\,{}^tB; \\ AR &: \Sigma = A^{-1}\Gamma\,{}^t(A^{-1}); \\ ARMA &: \Sigma = (A^{-1}B)\Gamma({}^t(A^{-1}B)). \end{aligned}$$

Suppose we choose ε to be a WWN with variance 1 ($\Gamma = I_n$) and denote $<$ an arbitrary total order enumerating the points of S. If X is centered with invertible covariance Σ, then X has a unique causal AR representation relative to ε and order $<$; this representation is associated with the lower triangular matrix A^* from the Cholesky factorization $\Sigma = {}^tA^*A^*$. The fact that A^*, like Σ, depends on $n(n+1)/2$ parameters confirms identifiability of the causal AR model. Equivalent AR representations, generally non-identifiable are written $\widetilde{A}X = \eta$, where for some orthogonal matrix P, $\eta = P\varepsilon$ (η still a WWN with variance 1) and $\widetilde{A} = PA^*$.

In practice, AR models are associated with not necessarily symmetric influence graphs $\mathscr{R}$: $s \to t$ is a (directed) edge of $\mathscr{R}$ if X_s influences X_t with some weight $a_{t,s}$ and the neighborhood of t is defined as $\mathscr{N}_t = \{s \in S : s \to t\}$.

Local one-parameter SAR representation

Let $W = (w_{t,s})_{t,s\in S}$ be a weights matrix or influence graph measuring the influence of s on t where, for each t, $w_{t,t} = 0$: for example, W could be the spatial contiguity matrix made up of ones where s has influence over t and zeros elsewhere. Other choices of W are presented in Cliff and Ord's book (45) (cf. §5.2 also). Once W has been chosen, a classical choice of spatial model for econometrics or spatial epidemiology is a SAR with parameter ρ. If $t \in S$ and if ε is a $SWN(\sigma_\varepsilon^2)$,

$$X_t = \rho \sum_{s:s\neq t} w_{t,s}X_s + \varepsilon_t, \text{ or } X = \rho WX + \varepsilon.$$

This model is well-defined as long as $A = I - \rho W$ is invertible.

Markov CAR representation

Once again, consider the centered vector X. CAR representations are written in terms of *linear conditional expectation* (conditional expectation if X is a Gaussian random field):

$$X_t = \sum_{s\in S:\, s\neq t} c_{t,s}X_s + e_t, \qquad \forall t \in S, \tag{1.25}$$

with $E(e_t) = 0$, $Var(e_t) = \sigma_t^2 > 0$ and $Cov(X_t, e_s) = 0$ when $t \neq s$. In this *intrinsic* representation, e is a *conditional residual.*

CAR representations are associated with a neighborhood graph $\mathscr{G}$ of S defined in the following way: $s \to t$ is an edge of $\mathscr{G}$ if $c_{t,s} \neq 0$. As we will see, $\mathscr{G}$ is symmetric. Denote C the matrix with coefficients $C_{s,s} = 0$ and $C_{t,s} = c_{t,s}$ when $s \neq t$ and let D be the diagonal matrix with coefficients $D_{t,t} = \sigma_t^2$. The Yule-Walker equations

$\Sigma = C\Sigma + D$ can be obtained by multiplying (1.25) by X_s for $s \in S$, then taking the expectation. Σ then satisfies:

$$(I - C)\Sigma = D.$$

Hence, (1.25) defines a CAR model with regular covariance matrix Σ iff $\Sigma^{-1} = D^{-1}(I - C)$ is symmetric and positive definite. In particular, representation (1.25) has to satisfy the constraints:

$$c_{t,s}\sigma_s^2 = c_{s,t}\sigma_t^2, \qquad \forall t \neq s \in S. \tag{1.26}$$

Hence, $c_{t,s} \neq 0$ when $c_{s,t} \neq 0$, implying that the CAR's graph $\mathscr{G}$ is symmetric. For algorithms that estimate CAR models, it is necessary to include these constraints. If X is stationary (for example on the finite torus $S = (\mathbb{Z}/p\mathbb{Z})^d$), we can reparametrize the model with $c_{t-s} = c_{t,s} = c_{s,t} = c_{s-t}$ for $t \neq s$. Under Gaussian hypotheses, (1.25) entirely characterizes this model.

We note that unlike stationary models on $\mathbb{Z}^d$ (cf. Prop. 1.8), when S is finite the family of SAR models is the same as the CAR one.

Markov Gaussian random fields

Suppose X is a Gaussian process on S, $X \sim \mathscr{N}_n(\mu, \Sigma)$ with invertible Σ and where S is associated with a symmetric graph $\mathscr{G}$ without loops. Let $\langle s,t \rangle$ mean s and t are neighbors in $\mathscr{G}$. We say that X is a $\mathscr{G}$-Markov random field if, noting $Q = (q_{s,t}) = \Sigma^{-1}$, $q_{st} = 0$ except when $\langle s,t \rangle$. In this case, we have for all $t \in S$,

$$\mathscr{L}(X_t | X_s, s \neq t) \sim \mathscr{N}(\mu_t - q_{t,t}^{-1} \sum_{s:\langle t,s \rangle} q_{t,s}(X_s - \mu_s), q_{t,t}^{-1})$$

and X follows a CAR model: for all $t \in S$,

$$X_t - \mu_t = -q_{t,t}^{-1} \sum_{s:\langle t,s \rangle} q_{t,s}(X_s - \mu_s) + e_t, \qquad Var(e_t) = q_{t,t}^{-1}. \tag{1.27}$$

Let $[Q]$ be the $n \times n$ matrix with diagonal 0 and $[Q]_{t,s} = q_{t,s}$ if $t \neq s$, and $\mathrm{Diag}(Q)$ the diagonal matrix whose diagonal is the same as that of Q. (1.27) can be written

$$X - \mu = -(\mathrm{Diag})^{-1}[Q](X - \mu) + e.$$

As we will see in Chapter 2, Gaussian CAR models are Gibbs models with quadratic potentials (189).

The Markov graph $\mathscr{G}$ of a SAR model

Let ε be a Gaussian WN with variance σ^2. The Gaussian SAR model $AX = \varepsilon$ exists if A^{-1} exists and has inverse covariance $\Sigma^{-1} = Q = \sigma^{-2}({}^tAA)$ and SAR graph: $\langle t,s \rangle_{\mathscr{R}} \Longleftrightarrow a_{t,s} \neq 0$. Its CAR representation (1.27) is:

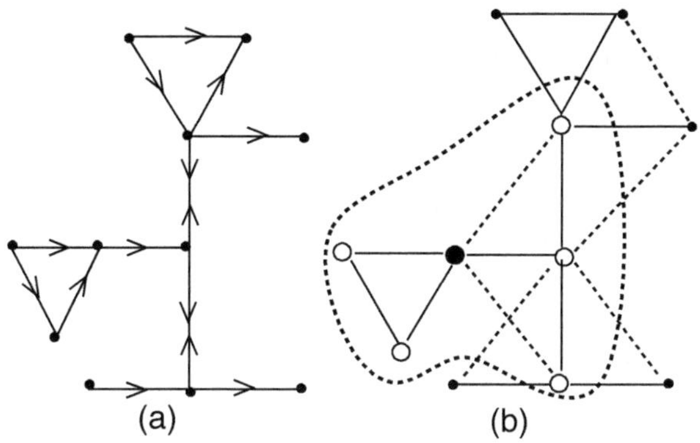

Fig. 1.8 (a) Directed graph $\mathscr{R}$ of a SAR; (b) Associated CAR model, graph $\mathscr{G}$ (new edges represented by dotted line) and conditional neighborhood (∘) of point (•).

1. CAR coefficients are: $c_{t,s} = -q_{t,s}/q_{t,t}$, where $q_{t,s} = \sum_{l \in S} a_{l,t} a_{l,s}$;
2. The graph $\mathscr{G}$ of the Markov CAR representation of X is:

$$\langle t,s \rangle_{\mathscr{G}} \Longleftrightarrow \begin{cases} \text{either } \langle t,s \rangle_{\mathscr{R}}, \\ \langle s,t \rangle_{\mathscr{R}} \\ \text{or } \exists l \in S \text{ s.t. } \langle l,t \rangle_{\mathscr{R}} \text{ and } \langle l,s \rangle_{\mathscr{R}}. \end{cases}$$

$\mathscr{G}$ is *undirected* with "double" the range of $\mathscr{R}$ (cf. Fig. 1.8).

Example 1.10. CAR representation of nearest neighbor SAR models.

Let $W = (w_{t,s})_{t,s \in S}$ be a weights matrix representing the influence of s on t with, for all t, $w_{t,t} = 0$; consider the SAR with one parameter ρ:

$$X = \rho W X + \varepsilon,$$

where ε is a $WWN(\sigma_\varepsilon^2)$. The CAR model associated with this SAR model is given by (1.27) with $\mu = 0$ and precision matrix $Q = \Sigma_X^{-1}$:

$$Q = \sigma_\varepsilon^{-2}(I - \rho(W + {}^tW) + \rho^2\, {}^tWW).$$

As for the best linear prediction of X, it is given by the vector

$$\widehat{X} = -(\mathrm{Diag})^{-1}[Q]X.$$

1.7.5 Autoregressive models with covariates

These types of models are especially used in spatial econometrics. Suppose that Z is a real-valued $n \times p$ matrix of observable exogenous conditions. SARX models (X for eXogenous) propose, in addition to regression of X onto Z, a weights structure W acting separately on the endogenous X and exogenous Z (7):

$$X = \rho WX + Z\beta + WZ\gamma + \varepsilon,\ \rho \in \mathbb{R},\ \beta \text{ and } \gamma \in \mathbb{R}^p. \tag{1.28}$$

X has three explicative factors: the usual regression variables ($Z\beta$), the endogenous (ρWX) and spatially lagged exogenous variables ($WZ\gamma$) with the same weights vector W but their own parameters.

A sub-model with common factors, also known as a *spatial Durbin model* is associated with the choice of constraint $\gamma = -\rho\beta$, i.e., with the regression model with SAR errors:

$$(I-\rho W)X = (I-\rho W)Z\beta + \varepsilon \text{ or } X = Z\beta + (I-\rho W)^{-1}\varepsilon. \tag{1.29}$$

The *spatial lag* sub-model corresponds to the choice $\gamma = 0$:

$$X = \rho WX + Z\beta + \varepsilon. \tag{1.30}$$

Note that these models offer three different ways to model the mean: respectively,

$$E(X) = (I-\rho W)^{-1}[Z\beta + WZ\gamma], \quad E(X) = Z\beta \quad \text{and} \quad E(X) = (I-\rho W)^{-1}Z\beta,$$

but the same covariance structure $\Sigma^{-1} \times \sigma^2 = (I - \rho\, {}^tW)(I - \rho W)$ if ε is a WWN with variance σ^2. An estimation of these models using Gaussian maximum likelihood can be obtained by expressing the mean and variance of X in terms of the model's parameters.

Variants of these models are possible, for example by choosing ε to be a SAR model associated with weights matrix H and some real-valued parameter α. We can also let different weights matrices be associated with the endogenous and exogenous variables.

1.8 Spatial regression models

We talk of *spatial regression* when the process $X = (X_s, s \in S)$ is made up of a deterministic part $m(\cdot)$ representing large scale variations, drift, trend or mean of X, and ε a centered residuals process:

$$X_s = m(s) + \varepsilon_s, \qquad E(\varepsilon_s) = 0.$$

Depending on the context of the study and available exogenous information, there are many ways to model $m(\cdot)$, whether it be by regression (linear or otherwise), analysis of variance (qualitative exogenous variables), analysis of covariance (exogenous variables with quantitative and qualitative values) or with generalized linear models:

Response surface: $m(s) = \sum_{l=1}^{p} \beta_l f_l(s)$ belongs to a linear space of known functions f_l. If $\{f_l\}$ is a polynomial basis, the spanned space is invariant with respect to the coordinate origin. If $s = (x, y) \in \mathbb{R}^2$, a quadratic model in these coordinates is associated with the monomials $\{f_l\} = \{1, x, y, xy, x^2, y^2\}$:

$$m(x,y) = \mu + ax + by + cx^2 + dxy + ey^2.$$

Exogenous dependency: $m(s,z) = \sum_{l=1}^{p} \alpha_l z_s^{(l)}$ is expressed in terms of observable exogenous variables z_s.

Analysis of variance: if $s = (i, j) \in \{1,2,\dots,I\} \times \{1,2,\dots,J\}$, we consider an "additive" model: $m(i, j) = \mu + \alpha_i + \beta_j$, where $\sum_i \alpha_i = \sum_j \beta_j = 0$.

Analysis of covariance: $m(\cdot)$ is a combination of regressions and components of analysis of variance:

$$m(s) = \mu + \alpha_i + \beta j + \gamma z_s, \qquad \text{with } s = (i, j).$$

Cressie (48) suggested the following decomposition of the residuals ε_s:

$$X_s = m(s) + \varepsilon_s = m(s) + W_s + \eta_s + e_s. \tag{1.31}$$

W_s is a "smooth" component modeled by an intrinsic process whose range is in the order of c times ($c < 1$) the maximum distance between the observation sites; η_s is a micro-scale component independent of W_s with a range in the order of c^{-1} times the minimum distance between observation sites and e_s is a measurement error or nugget component independent of W and η.

Generally speaking, if X is observed at n sites $s_i \in S$, linear models with linear spatial regressions are written:

$$X_{s_i} = {}^t z_{s_i} \delta + \varepsilon_{s_i}, \qquad i = 1, \dots, n, \tag{1.32}$$

where z_{s_i} and $\delta \in \mathbb{R}^p$, z_{s_i} is a covariate (qualitative, quantitative, mixed) observed at s_i and $\varepsilon = (\varepsilon_{s_i}, i = 1, \dots, n)$ spatially correlated centered residuals. Denoting $X = {}^t(X_{s_1}, \dots, X_{s_n})$, $\varepsilon = {}^t(\varepsilon_{s_1}, \dots, \varepsilon_{s_n})$ and $Z = {}^t(z_{s_1}, \dots, z_{s_n})$ the $n \times p$ matrix of exogenous variables, (1.32) can be written in matrix form as:

$$X = Z\delta + \varepsilon,$$

with $E(\varepsilon) = 0$ and $Cov(\varepsilon) = \Sigma$.

The second step consists of modeling Σ using a covariance function, variogram or spatial AR model.

Example 1.11. Rainfall in the State of Parana (`parana` dataset from the `geoR` package in *R* (181))

These data give the average rainfall during May-June over a number of years at 143 meteorological stations in the State of Parana, Brazil. The amount of rainfall can be influenced by various exogenous factors, climatic or otherwise, e.g., orography, though taking time-averages helps to diminish their effect. Upon examining

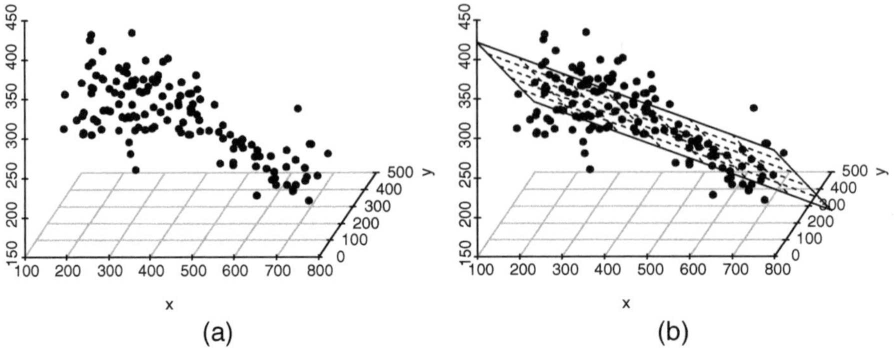

Fig. 1.9 (a) Rainfall data from 143 meteorological stations in the State of Parana, Brazil (`parana` dataset in `geoR`); (b) estimating a linear trend $m(s)$.

the cloud of points shown in Figure 1.9-a, we notice that the phenomena is not, on average, stationary and suggest that an affine model of the *response surface* $m(s) = \beta_0 + \beta_1 x + \beta_2 y$, $s = (x,y) \in \mathbb{R}^2$ is a reasonable choice. It then remains to suggest a covariance on $\mathbb{R}^2$ for the residuals that would allow us to quantify the covariance Σ of the 143 observations and validate the model in first and second-order (cf. §1.3.3).

Example 1.12. Modeling controlled experiments

The Mercer and Hall (156) dataset (cf. `mercer-hall` dataset on the website) gives the quantity of harvested wheat from an untreated field trial on a rectangular domain cut up into 20×25 parcels (i,j) of the same size $2.5\ m \times 3.3\ m$. A first glance at Fig. 1.10-a showing amounts of harvested wheat does not easily help us determine whether the mean $m(\cdot)$ is constant or not. To try to come to a decision, we can use the fact that the data is on a grid to draw boxplots by row and column and attempt to discover if there is a trend (or not) in either direction.

A graphical analysis (cf. Fig. 1.10-b and 1.10-c) suggests that there is no trend with respect to rows (i). We thus propose a model that only includes a column trend (j):

$$X_{i,j} = \beta_j + \varepsilon_{i,j}, \qquad i = 1,\dots,20,\ j = 1,\dots,25.$$

Example 1.13. Percentage of gross agricultural produce consumed locally

Cliff and Ord (45) analyzed spatial variability of the percentage X of gross agricultural output consumed where it was made in Ireland's 26 counties S (`eire` dataset in the `spdep` package). These percentages have been linked with an index z measuring the quality of road access of each county (cf. Fig. 1.11-a). The dispersion diagram (Fig. 1.11-b) shows that the linear model,

$$X_s = \beta_0 + \beta_1 z_s + \varepsilon_s, \qquad s \in S, \tag{1.33}$$

is a reasonable choice. A preliminary analysis of the residuals of (1.33) estimated by OLS shows that there is spatial correlation in the data. We model this using a

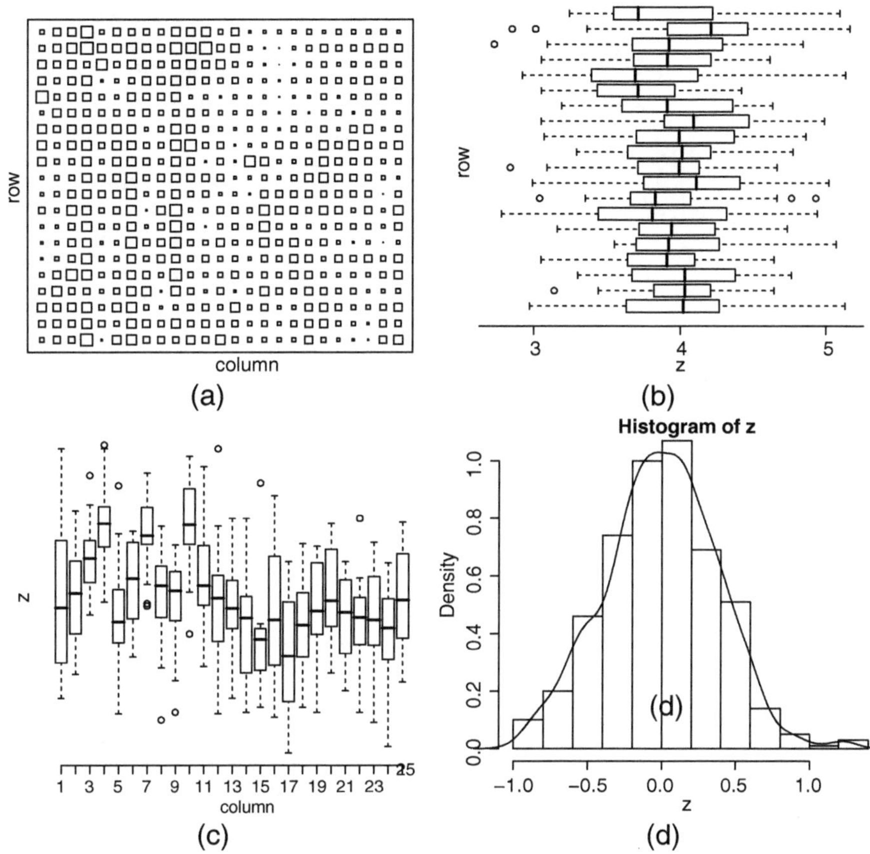

Fig. 1.10 (a) Mercer and Hall dataset: quantity of wheat harvested from a field divided into 20×25 parcels of the same size; the dimension of symbols is proportional to quantity harvested; (b) boxplots by row; (c) boxplots by column; (d) histogram of the data giving a nonparametric density estimation.

weights matrix $W = (w_{t,s})_{t,s \in S}$ with known weights representing the influence of s on t. Figure 1.11-d shows the influence graph associated with the symmetric binary specification and we choose $w_{t,s} = 1$ if s and t are neighbors, $w_{t,s} = 0$ otherwise.

A first choice of model is the spatial lag model (1.30):

$$X_s = \beta_0 + \beta_1 z_s + \gamma \sum_{t \in S} w_{s,t} X_t + \varepsilon_s.$$

A second possibility is to consider a regression with SAR residuals:

$$X_s = \beta_0 + \beta_1 z_s + \varepsilon_s, \qquad \varepsilon_s = \lambda \sum_{t \in S} w_{s,t} \varepsilon_t + \eta_s,$$

where η is a WWN. A model that generalized both choices is (cf. Ex. 1.21):

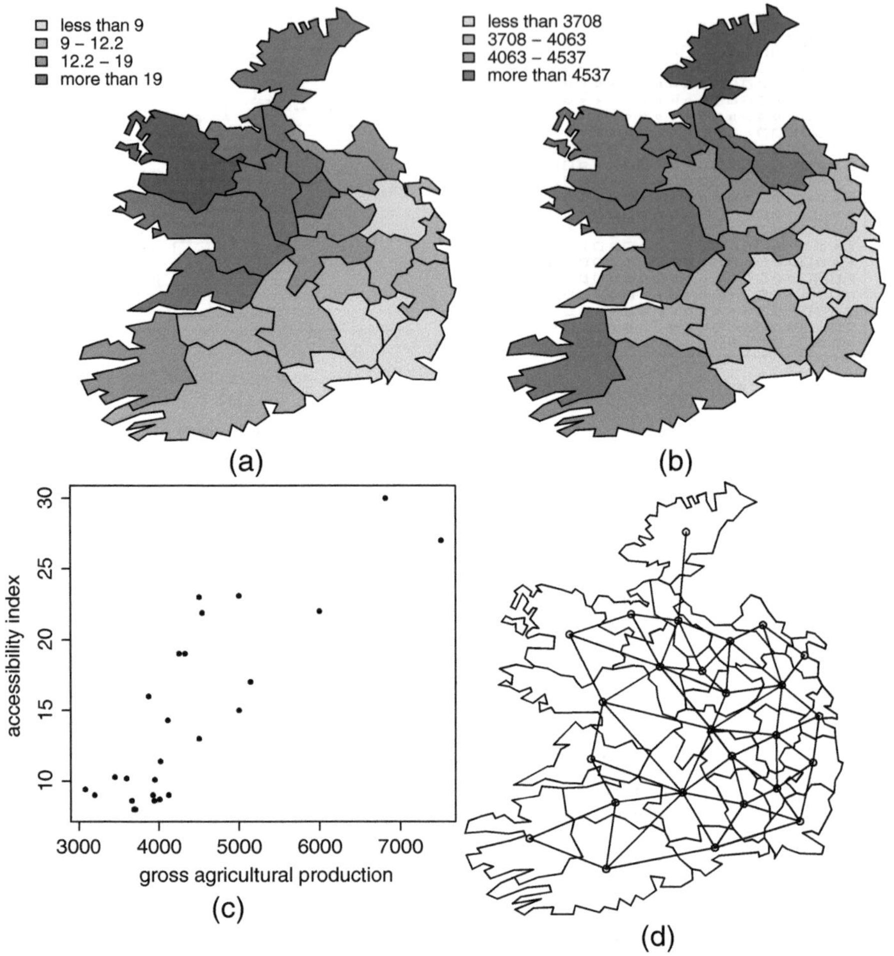

Fig. 1.11 (a) Percentage X of gross agricultural produce consumed in each of the 26 counties of Ireland where it was produced; (b) road access index Y; (c) diagram showing dispersion between X and Y; (d) influence graph associated with the binary specification.

$$X_s = \beta_0 + \beta_1 z_s + \gamma \sum_{t \in S} w_{s,t} X_t + \varepsilon_s,$$

where $\varepsilon_s = \lambda \sum_{t \in S} w_{s,t} \varepsilon_t + \eta_s$.

1.9 Prediction when the covariance is known

Our goal is to create a *prediction map* for X over all S when X is only observed at a finite number of points of S. *Kriging*, introduced by Matheron, deals with this prediction problem. It builds on the work of Krige (134), a South African mining engineer.

1.9.1 Simple kriging

Given n observations $\{X_{s_1},\dots,X_{s_n}\}$ in S, kriging aims to give a linear prediction of X_{s_0} at unobserved sites s_0. We suppose that the covariance (variogram) of X is known. If, as is the case in practice, it is not, it must be pre-estimated (cf. §5.1.3).

Denote: $X_0 = X_{s_0}$, $X = {}^t(X_{s_1},\dots,X_{s_n})$, $\Sigma = Cov(X)$, $\sigma_0^2 = Var(X_0)$ and $c = Cov(X,X_0)$, $c \in \mathbb{R}^n$ the second-order characteristics (known or estimated) of X and consider the linear predictor of X_0:

$$\widehat{X_0} = \sum_{i=1}^{n} \lambda_i X_{s_i} = {}^t\lambda X.$$

We keep the predictor that minimizes the mean of the square of errors $e_0 = X_0 - \widehat{X_0}$ (MSE),

$$MSE(s_0) = E\{(X_0 - \widehat{X_0})^2\}. \tag{1.34}$$

Simple kriging can be used when the mean $m(\cdot)$ of X is known. Without loss of generality, we suppose that X is centered.

Proposition 1.9. *Simple kriging: The linear predictor of X_0 minimizing (1.34) and the variance of the prediction error are, respectively:*

$$\widehat{X_0} = {}^tc\Sigma^{-1}X, \qquad \tau^2(s_0) = \sigma_0^2 - {}^tc\Sigma^{-1}c. \tag{1.35}$$

$\widehat{X_0}$ is the Best Linear Unbiased Predictor (BLUP) of X_0, i.e., the one having the smallest mean square error.

Proof.

$$MSE(s_0) = \sigma_0^2 - 2\,{}^t\lambda c + {}^t\lambda\Sigma\lambda = \Psi(\lambda);$$

the minimum is located at some λ for which the partial derivatives of Ψ are zero. We find $\lambda = \Sigma^{-1}c$ and it can be easily shown that it is a minimum. Substituting, we obtain the variance of the error given in (1.35). □

Remarks

The optimal value of λ is none other than $c = \Sigma\lambda$, i.e.,

$$Cov(X_{s_i}, X_0 - \widehat{X_0}) = 0 \qquad \text{for } i = 1,\dots,n.$$

These equations can be interpreted as showing that $\widehat{X_0}$ is the orthogonal projection (with respect to the scalar product of the covariance of X) of X_0 onto the space spanned by the variables $X_{s_1},\dots,X_{s_n}$. The predictor is identical to X_0 whenever s_0 is one of the observation sites. If X is a Gaussian process, $\widehat{X}_0$ is exactly the conditional expectation $E(X_0 \mid X_{s_1},\dots,X_{s_n})$; the distribution of this predictor is Gaussian and the error $X_0 - \widehat{X}_0 \sim \mathcal{N}(0,\tau^2(s_0))$.

1.9.2 Universal kriging

More generally, suppose that $X = Z\delta + \varepsilon$ follows a spatial linear regression model (1.32). Given z_0 and the covariance Σ of the residual ε (but not the mean parameter δ), we want to make an unbiased linear prediction of X_0, i.e., satisfying ${}^t\lambda Z = {}^t z_0$, which minimizes the mean square error (1.34). If Σ is unknown, it first must be estimated (cf. §5.1.3).

Proposition 1.10. *Universal kriging: the best unbiased linear predictor of X_0 is*

$$\widehat{X}_0 = \{{}^t c\Sigma^{-1} + {}^t(z_0 - {}^t Z\Sigma^{-1}c)({}^t Z\Sigma^{-1}Z)^{-1}{}^t Z\Sigma^{-1}\}X. \tag{1.36}$$

The variance of the prediction error is

$$\tau^2(s_0) = \sigma_0^2 - {}^t c\Sigma^{-1}c + {}^t(z_0 - {}^t Z\Sigma^{-1}c)({}^t Z\Sigma^{-1}Z)^{-1}(z_0 - {}^t Z\Sigma^{-1}c). \tag{1.37}$$

Proof. The MSE of the predictor ${}^t\lambda X$ is:

$$MSE(s_0) = \sigma_0^2 - 2\,{}^t\lambda c + {}^t\lambda\Sigma\lambda.$$

We consider the quantity:

$$\phi(\lambda,\nu) = \sigma_0^2 - 2\,{}^t\lambda c + {}^t\lambda\Sigma\lambda - 2\nu({}^t\lambda Z - {}^t z_0),$$

where ν is a Lagrange multiplier. The minimum of ϕ is found where the partial derivatives of ϕ at λ and ν are zero. For λ, we find $\lambda = \Sigma^{-1}(c + Z\nu)$. To obtain ν, we substitute λ into the unbiased condition and find

$$\nu = ({}^t Z\Sigma^{-1}Z)^{-1}(z_0 - {}^t Z\Sigma^{-1}c),$$
$$\lambda = \Sigma^{-1}c + \Sigma^{-1}Z({}^t Z\Sigma^{-1}Z)^{-1}(z_0 - {}^t Z\Sigma^{-1}c).$$

By substitution into $MSE(s_0)$, we obtain (1.36) and (1.37). □

An interpretation of the universal kriging prediction (1.36) is as follows: we rewrite (1.36) as

$$\widehat{X}_0 = {}^t z_0\widehat{\delta} + c\Sigma^{-1}(X - Z\widehat{\delta}), \qquad \text{where}$$
$$\widehat{\delta} = ({}^t Z\Sigma^{-1}Z)^{-1}{}^t Z\Sigma^{-1}X.$$

We will see in Chapter 5 that $\widehat{\delta}$ is the (optimal) generalized least squares (GLS) estimator of δ (cf. §5.3.4). Universal kriging of X is thus the sum of the (optimal) estimation ${}^t z_0 \widehat{\delta}$ of $E(X_0) = {}^t z_0 \delta$ and the simple kriging $c\Sigma^{-1}(X - Z\widehat{\delta})$ with residuals $\widehat{\varepsilon} = (X - Z\widehat{\delta})$ estimated by GLS.

When $X_s = m + \varepsilon_s$ with unknown but constant m, we say we are performing *ordinary kriging*.

Kriging formulae can be written analogously in terms of variograms if ε is an intrinsic process (cf. Ex. 1.10); in effect, stationarity plays no part in obtaining results (1.36) and (1.37).

Kriging is an *exact interpolator* as $\widehat{X}_{s_0} = X_{s_0}$ if s_0 is an observation site: in effect, if $s_0 = s_i$ and if c is the ith column of Σ, then $\Sigma^{-1}c = {}^t e_i$ where e_i is the ith vector of the canonical basis of $\mathbb{R}^n$ and ${}^t Z\Sigma^{-1}c = {}^t z_0$.

Regularity of kriging surfaces

Regularity at the origin of covariance C (variogram 2γ) determines the regularity of the kriging surface $s \mapsto \widehat{X}_s$, especially at the data sites (cf. Fig. 1.12):

1. For the nugget effect model, if $s_0 \neq s_i$, then $\Sigma^{-1}c = 0$, $\widehat{\delta} = n^{-1}\sum_{i=1}^n X_{s_i}$ and the prediction is none other than the arithmetic mean of the (X_{s_i}) if $s_0 \neq s_i$, with peaks $\widehat{X}_{s_0} = X_{s_i}$ when $s_0 = s_i$. More generally, if $C(h)$ is discontinuous at 0, $s \mapsto \widehat{X}_s$ is discontinuous at the data sites.
2. If $C(h)$ is linear at the origin, $s \mapsto \widehat{X}_s$ is everywhere continuous but not differentiable at the data sites.
3. If $C(h)$ is parabolic at 0, $s \mapsto \widehat{X}_s$ is everywhere continuous and differentiable.

If in dimension $d = 1$ kriging is done with cubic covariance (1.8), the interpolation function is a cubic spline. In higher dimensions and for separable cubic covariances, predictions are piecewise cubic in each variable (43, p. 272). Laslett (140) gives empirical comparisons between spline functions and kriging predictions.

Example 1.14. Kriging the rainfall data for the State of Parana (continued).

After preliminary analyses, we estimate (cf. §5.3.4) the affine regression model $m(s) = \beta_0 + \beta_1 x + \beta_2 y$, $s = (x,y) \in \mathbb{R}^2$ using a Gaussian covariance and a nugget effect for the residuals. Figure 1.13-a shows the prediction map using universal kriging with this model and Figure 1.13-b shows its standard error.

1.9.3 Simulated experiments

Simulated experiments are procedures aiming to learn a program (metamodel) $y = f(x)$ that associates input $x \in S = [0,1]^d$ to output $y \in \mathbb{R}^q$ (132; 193). Here, the "spatial" dimension d of input x is generally ≥ 3. Existing spatial statistics methods

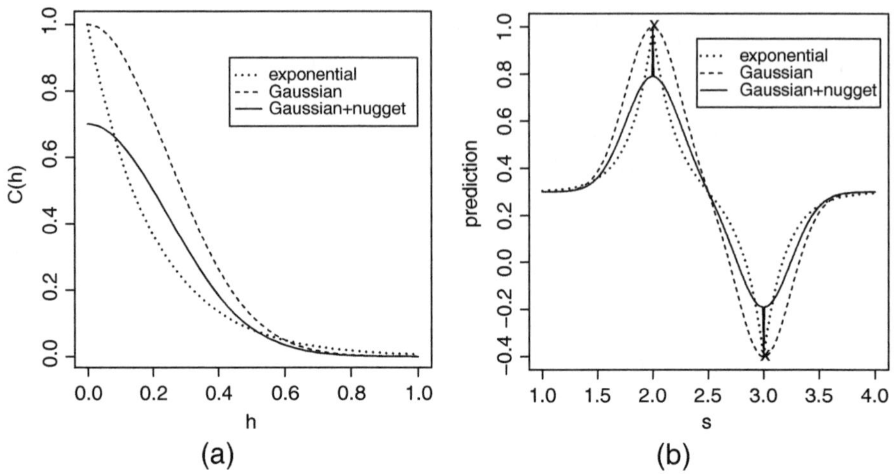

Fig. 1.12 Regularity (right) of the kriging surface (here, $S = \mathbb{R}, X_2 = 1, X_3 = -0.4$) as a function of regularity of the covariance at 0 (left); (i) exponential $C(h) = \exp(-|h|/0.2)$, (ii) Gaussian $C(h) = \exp(-0.6|h|^2/\sqrt{3})$, (iii) Gaussian + nugget effect $C(h) = \mathbf{1}_0(h) + 0.7\exp(-0.6\|h\|^2/\sqrt{3})(1 - \mathbf{1}_0(h))$.

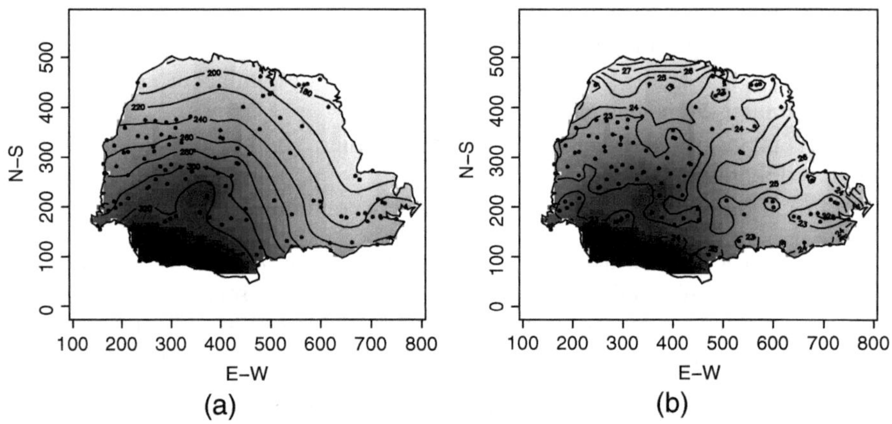

Fig. 1.13 Kriging of the rainfall data for the State of Parana showing in grayscale the prediction $\widehat{X}_s$ across the whole state: (a) contours associated with $\widehat{X}_s$; (b) contours associated with the standard deviation of the prediction error.

suppose that y is random and associated with x via a spatial model, for example a spatial regression

$$y = {}^t z(x)\beta + \varepsilon(x),$$

where $z(x)$ is known, $\beta \in \mathbb{R}^p$ unknown and ε is a stationary Gaussian process with covariance C. Usually, we choose a separable C to have quick algorithms. If we have observations (calculations) $x_i \mapsto y_i$ at points of some experimental design $\mathscr{O} = \{x_1, x_2, \dots, x_n\}$ of S, universal kriging $\widehat{y}_{\mathscr{O}} = \{\widehat{y}_{\mathscr{O}}(x), x \in S\}$ with covariance C gives a prediction of y over all of S. If the covariance depends on some unknown parameter θ, we must first pre-estimate θ and use universal kriging with $C_{\widehat{\theta}}$. For a fixed budget and given prediction criteria, our goal is to choose the optimal $\mathscr{O}$ that minimizes the criteria over all S.

Choosing a sampling scheme for selecting observation sites

Suppose that X is a random field with constant mean and that we want to choose an experimental design $\mathscr{O} = \{x_1, x_2, \dots, x_n\}$ of n points that minimizes the integral of the variance over a region of interest $A \subset \mathbb{R}^d$,

$$V(\mathscr{O}, A) = \int_A E[X_s - \widehat{X}_s(\mathscr{O})]^2 ds = \int_A \tau^2(s, \mathscr{O}) ds. \tag{1.38}$$

In this set-up, $\widehat{X}_s(\mathscr{O})$ and $\tau^2(s, \mathscr{O})$ are respectively the prediction and variance of the prediction using ordinary kriging. An approximation of (1.38) can be calculated by discretizing A at a finite subset $R \subset A$ of cardinality $M > m$. Minimizing (1.38) over R necessitates an exhaustive search over a set with $\binom{M}{m}$ elements. In practice, we use the following sequential algorithm:

1. Let $\mathscr{O}_{k-1} = \{x_1^*, \dots, x_{k-1}^*\}$ be the first $k-1$ chosen points and $R_k = R \backslash \mathscr{O}_{k-1}$. Choose $x_k^* = \arg\min_{x \in R_k} V(\mathscr{O}_{k-1} \cup s, A)$.
2. Repeat until $k = m$.

This V criteria can be generalized to non-stationary mean processes. Other criteria may also be considered.

Exercises

1.1. Effect of spatial correlations on the variance of empirical means.
Suppose $X = \{X_s, s \in \mathbb{Z}^d\}$ is a stationary random field on $\mathbb{Z}^d$ with mean m and covariance $C(h) = \sigma^2 \rho^{\|h\|_1}$, where $\|h\|_1 = \sum_{i=1}^d |h_i|$ and $|\rho| < 1$.

1. For $d = 1$ and $\overline{X} = \sum_{t=1}^9 X_t / 9$, show that:

$$\mathrm{V}_1(\rho) = \mathrm{Var}\{\overline{\mathrm{X}}\} = \frac{\sigma^2}{81}\{9 + 2\sum_{\mathrm{k}=1}^{8}(9-\mathrm{k})\rho^{\mathrm{k}}\}.$$

2. For $d = 2$ and $\overline{X} = \sum_{s=1}^3 \sum_{t=1}^3 X_{s,t} / 9$, show that:

$$\mathrm{V}_2(\rho) = \mathrm{Var}(\overline{\mathrm{X}}) = \frac{\sigma^2}{81}\{9 + 24\rho + 28\rho^2 + 16\rho^3 + 4\rho^4\}.$$

Compare $V_1(0)$, $V_1(\rho)$ and $V_2(\rho)$. Show that for $\rho = 1/2$, these three values are respectively proportional to 9, 15.76 and 30.25.

3. Denote $N = n^d$ and $V_d = Var(\overline{X}_N)$ where $\overline{X}_N = N^{-1}\sum_{t\in\{1,2,\dots,n\}^d} X_t$. Show that:

$$V_d(\rho) = Var(\overline{X}_N) = \frac{\sigma^2}{N^2}\left\{n\frac{1+\rho}{1-\rho} - \frac{2\rho}{1-\rho^2} + o_n(1)\right\}^d.$$

Compare the width of the confidence intervals for m under ρ-correlation and under independence.

1.2. Three methods of prediction for factorizing ARs on $\mathbb{Z}^2$.
Let X be a stationary centered Gaussian process on $\mathbb{Z}^2$ with covariance $C(h) = \sigma^2\rho^{\|h\|_1}$. Consider the following three predictions of X_0:

1. The predictor via $0 = E(X_0)$.
2. The optimal causal SAR predictor using $X_{1,0}$, $X_{0,1}$ and $X_{1,1}$.
3. The optimal CAR predictor using $\{X_t, t \neq 0\}$.

Give explicit representations of the last two predictors. Show that when $\rho = 1/2$, the variances of the prediction errors are respectively proportional to 1, 0.5625 and 0.36.

1.3. Krige's formula.
Let $X = \{X_t, t \in \mathbb{R}^d\}$ be a centered stationary process in L^2 with covariance C. For a bounded Borel set $V \in \mathscr{B}_b(\mathbb{R}^d)$, note:

$$X(V) = \frac{1}{\nu(V)}\int_V X(z)dz \text{ and } C(u,U) = \frac{1}{\nu(u)\nu(U)}\int_u\int_U C(y-z)dydz,$$

where $u, U \in \mathscr{B}_b(\mathbb{R}^d)$ have volumes $\nu(u) > 0$ and $\nu(U) > 0$.

1. The *extension variance* of X from v to V is defined as $\sigma_E^2(v,V) = Var(X(v) - X(V))$: i.e., it is the variance of the prediction error of $X(V)$ when using $\widehat{X}(V) = X(v)$. Show that $\sigma_E^2(v,V) = C(v,v) + C(V,V) - 2C(v,V)$.
2. Suppose that $D \subset \mathbb{R}^d$ is partitioned into I subdomains V_i and each V_i in turn is divided into J equally sized subdomains v_{ij} in such a way that we pass from partition V_i to V_j via a translation. We denote by v the shared generating form of the v_{ij} and V that of the V_i. Noting $X_{ij} = X(v_{ij})$ and $X_{i\cdot} = \frac{1}{J}\sum_{j=1}^J X_{ij}$, we define the *empirical dispersion variance* and *dispersion variance* of X for v in V, $v \subset V$, by:

$$s^2(v \mid V) = \frac{1}{J}\sum_{j=1}^J (X_{ij} - X_{i\cdot})^2 \text{ and } \sigma^2(v \mid V) = E\{s^2(v \mid V)\}).$$

   a. Show that $\sigma^2(v \mid V) = C(v,v) - C(V,V)$.
   b. Show that $\sigma^2(v \mid D) = \sigma^2(v \mid V) + \sigma^2(V \mid D)$.

1.4. A sufficient p.s.d. condition for matrices.

1. Show that C is p.s.d if C is diagonally dominant, i.e., if for all i: $C_{ii} \geq \sum_{j:j\neq i}|C_{ij}|$.

2. Investigate this condition for the two covariances:

   a. $C(i,j) = \rho^{|i-j|}$ on $\mathbb{Z}$;
   b. $C(i,j) = 1$ if $i = j$, ρ if $\|i-j\|_\infty = 1$ and 0 otherwise, on $\mathbb{Z}^d$.

   Show that the converse of 1 is not true.

1.5. Product covariance, restriction and extension of covariances.

1. Show that if $C_k(\cdot)$ are stationary covariances on $\mathbb{R}^1$, the function $C(h) = \prod_{k=1}^d C_k(h_k)$ is a covariance on $\mathbb{R}^d$.
2. Show that if C is a covariance on $\mathbb{R}^d$, then C is a covariance on any vectorial subspace of $\mathbb{R}^d$.
3. Consider the function $C_0(h) = (1 - |h|/\sqrt{2})\mathbf{1}\{|h| \leq \sqrt{2}\}$ on $\mathbb{R}$.

   a. Show that C_0 is a covariance on $\mathbb{R}$.
   b. For $(i,j) \in A = \{1,2,\ldots,7\}^2$, suppose $s_{ij} = (i,j)$ and $a_{ij} = (-1)^{i+j}$. Show that $\sum_{(i,j),(k,l)\in A} a_{ij}a_{kl}C_0(\|s_{ij} - s_{kl}\|) < 0$. Deduce that $C(h) = C_0(\|h\|)$ is not a covariance on $\mathbb{R}^2$.
   c. Show that $\sum_{u,v\in\{0,1\}^d}(-1)^{\|u\|_1+\|v\|_1}C_0(\|u-v\|) < 0$ when $d \geq 4$. Deduce that $C(h) = C_0(\|h\|)$ is not a covariance on $\mathbb{R}^d$ when $d \geq 4$.
   d. Show that $C(h) = C_0(\|h\|)$ is not a covariance on $\mathbb{R}^3$. Deduce that no isotropic extension of C_0 is a covariance on $\mathbb{R}^d$ if $d \geq 2$.

1.6. χ^2 random fields.

1. If (X,Y) is a pair of Gaussian variables, show that:

$$Cov(X^2,Y^2) = 2\{Cov(X,Y)\}^2.$$

   Hint: if (X,Y,Z,T) is a Gaussian vector in $\mathbb{R}^4$, then

$$E(XYZT) = E(XY)E(ZT) + E(XZ)E(YT) + E(XT)E(YZ).$$

2. Suppose that $X^1,\ldots,X^n$ are n centered i.i.d. Gaussian processes on $\mathbb{R}^d$, each with covariance C_X. Show that the random field Y defined by

$$Y = \{Y_s = \sum_{i=1}^n [X_s^i]^2,\ s \in \mathbb{R}^d\}$$

   is stationary with covariance $C_Y(h) = 2nC_X(h)^2$.

1.7. Markov property of exponential covariances.
Consider a stationary process X on $\mathbb{R}$ with covariance $C(t) = \sigma^2\rho^{|t|}$, with $|\rho| < 1$. X is observed at n sites $\{s_1 < s_2 < \ldots < s_n\}$.

1. Show that if $s_0 < s_1$, the kriging of X at s_0 is $\widehat{X}_{s_0} = \rho^{s_1 - s_0}X_{s_1}$.
2. Show that if $s_k < s_0 < s_{k+1}$, the kriging $\widehat{X}_{s_0}$ only depends on X_{s_k} and $X_{s_{k+1}}$ and give an explicit formulation of this kriging.

1.8. For a stationary process X on $\mathbb{R}$ with covariance C, we observe: $X_0 = -1$ and $X_1 = 1$. Show that simple kriging gives

$$\widehat{X}_s = \frac{C(s-1)-C(s)}{C(0)-C(1)}$$

and that the variance of the prediction error at s is

$$\tau^2(s) = C(0)\left(1 - \frac{(C(s)+C(s-1))^2}{C(0)^2 - C(1)^2}\right) + 2\frac{C(s)C(s-1)}{C(0)-C(1)}.$$

Draw the graphs of $s \mapsto \widehat{X}_s$ and $s \mapsto \tau^2(s)$ for $s \in [-3,3]$ when C is Matérn's covariance with parameters $C(0)=1$, $a=1$, $\nu = 1/2$ (resp. $\nu = 3/2$). Comment on these graphs.

1.9. Models with zero correlation for distances > 1.
Consider the stationary process $X = (X_t, t \in \mathbb{Z})$ with correlation ρ at a distance 1 and correlation 0 at distances > 1. Denote $X(n) = {}^t(X_1, X_2, \dots, X_n)$ and $\Sigma_n = Cov(X(n))$.

1. Under what condition on ρ is Σ_3 p.s.d.? Same question for Σ_4.
   Find the correlation function of the $MA(1)$ process: $X_t = \varepsilon_t + a\varepsilon_{t-1}$, where ε is a SWN. Deduce that the condition $|\rho| \leq 1/2$ ensures that for all $n \geq 1$, Σ_n is p.s.d.
2. Calculate the kriging of X_0 using X_1 and X_2. Same question when:
   a. $E(X_t) = m$ is unknown.
   b. $E(X_t) = at$.
3. Try to answer similar questions when $X = (X_{s,t}, (s,t) \in \mathbb{Z}^2)$ is a stationary random field with correlation ρ at (Euclidean) distance 1 and 0 at distances > 1.

1.10. If $X_s = m + \varepsilon_s$, where $\{\varepsilon_s, s \in S\}$ is an intrinsic process with variogram 2γ, give the kriging predictions (1.36) and kriging variances (1.37) as a function of γ.

1.11. Consider the process $X_s = cos(U + sV)$, $s \in \mathbb{R}$, where U is the uniform distribution $\mathscr{U}(0, 2\pi)$, V a Cauchy variable on $\mathbb{R}$ (with density $1/\pi(1+x^2)$) and U and V independent. Show that $E(X_s) = 0$ and $Cov(X_s, X_t) = 2^{-1}\exp\{-|s-t|\}$. Deduce that the trajectories of X are infinitely differentiable but that X is not L^2 differentiable.

1.12. "Boundary" CAR models.
Show that the equation $X_t = \frac{1}{d}\sum_{s:\|s-t\|_1=1} X_s + e_t$ defines a CAR model on $\mathbb{Z}^d$ if and only if $d \geq 3$.

1.13. Give explicit CAR representations of the following SAR models (graph, coefficients, variance ratio κ^2):

1. $X_{s,t} = aX_{s-1,t} + bX_{s,t-1} + cX_{s+1,t-1} + \varepsilon_{s,t}$, $(s,t) \in \mathbb{Z}^2$.
2. $X_{s,t} = a(X_{s-1,t} + X_{s+1,t}) + b(X_{s,t-1} + X_{s,t+1}) + c(X_{s-1,t-1} + X_{s+1,t+1}) + \varepsilon_{s,t}$, $(s,t) \in \mathbb{Z}^2$.

3. $X_t = aX_{t-1} + bX_{t+2} + \varepsilon_t$, $(s,t) \in \mathbb{Z}$.

1.14. Simulating factorizing SAR models on $\mathbb{Z}^2$.
Consider the factorizing centered Gaussian SAR model:

$$X_{s,t} = \alpha X_{s-1,t} + \beta X_{s,t-1} - \alpha\beta X_{s-1,t-1} + \varepsilon_{s,t}, \qquad 0 < \alpha, \beta < 1.$$

If $Var(\varepsilon) = \sigma_\varepsilon^2 = (1-\alpha^2)(1-\beta^2)$, the covariance of X is $C(s,t) = \alpha^{|s|}\beta^{|t|}$. We propose three ways to simulate X on the rectangle $S = \{0,1,2,\ldots,n-1\} \times \{0,1,2,\ldots,m-1\}$, of which the first and third are exact simulations:

1. Using the Cholesky decomposition of $\Sigma = \Sigma_1 \otimes \Sigma_2$, give this decomposition and the associated simulation algorithm.
2. Using Gibbs sampling with the associated CAR model (cf. §4.2; give the associated CAR model and the simulation algorithm).
3. Using a recursive formulation that exactly defines X on S: let W be the Brownian sheet on $(\mathbb{R}^+)^2$ (cf. (1.1)) and define variables

$$Z_{s,0} = \alpha^s \times W([0,\alpha^{-2s}] \times [0,1]) \qquad \text{for } s = 0,\ldots,n-1 \text{ and}$$
$$Z_{0,t} = \beta^t \times W([0,1] \times [0,\beta^{-2t}]) \qquad \text{for } t = 0,\ldots,m-1.$$

   Show that Z has the same covariance as X in the directions of the two axes. Deduce an exact simulation method for X on a subgrid of S.

1.15. The Yule-Walker equations.
Give the Yule-Walker equations for the stationary models with spectral density:

1. $f(u_1,u_2) = \sigma^2(1 - 2a\cos u_1 - 2b\cos u_2)$.
2. $g(u_1,u_2) = (1 + 2\cos u_1) f(u_1,u_2)$.

1.16. Identify in series form the variance of the isotropic 4-NN CAR model on $\mathbb{Z}^2$ with parameter a, $|a| < 1/4$. Determine the correlation at distance 1 of this model and draw the correlation graph $a \to \rho(a)$. Same question in dimension $d = 1$ with $|a| < 1/2$.

1.17. Behavior of CAR models at their parametric boundary.

1. Consider the isotropic 4-NN CAR model on $\mathbb{Z}^2$ with parameter $a = 1/4 - \varepsilon$, with $\varepsilon \downarrow 0$. What equation satisfies $\rho_\varepsilon = \rho_X(1,0)$? Show that $1 - \rho_\varepsilon \sim -(\pi/2)(\log\varepsilon)^{-1}$ when ε is close to 0. For what values of a do we get $\rho_\varepsilon = 0.9$, 0.95 and 0.99?
2. Attempt the same questions for the (isotropic) 2-NN model on $\mathbb{Z}^1$. Compare the behavior of $\rho(\varepsilon)$ when $d = 1$ and $d = 2$ for small ε.

1.18. The restriction of a Markov random field in $\mathbb{Z}^2$ to $\mathbb{Z}^1$.
Suppose X is the isotropic 4-NN CAR model on $\mathbb{Z}^2$. What is the spectral density of the one parameter process $\{X_{s,0},\ s \in \mathbb{Z}\}$? In $\mathbb{Z}$, is this model still a Markov random field?

1.19. Exchangeable Gaussian models on $S = \{1, 2, \ldots, n\}$.
Consider the following SAR model on $\mathbb{R}^n$:

$$X = \alpha JX + \varepsilon,$$

where ε is a Gaussian WN and J the $n \times n$ matrix with coefficients $J_{ij} = 1$ if $i \neq j$, $J_{ii} = 0$ otherwise. We say that the set $\{aI + bJ,\ a, b \in \mathbb{R}\}$ is stable under multiplication and inversion if $aI + bJ$ is invertible.

1. Under what condition on α is the model well-defined? Show that $Cov(X) = r_0 I + r_1 J$ and find r_0 and r_1.
2. After identification of β and $Cov(e)$, show that X can be written in the following CAR form:
$$X = \beta JX + e.$$

1.20. Suppose that X is a non-stationary Gaussian SAR model over the sites $S = \{1, 2, \ldots, n\}$. Give its CAR representation: graph, joint distribution, conditional distributions and prediction of X_i using the other observations.

1.21. Two SARX models with covariates.
Let Y be an $n \times p$ matrix of deterministic covariates that influence some spatial variable $X \in \mathbb{R}^n$, W a spatial contiguity matrix on $S = \{1, 2, \ldots, n\}$ and η a Gaussian WN with variance σ^2. Suppose we are interested in the following two models: first, the *spatial lag model* defined as the SAR model: $X = Y\beta + \alpha WX + \eta$ with exogenous Y; second, the *Durbin model*, defined as the SAR model on the residuals $X - Y\beta = \alpha W(X - Y\beta) + \eta$. If $I - \alpha W$ is invertible, calculate the distribution of X for each model. How do the results change if η is itself a SAR model, $\eta = \rho \Delta \eta + e$, where Δ is a known proximity matrix and e a Gaussian WN with variance σ_e^2? Calculate the log-likelihood of each model.

Chapter 2
Gibbs-Markov random fields on networks

Throughout this chapter, $X = (X_i,\, i \in S)$ will denote a random field defined on a *discrete set* of sites S with values in $\Omega = E^S$, where E is some *general state space*. The discrete set S may be finite or infinite, regular ($S \subseteq \mathbb{Z}^2$ in imaging and radiography) or irregular (regions in epidemiology). We denote by π either the distribution of X or its density. Instead of describing X using some of its global characteristics (e.g., mean or covariance), we are interested here in characterizing π using its conditional distributions, useful when the observed phenomenon is given in terms of its local conditional behavior. In particular, we attempt to answer the following question: given a family $\{\nu_i(\cdot|x^i),\, i \in S\}$ of distributions on E indexed by the configuration x^i of x outside of i, under what conditions do these distributions represent conditional distributions of a joint distribution π?

Being able to reply in some way to this question means being able to specify either wholly or partially the joint distribution π using its conditional distributions. Furthermore, if $\nu_i(\cdot|x^i)$ is only locally dependent, the complexity of the model is greatly reduced. In this case we say that X is a *Markov random field.*

Without additional hypotheses, conditional distributions $\{\nu_i\}$ are not generally compatible in this way. In this chapter, we will begin by describing a general family of conditional distributions called *Gibbs specifications* that are compatible without further conditions; Gibbs specifications are characterized by potentials. Their importance is enhanced by the Hammersley-Clifford theorem showing that Markov random fields are Gibbs random fields with local potentials. Besag's *auto-models* are a particularly simple subclass of Markov random fields that are useful in spatial statistics.

Unlike second-order models for which $E = \mathbb{R}$ or $\mathbb{R}^p$ (cf. Ch. 1), the state space E of a Gibbs-Markov random field can be general, quantitative, qualitative or mixed. For example, $E = \mathbb{R}^+$ for a positive-valued random field (exponential or Gamma random field), $E = \mathbb{N}$ for count variables (Poisson random fields in epidemiology), $E = \{a_0, a_1, \ldots, a_{m-1}\}$ for categorical random fields (spatial pattern of m plant types in ecology), $E = \{0, 1\}$ for binary random fields (presence/absence of an illness or species), $E = \Lambda \times \mathbb{R}^p$ for random fields combining categorical labels ν with quantitative multivariate values x (in remote sensing, ν represents a landscape texture

C. Gaetan, X. Guyon, *Spatial Statistics and Modeling,* Springer Series in Statistics,
DOI 10.1007/978-0-387-92257-7_2, © Springer Science+Business Media, LLC 2010

and x a multispectral signature), $E = \{0\} \cup]0, +\infty[$ for mixed states (in pluviometry, $X = 0$ if it does not rain, $X > 0$ otherwise), or even, as before, $E = \mathbb{R}$ or $E = \mathbb{R}^p$ for Gaussian random fields. We suppose that measurable state spaces $(E, \mathscr{E})$ are associated with some reference measure $\lambda > 0$: $\mathscr{E}$ the Borel σ-algebra and λ the Lebesgue measure when $E \subset \mathbb{R}^p$, or $\mathscr{E}$ the set of subsets of E and λ the counting measure when E is discrete. Definitions and results from this chapter can be easily extended to cases where the state space E_i may depend on $i \in S$.

2.1 Compatibility of conditional distributions

Without supplementary hypotheses, conditional distributions $\{\nu_i\}$ are generally not compatible. Exercise 2.3 uses a parametric dimension argument to show simply why this is the case. A different approach involves examining the condition of Arnold, Castillo and Sarabia (8) that guarantees two families of conditional distributions $(X|y)$ and $(Y|x)$ with states $x \in S(X)$ and $y \in S(Y)$ are compatible: let μ and ν be two reference measures on $S(X)$ and $S(Y)$, $a(x,y)$ the density of $(X = x|y)$ with respect to μ, $b(x,y)$ that of $(Y = y|x)$ with respect to ν, $N_a = \{(x,y) : a(x,y) > 0\}$ and $N_b = \{(x,y) : b(x,y) > 0\}$; then, the two families are compatible iff $N_a = N_b (= N)$ and if:

$$a(x,y)/b(x,y) = u(x)v(y), \qquad \forall (x,y) \in N, \tag{2.1}$$

with $\int_{S(X)} u(x)\mu(dx) < \infty$.

This factorization condition is necessary because the conditional densities are written $a(x,y) = f(x,y)/h(x)$ and $b(x,y) = f(x,y)/k(y)$, with f, h and k the densities of (X,Y), X and Y. Conversely, it is sufficient to show that $f^*(x,y) = b(x,y)u(x)$ is, up to a multiplicative constant, a density whose conditional densities are a and b.

Hence, Gaussian distributions $(X|y) \sim \mathscr{N}(a+by, \sigma^2 + \tau^2 y^2)$ and $(Y|x) \sim \mathscr{N}(c + dx, \sigma'^2 + \tau'^2 x^2)$, σ^2 and $\sigma'^2 > 0$ are not compatible if $\tau\tau' \neq 0$. If $\tau = \tau' = 0$, the distributions are compatible if $d\sigma^2 = b\sigma'^2$ as the joint distribution is Gaussian. We will come back to this example, which belongs to the class of *Gaussian auto-models* (cf. §2.4.2) and also to the CAR class studied in Chapter 1 (cf. §1.7.3). An example of compatible Gaussian conditional distributions that lead to compatible non-Gaussian joint distributions is examined in Exercise 2.4 (cf. (8)).

Another example of compatible conditional distributions is that of the auto-logistic distributions on $E = \{0,1\}$. These distributions,

$$\nu_i(x_i|x^i) = \frac{\exp x_i(\alpha_i + \sum_{j \neq i} \beta_{ij} x_j)}{1 + \exp(\alpha_i + \sum_j \beta_{ij} x_j)}, \qquad i \in S, x_i \in \{0,1\}$$

are compatible on $\Omega = \{0,1\}^S$ if $\beta_{ij} = \beta_{ji}$ for all $i \neq j$. In this case, the associated joint distribution is that of an *auto-logistic model* (cf. §2.4) with joint energy:

$$U(x) = C \exp\{\sum_{i \in S} \alpha_i x_i + \sum_{i,j \in S, i<j} \beta_{ij} x_i x_j\}.$$

Defining conditions that ensure compatibility of conditional distributions is not easy (8). To see why, set $x^* = (x_1^*, x_2^*, \ldots, x_n^*)$ as a reference state of E^S and suppose that π has conditional distributions $\nu = \{\nu_i\}$. Then (Brook's lemma (35)):

$$\frac{\pi(x_1,\ldots,x_n)}{\pi(x_1^*,\ldots,x_n^*)} = \prod_{i=0}^{n-1} \frac{\pi(x_1^*,\ldots,x_i^*,x_{i+1},x_{i+2},\ldots,x_n)}{\pi(x_1^*,\ldots,x_i^*,x_{i+1}^*,x_{i+2},\ldots,x_n)}$$
$$= \prod_{i=0}^{n-1} \frac{\nu_i(x_{i+1}|x_1^*,\ldots,x_i^*,x_{i+2},\ldots,x_n)}{\nu_i(x_{i+1}^*|x_1^*,\ldots,x_i^*,x_{i+2},\ldots,x_n)}.$$

This shows that if the distributions ν_i are compatible for π, then π can be reconstructed from the ν_i. However, this reconstruction has to be invariant, both by coordinate permutation $(1,2,\ldots,n)$ and with respect to the choice of reference state x^*: these invariances represent constraints on the $\{\nu_i\}$ ensuring coherency of the conditional distributions.

Let us now begin our study of Gibbs random fields, which are defined by conditional distributions without compatibility constraints.

2.2 Gibbs random fields on S

We suppose in this section that the set of sites S is countable, typically $S = \mathbb{Z}^d$ if S is regular. Denote $\mathscr{S} = \mathscr{P}_F(S)$ the family of finite subsets of S, $x_A = (x_i, i \in A)$ the configuration of x on A, $x^A = (x_i, i \notin A)$ the exterior of A, $x^i = x^{\{i\}}$ the exterior of site i, $\Omega_A = \{x_A, y \in \Omega\}$ and $\Omega^A = \{x^A, x \in \Omega\}$. Furthermore, let $\mathscr{F} = \mathscr{E}^{\otimes S}$ be the σ-field on Ω and dx_Λ the measure $\lambda^{\otimes \Lambda}(dx_\Lambda)$ on $(E^\Lambda, \mathscr{E}^{\otimes \Lambda})$ restricted to Ω_Λ.

2.2.1 Interaction potential and Gibbs specification

Gibbs random fields are associated with families π^Φ of conditional distributions defined with respect to interaction potentials Φ.

Definition 2.1. Potential, admissible energy and Gibbs specification

1. An interaction potential is a family $\Phi = \{\Phi_A, A \in \mathscr{S}\}$ of measurable mappings $\Phi_A : \Omega_A \longmapsto \mathbb{R}$ such that, for every subset $\Lambda \in \mathscr{S}$, the following sum exists:

$$U_\Lambda^\Phi(x) = \sum_{A \in \mathscr{S}: A \cap \Lambda \neq \emptyset} \Phi_A(x). \tag{2.2}$$

   U_Λ^Φ is the energy of Φ on Λ, Φ_A the potential on A.
2. Φ is admissible if for all $\Lambda \in \mathscr{S}$ and $y^\Lambda \in \Omega^\Lambda$,

$$Z_\Lambda^\Phi(x^\Lambda) = \int_{\Omega_\Lambda} \exp U_\Lambda^\Phi(x_\Lambda, x^\Lambda) dx_\Lambda < +\infty.$$

3. If Φ is admissible, the Gibbs specification π^{Φ} associated with Φ is the family $\pi^{\Phi} = \{\pi_{\Lambda}^{\Phi}(\cdot|x^{\Lambda}); \Lambda \in \mathscr{P}_F(S), x^{\Lambda} \in \Omega^{\Lambda}\}$ with conditional distributions $\pi_{\Lambda}^{\Phi}(\cdot|x^{\Lambda})$ of density with respect to $\lambda^{\otimes \Lambda}$:

$$\pi_{\Lambda}^{\Phi}(x_{\Lambda}|x^{\Lambda}) = \{Z_{\Lambda}^{\Phi}(x^{\Lambda})\}^{-1} \exp U_{\Lambda}^{\Phi}(x),\ \Lambda \in \mathscr{S}.$$

The family π^{Φ} is *well-defined* because if $\Lambda \subset \Lambda^*$, the conditional distribution π_{Λ} can be obtained by restricting the conditional distribution π_{Λ^*} to Λ. Summability (2.2) is guaranteed if (S,d) is a metric space and if the potential Φ has *finite range*, i.e., if there exists some $R > 0$ such that $\Phi_A \equiv 0$ as soon as the diameter of A, $\delta(A) = \sup_{i,j \in A} d(i,j)$ is larger than R. In this case only a finite number of potentials $\Phi_A \neq 0$ contribute to the energy U_{Λ}^{Φ}.

Conditional specifications π^{Φ} and Gibbs measures $\mathscr{G}(\Phi)$

A *Gibbs measure* associated with the potential Φ is a distribution μ on $(\Omega, \mathscr{F})$ whose conditional distributions coincide with π^{Φ}: we write $\mu \in \mathscr{G}(\Phi)$. If S is finite, $\pi(x) = \pi_S(x)$ and $\mathscr{G}(\Phi) = \{\pi\}$. If S is infinite the question arises as to whether there is existence and/or uniqueness of a global distribution with specification π^{Φ}. This is not necessarily the case without additional hypotheses and answering this question is one of the goals of statistical mechanics (cf. (85)). For a large class of spaces E (E Polish, for example a Borel set in $\mathbb{R}^d$, a compact set of a metric space or a finite set) and if the measure λ is finite, Dobrushin (66) showed that $\mathscr{G}(\Phi) \neq \emptyset$; thus there exists at least one global distribution with specification π^{Φ}. However, this distribution is not necessarily unique: if $\sharp\mathscr{G}(\Phi) > 1$, we say that there is a phase transition, namely two different distributions could represent the same conditional distributions. Thus, on an infinite network S, a Gibbs specification does not necessarily specify a joint distribution model on S. This situation shows one of the difficulties encountered in the study of statistical asymptotics of Gibbs random fields. Dobrushin (66) gives a sufficient condition ensuring uniqueness of the Gibbs measure associated with π^{Φ} (also cf. (85) and §B.2).

Identifiability of potential Φ

Without additional constraints, the mapping $\Phi \mapsto \pi^{\Phi}$ is not identifiable: for example, for any constant c, if a potential Φ_A is modified to $\widetilde{\Phi}_A = \Phi_A + c$, then $\pi^{\Phi} \equiv \pi^{\widetilde{\Phi}}$. In effect, increasing the energy by a constant c does not modify the conditional distribution π_{Λ}^{Φ}. One way to make this mapping identifiable is as follows: fix a reference state τ of E; then, the following constraints make Φ identifiable:

$$\forall A \neq \emptyset,\ \Phi_A(x) = 0 \qquad \text{if for some } i \in A,\ x_i = \tau. \tag{2.3}$$

This result follows from the Moëbius inversion formula:

$$\Phi_A(x_A) = \sum_{V \subseteq A} (-1)^{\sharp(A \setminus V)} U(x_A, \tau^A), \tag{2.4}$$

a formula that uniquely associates potentials satisfying (2.3) with the energy U. Following analysis of variance terminology we say that the Φ_A are the A-interactions in the decomposition of U; for example, for $S = \{1,2\}$, this decomposition is written $U - U(\tau,\tau) = \Phi_1 + \Phi_2 + \Phi_{12}$, where

$$\Phi_1(x) = U(x,\tau) - U(\tau,\tau), \qquad \Phi_2(y) = U(\tau,y) - U(\tau,\tau),$$
$$\Phi_{1,2}(x,y) = U(x,y) - U(x,\tau) - U(\tau,y) + U(\tau,\tau).$$

Φ_1 is the main effect of the first factor (the potential associated with $\{1\}$), Φ_2 that of the second factor (the potential associated with $\{2\}$) and $\Phi_{1,2}$ the second-order interaction effect.

2.2.2 Examples of Gibbs specifications

Modeling random fields via potentials $\Phi = \{\Phi_A, A \in \mathscr{C}\}$, where $\mathscr{C}$ is a family of finite subsets of S, needs specialist knowledge. The model's parameters are made up of the subsets $A \in \mathscr{C}$ that index the potential Φ and the potential functions Φ_A for $A \in \mathscr{C}$. If the Φ_A have parametric form $\Phi_A(x) = \theta_A \phi_A(x)$ where the ϕ_A are known real-valued functions, π belongs to the *exponential family*

$$\pi(x) = Z^{-1}(\theta) \exp {}^t\theta T(x),$$

with parameter $\theta = (\theta_A, A \in \mathscr{C})$ and sufficient statistic $T(x) = (\phi_A(x), A \in \mathscr{C})$. The explicit forms of the conditional distributions allow us to:

1. Use Markov Chain Monte Carlo algorithms such as Gibbs sampling and the Metropolis algorithm (cf. Ch. 4).
2. Use conditional pseudo-likelihood (CPL) estimation methods which are easy to implement and which retain good asymptotic properties in situations where maximum likelihood (ML) is more difficult to implement (cf. Ch. 5 and §5.4.2).

Ising models on $S \subset \mathbb{Z}^2$

Introduced by physicists to model spin configurations $E = \{-1,+1\}$ on networks, these binary-state models are also widely used in imaging and in statistics with state space $E^* = \{0,1\}$. They can be defined both on regular and irregular networks.

Isotropic 4-nearest neighbor model on $\mathbb{Z}^2$

The only non-zero potentials are the singleton potentials $\Phi_{\{i\}}(x) = \alpha x_i$, $i \in S$ and the pair potentials $\Phi_{\{i,j\}}(x) = \beta x_i x_j$ when i and j are neighbors at a distance of 1,

$\|i-j\|_1 = |i_1 - j_1| + |i_2 - j_2| = 1$. Denoting this relationship by $\langle i,j \rangle$, the conditional specification on finite Λ has energy:

$$U_\Lambda(x_\Lambda|x^\Lambda) = \alpha \sum_{i\in\Lambda} x_i + \beta \sum_{i\in\Lambda,\, j\in S:\, \langle i,j\rangle} x_i x_j.$$

The conditional distribution π_Λ^Φ is given by:

$$\pi_\Lambda(x_\Lambda|x^\Lambda) = Z_\Lambda^{-1}(\alpha,\beta;x^\Lambda)\exp U_\Lambda(x_\Lambda;x^\Lambda),$$

where

$$Z(\alpha,\beta;x^\Lambda) = \sum_{x_\Lambda\in E^\Lambda} \exp U_\Lambda(x_\Lambda;x^\Lambda).$$

The parameters can be interpreted in the following way: α controls the marginal distribution and β the spatial correlation. If $\alpha = 0$, the marginal configurations $\{X_i = +1\}$ and $\{X_i = -1\}$ are equiprobable whereas $\alpha > 0$ increases the probability of $+1$ appearing and $\alpha < 0$ the probability of -1. β is a parameter related to spatial dependency: $\beta = 0$ corresponds to independent X_i, $\beta > 0$ encourages neighbors to be equal and $\beta < 0$ encourages opposite-valued neighbors. For example, generating a model with parameters $\alpha = 0$ and a fairly large positive value of β leads to geometrically regular groupings of $+1$s and -1s, increasingly regular as the value of β increases. On the other hand, models generated with large negative values of β lead to an alternating chequered pattern of $+1$s and -1s, more and more so as $|\beta|$ increases.

The normalization constant $Z_\Lambda(\alpha,\beta;x^\Lambda)$, a sum of $2^{\sharp\Lambda}$ terms, is impossible to calculate when Λ is large. This poses a problem both for simulating Y on Λ and model estimation by maximum likelihood because the constant Z_Λ depends on the parameter (α,β) being estimated. For example, if Λ is the (small) 10×10 grid, then there are $2^{100} \simeq 1.27\times 10^{30}$ terms contributing to Z_Λ! Nevertheless, the conditional distribution at a site i, depending on the configuration of the 4-NN, is easy to calculate:

$$\pi_i(x_i|x^i) = \pi(x_i|x^i) = \frac{\exp x_i(\alpha+\beta v_i(x))}{2ch(\alpha+\beta v_i(x))},$$

where $v_i(x) = \sum_{j:\langle i,j\rangle} x_j$. It is simple to show that the parametrization of $\pi(\cdot|\cdot)$ by (α,β) is well-defined, that is, two different sets of parameters lead to two different families $\pi(\cdot|\cdot)$. As the state space is finite, $\mathscr{G}(\Phi) \neq \emptyset$. Thus, there always exists a global distribution with specification $\pi(\Phi)$. However, this distribution is not always unique: for example, with $\alpha = 0$ there is uniqueness only if $\beta < \beta_c = \log(1+\sqrt{2})/4$ (Onsager, (166); (85)). When $\alpha = 0$ and $\beta > \beta_c$, there are several distributions with the same conditional specifications. There is thus a phase transition.

Ising model on finite subsets $S \subset \mathbb{Z}^2$

For a given potential, a global model π always exists and is unique. The model on the *torus* $S = T^2 = \{1,2,\ldots,m\}^2$ can be obtained by considering the neighbor relation as being defined modulo m, i.e., $\langle i,j\rangle \iff |i_1 - j_1| + |i_2 - j_2| \equiv 1$

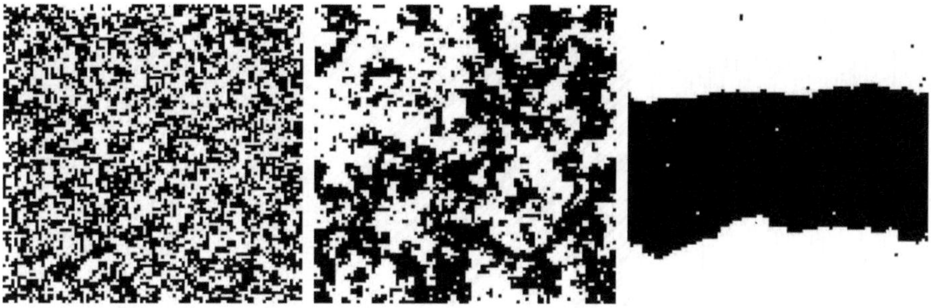

Fig. 2.1 Simulation of an isotropic 4-NN Ising model on $\{1,\ldots,100\}^2$ shown after 3000 iterations using Gibbs sampling: $\alpha = 0$ and (a) $\beta = 0.2$, (b) $\beta = 0.4$, (c) $\beta = 0.8$. Simulations were performed using the program *AntsInFields*.

(modulo m). Figure 2.1 gives three simulation results using an isotropic 4-NN Ising model on the 100×100 torus: $\alpha = 0$ ensures that the two states $+1$ and -1 are equiprobable. Parameters in Figs 2.1-b and -c correspond, for infinite networks, to phase transitions. These simulations were performed using the program *AntsInFields* (cf. Winkler (224) and `www.antsinfields.de`).

The binary space $E^* = \{0,1\}$ corresponds to *absence* ($x_i = 0$) and *presence* ($x_i = 1$) of a species (or illness) at site i. The mapping $y_i \mapsto 2x_i - 1$ from E^* to $E = \{-1,+1\}$ and the mapping $a = 2\alpha - 8\beta$ and $b = 4\beta$ associate with the Ising model with $y \in E$ the model with $x \in E^*$:

$$\pi_\Lambda(x_\Lambda|x^\Lambda) = Z^{-1}(a,b,x^\Lambda)\exp\{a \sum_{i\in\Lambda x_i} + b \sum_{i\in\Lambda,\, j\in S:\,\langle i,j\rangle} x_i x_j\}.$$

States 0 and 1 are equiprobable when $a + 2b = 0$, whereas $a + 2b > 0$ favors state 1 and $a + 2b < 0$ state 0. b can be interpreted in the same way as β in Ising models, $b > 0$ corresponds to "cooperation", $b < 0$ to "competition" and $b = 0$ to spatial independence.

Generalizations of the Ising model

Generalizations of the Ising model cover a large number of real-world situations.

1. We can introduce *anisotropy* by choosing a parameter β_H for horizontal neighbors and β_V for vertical ones.
2. *Non-stationary* models; these have potentials:

$$\Phi_{\{i\}}(x) = \alpha_i x_i, \qquad \Phi_{\{i,j\}}(x) = \beta_{\{i,j\}} x_i x_j,$$

   with α_i and $\beta_{\{i,j\}}$ depending on i and (i,j) and/or being defined with respect to known weights and/or covariates. An example on $\mathbb{Z}^2$ is the log-linear model with potentials

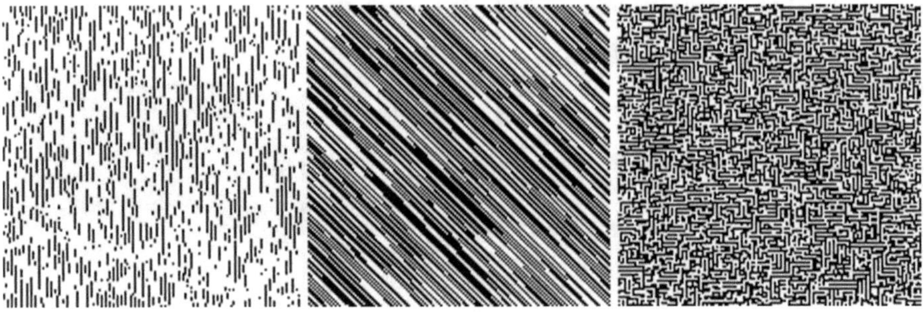

0.0	1.0	0.0		3.0	1.0	0.0		−0.4	0.4	−0.4
−1.0	2.0	−1.0		−1.0	0.0	−1.0		0.4	0.0	0.4
0.0	1.0	0.0		0.0	1.0	3.0		−0.4	0.4	−0.4

Fig. 2.2 Three 8-NN binary textures on $\{1,\ldots,150\}^2$. Simulations were performed with 3000 iterations of Gibbs sampling. Below each pattern we show the local model relative to the central pixel and given parameters. Simulations were performed using the *AntsInFields* program.

$$\Phi_{\{i\}}(x) = \alpha + \gamma_1 i_1 + \gamma_2 i_2 + {}^t\delta z_i, \qquad \Phi_{\{i,j\}}(x) = \beta x_i x_j,$$

where $i = (i_1, i_2)$.

3. The *neighbor relation* $\langle i,j \rangle$ can be associated with a general symmetric graph $\mathcal{G}$ without loops: for example, on $\mathbb{Z}^2$, an 8-NN graph can be defined with the neighbor relation: $\langle i,j \rangle$ if $\|i-j\|_\infty \le 1$, where $\|(u,v)\|_\infty = \max\{|u|,|v|\}$. Fig. 2.2 gives results of simulations of three 8-NN binary textures for various local parameters.
4. Potentials involving more than pairs can also be considered: for example, for an 8-NN Gibbs random field, it is possible to introduce potentials over triplets $\{i,j,k\}$ or quadruplets $\{i,j,k,l\}$ for sets of neighboring sites.
5. The number of states can be increased, either qualitatively (Potts model) or quantitatively (grayscale for imaging).
6. Lastly, these types of models can be defined over general networks S without regularity conditions as long as S is associated with a graph $\mathcal{G}$ defining a neighbor relation.

Other examples on real-valued textures (estimation and simulation) are given in Chapter 5 (cf. Fig. 5.13). These show the significant variety of configurations that these models are able to generate and their usefulness in representing real-valued patterns (cf. (224)).

Potts model

This model, also known as a Strauss model (207) generalizes the binary Ising model to situations with $K \ge 2$ qualitative states $E = \{a_0, a_1, \ldots, a_{K-1}\}$ (colors, textures, features). Its potentials are:

$$\Phi_{\{i\}}(x) = \alpha_k, \qquad \text{if } x_i = a_k,$$
$$\Phi_{\{i,j\}}(x) = \beta_{k,l} = \beta_{l,k}, \text{ if } \{x_i, x_j\} = \{a_k, a_l\} \text{ for neighbors } i \text{ and } j.$$

These models are particularly useful as prior models in Bayesian image segmentation. If S is finite, the energy associated with Φ is

$$U(x) = \sum_k \alpha_k n_k + \sum_{k<l} \beta_{kl} n_{kl}, \tag{2.5}$$

where n_k is the number of sites with state a_k and n_{kl} the number of neighboring sites with states $\{a_k, a_l\}$. For these types of models, the larger the value of α_k, the greater the marginal probability of state a_k; β_{kl} in turn controls the likelihood of configuration $\{a_k, a_l\}$ at neighboring sites. For example, to forbid the configuration $\{a_k, a_l\}$ at neighboring sites, we choose a large $-\beta_{kl} > 0$; if we want the marginal probability of a_k to be bigger than that of a_l, we set $\alpha_k > \alpha_l$.

As they stand, the parameters (α, β) are not identifiable. By taking $\tau = a_0$ as a reference state, constraints: $\alpha_0 = 0$ and for all k, $\beta_{0,k} = 0$, the potential Φ is identifiable. If the K states are exchangeable, i.e., for all $k \neq l$, $\alpha_k \equiv \alpha$ and $\beta_{kl} \equiv \beta$, the model depends only on the interaction parameter β:

$$\pi(x) = Z^{-1} \exp\{-\beta n(x)\}, \tag{2.6}$$

where $n(x)$ is the number of pairs of neighboring sites with the same state. In effect, $\sum_k \alpha_k n_k \equiv \alpha n$ is a constant independent of x and $\sum_{k<l} \beta_{kl} n_{kl} \equiv \beta(N - n(x))$, where N, the total number of edges in the neighbor graph, does not depend on x. In Bayesian image reconstruction, β plays the role of regularization parameter: the larger $-\beta$ is, the more the reconstructed regions with constant label are geometrically regular (cf. Fig. 2.3).

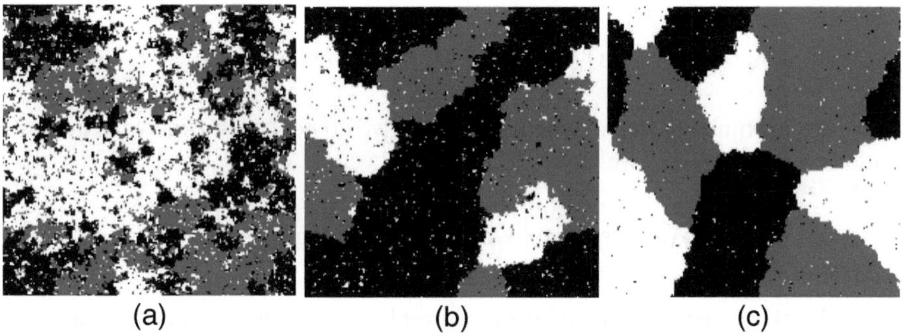

Fig. 2.3 3-level exchangeable grayscale Potts model on $\{1,\ldots,200\}^2$: (a) $-\beta = 0.5$, (b) $-\beta = 0.6$, (c) $-\beta = 0.7$. Simulations were performed using Gibbs sampling of 5000 iterations with *AntsInFields*.

Example 2.1. The role of regularization parameter β in Bayesian imaging.

A central goal in image processing or signal processing is to reconstruct an object x based on a noisy observation $y = \Phi(x,e)$. Bayesian methods involve prior modeling of x and then propose to reconstruct $\widehat{x}$ from its posterior distribution $\pi(\cdot|y)$ (cf. the pioneering article of Geman and Geman (82) and (96; 42; 224)). Several approaches are possible. The MAP or *maximum a posteriori* selects

$$\widehat{x} = \operatorname*{argmax}_{x\in\Omega} \pi(x|y),$$

where $\pi(\cdot|y)$ is the conditional distribution of X given y. The distribution of Y given x is known if the degradation process is well-defined. However, that of X, indispensable for calculating the posterior distribution $\pi(\cdot|y)$, is generally unknown. This is why we have to choose a *prior distribution* $\pi(x)$ for X. This step, partially *ad hoc* requires specialist knowledge of the problem being considered. Once π is chosen, the joint distribution (X,Y) and the conditional distribution of $(X|y)$ are given by

$$\pi(x,y) = \pi(y|x)\pi(x) \quad \text{and} \quad \pi(x|y) = \frac{\pi(y|x)\pi(x)}{\pi(y)},$$

and the MAP reconstruction of x is:

$$\widehat{x}_{MAP} = \operatorname*{argmax}_{x\in\Omega} \pi(y|x)\pi(x).$$

Figure 2.4 shows a simulated example of such a reconstruction: (a) is the binary 64×64 image x to be reconstructed; (b) is the observed image y of x degraded by an i.i.d. channel noise $P(Y_i = X_i) = 1 - P(Y_i \neq X_i) = p = 0.25$. The distribution y is thus

$$\pi(y|x) = c\exp\left\{n(y,x)\log\frac{1-p}{p}\right\},$$

where $n(y,x) = \sum_{i\in S}\mathbf{1}(y_i = x_i)$. We choose the Potts distribution (2.6) with two interchangeable states and parameter β as the prior distribution on x. We obtain:

$$\pi(x|y) = c(y)\exp\left\{-\beta n(x) + n(y,x)\log\frac{1-p}{p}\right\}.$$

As we are maximizing $\pi(x|y)$ with respect to x, we do not need to know the normalization constant $c(y)$, so we have:

$$\widehat{x}_{MAP} = \operatorname*{argmax}_{x\in\{0,1\}^S}\left\{-\beta n(x) + n(y,x)\log\frac{1-p}{p}\right\}.$$

To resolve this combinatorial optimization problem, (94) use an exact Ford-Fulkerson algorithm, though it is also possible to use simulated annealing (82; 96; 224). Images 2.4-c, 2.4-d and 2.4-e give respectively exact MAP reconstructions for increasing values $\beta = 0.3$, 0.7 and 1.1. Increasing the value of the regularization parameter

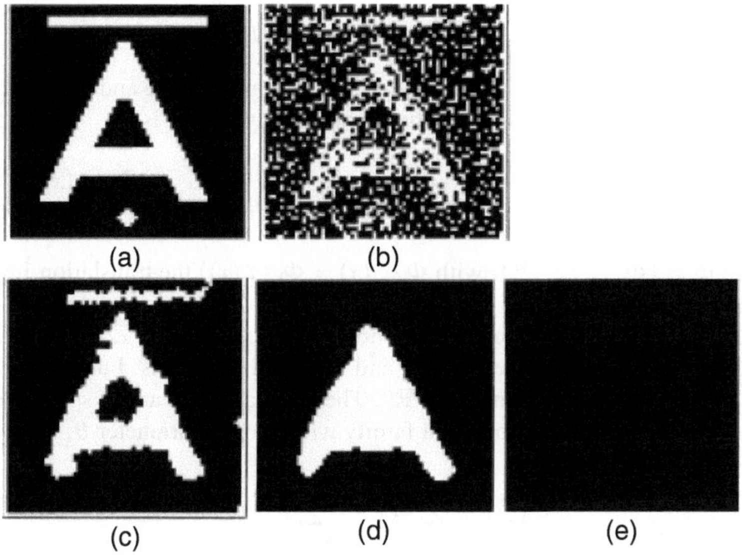

Fig. 2.4 Bayesian reconstruction of a noisy image: (a) original image (b) image with 25 % noise; exact MAP reconstructions with: (c) $\beta = 0.3$, (d) $\beta = 0.7$, (e) $\beta = 1.1$. Source: Greig, Porteous and Seheult (94), reproduced with permission of Blackwell Publishing.

β increases the regularity of the reconstructed zones of each state up to the point where, for $\beta = 1.1$, only the black zones remain!

Gaussian specification on $S = \{1,2,\ldots,n\}$

If $\Sigma^{-1} = Q$ exists, the Gaussian distribution $X = (X_i, i \in S) \sim \mathcal{N}_n(\mu, \Sigma)$ is a Gibbs random field with energy

$$U(x) = \frac{1}{2}\,{}^t(x-\mu)Q(x-\mu),$$

with singleton potentials $\Phi_{\{i\}}$ and pair potentials $\Phi_{\{i,j\}}$ given by

$$\Phi_{\{i\}}(x) = x_i \sum_{j:j\neq i} q_{ij}\mu_j - \frac{1}{2}q_{ii}x_i^2 \quad \text{and} \quad \Phi_{\{i,j\}}(x) = -q_{ij}x_i x_j \quad \text{if} \quad i \neq j.$$

It is easy to get the conditional distributions $\mathcal{L}_A(X_A|x^A)$ by fixing the conditional energy $U_A(\cdot|x^A)$ on A. X is a $\mathcal{G}$-Markov random field if, for all $i \neq j$: $q_{ij} \neq 0 \Longleftrightarrow \langle i,j \rangle$ is an edge in the graph $\mathcal{G}$ (cf. §2.3).

Translation-invariant potential

For $S = \mathbb{Z}^d$ and $\Omega = E^{\mathbb{Z}^d}$, an i-translation τ_i on Ω is defined as:

$$(\tau_i(x))_j = x_{i+j}, \qquad \forall j \in \mathbb{Z}^d.$$

Let V be a finite subset of $\mathbb{Z}^d$ and $\Phi_V : E^V \to \mathbb{R}$ a measurable and bounded mapping. The translation-invariant specification associated with Φ_V is

$$\pi_\Lambda^{\Phi_V}(x_\Lambda | x^\Lambda) = \{Z_\Lambda^{\Phi_V}(x^\Lambda)\}^{-1} \exp\{\sum_{i \in \mathbb{Z}^d : \{i+V\} \cap \Lambda \neq \emptyset} \Phi_V(\tau_i(x))\},$$

where $\Phi = \{\Phi_{V+i}, i \in \mathbb{Z}^d\}$, with $\Phi_{V+i}(x) = \Phi_V(\tau_i(x))$ the translation-invariant potential defining π.

Let V_k, $k = 1, \ldots, p$ be p finite non-empty subsets of $\mathbb{Z}^d$ and suppose $\phi_k : E^{V_k} \to \mathbb{R}$ are p known, measurable and bounded potentials. Furthermore, let $\theta = {}^t(\theta_1, \theta_2, \ldots, \theta_p)$ be a parameter in $\mathbb{R}^p$. The translation-invariant specification associated with (ϕ_k) is the exponential family with energy parameter θ,

$$U_\Lambda(x) = \sum_{k=1}^{p} \theta_k \{\sum_{i : \{i+V_k\} \cap \Lambda \neq \emptyset} \phi_k(\tau_i(x))\}. \tag{2.7}$$

For example, the translation-invariant $2d$-NN Ising model on $\mathbb{Z}^d$ is associated with the $p = d+1$ potentials $\phi_0(x) = x_0$ ($V_0 = \{0\}$), $\phi_k(x) = x_0 x_{e_k}$ ($V_k = \{0, e_k\}$), where e_k is the kth unitary vector of the canonical basis of $\mathbb{R}^d$, $k = 1, \ldots, d$. The parameter θ in this representation is identifiable. The isotropic sub-model has potentials ϕ_0 and $\phi^* = \sum_{k=1}^d \phi_k$.

Hierarchical model on product spaces $E = \Lambda \times \mathbb{R}^p$

Imaging and remote sensing are examples where product spaces are useful: $Y = (X, Z)$ where $X_i \in \Lambda$ is a texture label at i (forest, cultivated fields, wasteland, water, etc.) and $Z_i \in \mathbb{R}^p$ a quantitative multispectral measure. A hierarchical model can be obtained for example by modeling X using a Potts model, then, conditionally on X, modeling Z as a Gaussian multispectral grayscale texture (cf. Ex. 2.6).

2.3 Markov random fields and Gibbs random fields

Gibbs random fields are useful because they are able to represent simply coherent conditional specifications and also because they have the Markov random field property. We now give some relevant definitions.

2.3.1 Definitions: cliques, Markov random field

Suppose that $S = \{1, 2, \ldots, n\}$ has a symmetric neighbor graph $\mathcal{G}$ without loops. Two sites $i \neq j$ are neighbors if (i, j) is an edge of $\mathcal{G}$, noted $\langle i, j \rangle$; the neighborhood boundary of A is

$$\partial A = \{i \in S, i \notin A : \exists j \in A \text{ s.t. } \langle i, j\rangle\}.$$

We note $\partial i = \partial\{i\}$.

Definition 2.2. Markov random fields and graph cliques

1. X is a Markov random field on S for the graph $\mathscr{G}$ if, for each $A \subset S$ and $x^A \in \Omega^A$, the distribution of X on A conditional on x^A only depends on $x_{\partial A}$, the configuration of x on the neighborhood boundary of A, i.e.,

$$\pi_A(x_A|x^A) = \pi_A(x_A|x_{\partial A}).$$

2. A non-empty subset C of S is a clique of the graph $\mathscr{G}$ if C is a singleton or if all pairs of elements of C are neighbors in $\mathscr{G}$. The set of cliques of $\mathscr{G}$ is denoted $\mathscr{C}(\mathscr{G})$.

For example, for the 4-NN graph on $\mathbb{Z}^2$, cliques are either singletons $\{i\}$ or subsets $\{i, j\}$ with $\|i-j\|_1 = 1$. In the 8-NN graph, cliques also include triplets $\{i, j, k\}$ and quadruplets $\{i, j, k, l\}$ of sites u and v such that $\|u-v\|_\infty \leq 1$.

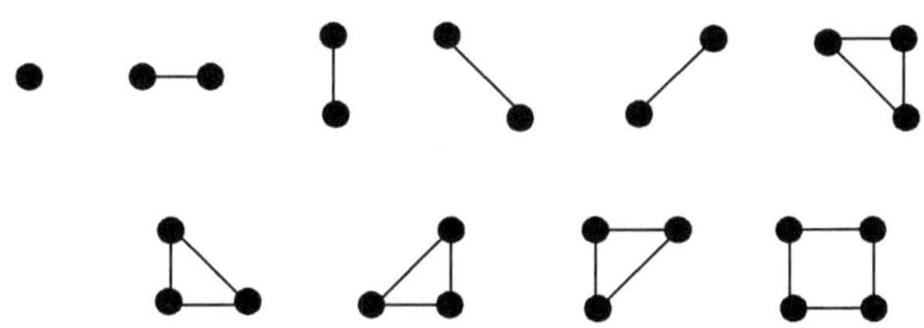

Fig. 2.5 The 10 types of clique for the 8-NN graph in $\mathbb{Z}^2$.

2.3.2 The Hammersley-Clifford theorem

To a family $\mathscr{C}$ of subsets of S containing all singletons we associate the neighbor graph $\mathscr{G}(\mathscr{C})$ defined as: sites $i \neq j$ are neighbors in $\mathscr{G}(\mathscr{C})$ if there exists a $C \in \mathscr{C}$ s.t. $\{i, j\} \subset C$. The following result relates Markov random fields to Gibbs random fields.

Theorem 2.1. *Hammersley-Clifford theorem (Besag, (25)).*

1. Let π be a Markov $\mathscr{G}$-random field on E satisfying:

$$\pi_A(x_A|x^A) > 0, \qquad \forall A \subset S \text{ and } x \in E^S. \tag{2.8}$$

Then there exists a potential $\Phi = \{\Phi_A, A \in \mathscr{C}\}$ defined on the set of cliques $\mathscr{C}$ of the graph $\mathscr{G}$ such that π is a Gibbs random field with potential Φ.

2. Conversely, let $\mathscr{C}$ be a family of subsets of S containing all singletons. Then a Gibbs random field with potentials $\Phi = \{\Phi_A, A \in \mathscr{C}\}$ is a Markov random field for the neighbor graph $\mathscr{G}(\mathscr{C})$.

Proof:

1. Denote 0 some reference state of E. We want to characterize the potentials Φ_A using the function $U(x) = \log\{\pi(x)/\pi(0)\}$ and then check that $\Phi_A \equiv 0$ whenever $A \notin \mathscr{C}$. The Moëbius formula (2.4) says that the energy $U(x)$ is:

$$U(x) = \sum_{A \subseteq S} \Phi_A(x), \quad \text{where } \Phi_A(x) = \sum_{B \subset A} (-1)^{\sharp(A \setminus B)} U(x_A, 0).$$

We show that $\Phi_A = 0$ when $A \supseteq \{i, j\}$ and $i \neq j$ are not neighbors. To begin, choose $\{i, j\} = \{1,2\}$, $A = \{1,2\} \cup C$ with $C \cap \{1,2\} = \varnothing$: Φ_A can again be written:

$$\begin{aligned}\Phi_A(x) = \sum_{B \subset C} (-1)^{\sharp(C \setminus B)} \{&U(x_1, x_2, x_C, 0) - U(x_1, 0, x_C, 0) \\ &- U(0, x_2, x_C, 0) + U(0, 0, x_C, 0)\}.\end{aligned}$$

As sites 1 and 2 are not neighbors,

$$\begin{aligned}U(x_1, x_2, x_C, 0) - U(x_1, 0, x_C, 0) &= \log \frac{\pi(x_1, x_2, x_C, 0)}{\pi(x_1, 0, x_C, 0)} = \log \frac{\pi(x_2 | x_1, x_C, 0)}{\pi(0 | x_1, x_C, 0)} \\ &= \log \frac{\pi(x_2 | x_C, 0)}{\pi(0 | x_C, 0)}.\end{aligned}$$

Then, as the resulting quantity does not depend on x_1, we have that $\Phi_A \equiv 0$. Thus, $\Phi_A \equiv 0$ if A is not a clique in $\mathscr{G}$.

2. The distribution π conditional on x^A is given by:

$$\pi_A(x_A | x^A) = Z_A^{-1}(x^A) \exp U_A(x),$$

where

$$U_A(x) = \sum_{\mathscr{C}: \mathscr{C} \cap A \neq \emptyset} \Phi_C(x),$$

$U_A(x)$ depending only on $\{x_i : i \in A \cup \partial A\}$ as π is a Markov $\mathscr{G}(\mathscr{C})$-random field. □

Denote $\pi_{\{i\}}(\cdot)$ the marginal distribution of π at i. Condition (2.8) of the theorem can be weakened to a condition known as "π positivity":

"if for $x = (x_i, i \in S)$, we have $\forall i \in S$: $\pi_{\{i\}}(x_i) > 0$, then $\pi(x) > 0$." (2.9)

Exercise 2.8 gives an example of a distribution π which does not satisfy this positivity condition.

2.4 Besag auto-models

These Markov models are characterized by conditional densities belonging to a certain exponential family (25). A little further on, we will present the result that this definition is based on.

2.4.1 Compatible conditional distributions and auto-models

Let X be a Markov random field on $S=\{1,2,\ldots,n\}$ with *pair potentials*:

$$\pi(x) = C\exp\{\sum_{i\in S}\Phi_i(x_i) + \sum_{\{i,j\}}\Phi_{ij}(x_i,x_j)\}. \tag{2.10}$$

For each $x\in E^n$, $\pi(x)>0$. Suppose furthermore that the identifiability constraints (2.3) are satisfied, denoting 0 the reference state of E. The following property allows us to find the potentials of X from its conditional distributions.

Theorem 2.2. *(Besag (25)) Suppose that each conditional distribution $\pi_i(\cdot|x^i)$ of π belongs to the exponential family:*

$$\log\pi_i(x_i|x^i) = A_i(x^i)B_i(x_i) + C_i(x_i) + D_i(x^i), \tag{2.11}$$

where $B_i(0)=C_i(0)=0$. Then:

1. *For any $i,j\in S$, $i\neq j$, there exists α_i and $\beta_{ij}=\beta_{ji}$ such that:*

$$A_i(x^i) = \alpha_i + \sum_{j\neq i}\beta_{ij}B_j(x_j), \tag{2.12}$$

$$\Phi_i(x_i) = \alpha_i B_i(x_i) + C_i(x_i), \qquad \Phi_{ij}(x_i,x_j) = \beta_{ij}B_i(x_i)B_j(x_j). \tag{2.13}$$

2. *Conversely, conditional distributions satisfying (2.11) and (2.12) are compatible for a joint distribution that is a Markov random field with potentials (2.13).*

Proof:

1. Denoting 0_i the state 0 at i, we have for the random field (2.10):

$$U(x) - U(0_i,x^i) = \Phi_i(x_i) + \sum_{j\neq i}\Phi_{ij}(x_i,x_j) = \log\frac{\pi_i(x_i|x^i)}{\pi_i(0_i|x^i)}.$$

Setting $x^i=\mathbf{0}$, i.e., the configuration 0 everywhere on $S\backslash\{i\}$, this equation gives $\Phi_i(x_i) = A_i(0)B_i(x_i)+C_i(x_i)$. Choose x so that $x^{\{i,j\}}=\mathbf{0}$ for $i=1$, $j=2$. A closer look at $U(x)-U(0_1,x^1)$ and $U(x)-U(0_2,x^2)$ gives:

$$\Phi_1(x_1)+\Phi_{12}(x_1,x_2) = A_1(0,x_2,0,\ldots,0)B_1(x_1),$$

$$\Phi_2(x_2) + \Phi_{12}(x_1,x_2) = A_2(x_1,0,\ldots;0)B_2(x_2),$$
$$\Phi_{12}(x_1,x_2) = [A_1(0,x_2,0) - A_1(0)]B_1(x_1) = [A_2(x_1,0) - A_2(0)]B_2(x_2),$$

and

$$A_2(x_1,0) - A_2(0) = \frac{A_1(0,x_2,0) - A_1(0)}{B_2(x_2)} B_1(x_1) \qquad \text{if } B_2(x_2) \neq 0.$$

We deduce that $B_2(x_2)^{-1}[A_1(0,x_2,0) - A_1(0)]$ is constant with respect to x_2 and equal to β_{21}. By permuting indices 1 and 2, we can analogously define β_{12} and show that $\beta_{12} = \beta_{21}$ and $\Phi_{12}(x_1,x_2) = \beta_{12}B_1(x_1)B_2(x_2)$. Equating the conditional distributions then gives (2.12) with $\alpha_i = A_i(0)$.

2. It suffices to show that the random field of potentials (2.13) has itself the conditional distributions associated with (2.11) and (2.12). We have that

$$\begin{aligned} U(x_i,x^i) - U(0_i,x^i) &= \Phi_i(x_i) + \sum \Phi_{ij}(x_i,x_j) \\ &= \alpha_i B_i(x_i) + C_i(x_i) + \sum \beta_{ij} B_i(x_i) B_j(x_j) \\ &= A_i(x^i)B_i(x_i) + C_i(x_i)\rangle = \log \frac{\pi_i(x_i|x^i)}{\pi_i(0_i|x^i)}. \end{aligned}$$

□

An important class of Markov random fields is that of Gibbs random fields with values in $E \subseteq \mathbb{R}$ and with potentials over at most pairs of points, with the pair potentials being quadratic.

Definition 2.3. Besag auto-models
X is an auto-model if X is real-valued and if its distribution π is given by:

$$\pi(x) = Z^{-1} \exp\{\sum_{i \in S} \Phi_i(x_i) + \sum_{\langle i;j \rangle} \beta_{ij} x_i x_j\}, \tag{2.14}$$

with $\beta_{ij} = \beta_{ji}$, $\forall i, j$.

We now give an important corollary of the previous theorem that allows us to define a joint model starting from conditional distributions whose compatibility is automatically assured.

Corollary 2.3 *Let $\nu = \{\nu_i(\cdot|x^i), i \in S\}$ be a family of real-valued conditional distributions satisfying (2.11) such that for each $i \in S$, $B_i(x_i) = x_i$. Then, these distributions are compatible as an auto-model with distribution π given by (2.14) if, whenever $i \neq j$, $\beta_{ij} = \beta_{ji}$.*

2.4.2 Examples of auto-models

Logistic auto-model: $E = \{0,1\}$

For each i, the conditional distribution $\pi_i(\cdot|x^i)$ is a *logit model* with parameter $\theta_i(x_i)$:

$$\theta_i(x_i) = \{\alpha_i + \sum_{j:\langle i,j\rangle} \beta_{ij} x_j\},$$

$$\pi_i(x_i|x^i) = \frac{\exp x_i\{\alpha_i + \sum_{j:\langle i,j\rangle} \beta_{ij} x_j\}}{1 + \exp\{\alpha_i + \sum_{j:\langle i,j\rangle} \beta_{ij} x_j\}}.$$

If, whenever $i \neq j$, $\beta_{ij} = \beta_{ij}$, these conditional distributions are compatible with the joint distribution π with energy U:

$$U(x) = \sum_i \alpha_i x_i + \sum_{\langle i,j\rangle} \beta_{ij} x_i x_j.$$

Binomial auto-model: $E = \{0, 1, 2, \ldots, N\}$

Let us now consider a family of conditional binomial distributions $\pi_i(\cdot|x_i) \sim \mathscr{B}in(N, \theta_i(x_i))$ for which the $\theta_i(x_i)$ satisfy:

$$A_i(x^i) = \log\{\theta_i(x_i)/(1-\theta_i(x_i))\} = \alpha_i + \sum_{j:\langle i,j\rangle} \beta_{ij} x_j.$$

Then, if for each $i \neq j$, $\beta_{ij} = \beta_{ji}$, these distributions are compatible with the joint distribution π with energy U:

$$U(x) = \sum_i (\alpha_i x_i + \log \tbinom{N}{x_i}) + \sum_{\langle i,j\rangle} \beta_{ij} x_i x_j.$$

The conditional binomial parameter is given by

$$\theta_i(x_i) = [1 + \exp -\{\alpha_i + \sum_{j\in S:\langle i,j\rangle} \beta_{ij} x_j\}]^{-1}.$$

In these two preliminary examples, as E is finite, U is always admissible. This is no longer the case in the two next examples.

Poisson auto-model: $E = \mathbb{N}$

Suppose the conditional distributions $\pi_i(\cdot|x^i)$ are Poisson $\mathscr{P}(\lambda_i(x^i))$, $i \in S$ with parameters satisfying log-linear models:

$$\log \lambda_i(x^i) = A_i(x^i) = \alpha_i + \sum_{j:\langle i,j\rangle} \beta_{ij} x_j.$$

If $\beta_{ij} = \beta_{ji} \leq 0$ when $i \neq j$, define the associated joint energy by

$$U(x) = \sum_i (\alpha_i x_i + \log(x_i!)) + \sum_{j:\langle i,j\rangle} \beta_{ij} x_i x_j.$$

U is admissible iff $\beta_{ij} \leq 0$ when $i \neq j$. In effect, in this case,

$$\exp U(x) \leq \prod_{i \in S} \frac{\exp(\alpha_i x_i)}{x_i!}$$

and $\sum_{\mathbb{N}^S} \exp U(x)$ converges. Otherwise, if for example $\beta_{1,2} > 0$, we have for x_1 and x_2 large enough,

$$\exp U(x_1, x_2, 0, \ldots, 0) = \frac{\exp\{\alpha_1 x_1 + \alpha_2 x_2 + \beta_{1,2} x_1 x_2\}}{x_1! x_2!} \geq \frac{\exp\{\beta_{1,2} x_1 x_2 / 2\}}{x_1! x_2!}.$$

As the lower bound is a divergent series, U is not admissible.

This admissibility condition $\beta_{ij} \leq 0$ can be seen as *competition* between neighboring sites. If we want to allow *cooperation* between sites, we can do one of the following:

1. Bound the state space E by some value $K < \infty$, $E = \{0, 1, , \ldots, K\}$, for example by considering the right-censored variables $Z_i = \inf\{X_i, K\}$.
2. Restrict the conditional Poisson distributions to $\{X \leq K\}$.

Then, as the state spaces are now finite, these models are always admissible and allow us to deal with cooperation as well as competition (127; 9).

Exponential auto-model: $E =]0, +\infty[$

These models have conditional exponential distributions:

$$\pi_i(x_i | x^i) \sim \mathscr{E}xp(\mu_i(x^i)), \qquad \mu_i(x^i) = \{\alpha_i + \sum_{j:\langle i,j \rangle} \beta_{ij} x_j\}.$$

If for all $i \neq j$, $\alpha_i > 0$ and $\beta_{ij} = \beta_{ji} \geq 0$, these distributions are compatible with a joint distribution with admissible energy:

$$U(x) = -\sum_i \alpha_i x_i - \sum_{\langle i,j \rangle} \beta_{ij} x_i x_j.$$

$\beta_{ij} \geq 0$ represents competition between sites i and j. We can allow cooperation between neighboring sites ($\beta_{ij} < 0$) either by truncating E to $[0, K]$ or by restricting the conditional exponential distributions to $X \leq K$.

Arnold, Castillo and Sarabia (8) have generalized these models to cliques of more than 2 points: the joint distribution

$$\pi(x) = Z^{-1} \exp -\{\sum_{A \in \mathscr{C}} \beta_A \prod_{l \in A} x_l\}$$

is admissible if, for each $i \in S$, $\beta_{\{i\}} > 0$ and if $\beta_A \geq 0$ whenever $\sharp A \geq 2$. It is easy to show that the conditional distributions at each site are exponential. (8) also generalize to potentials over more than pairs for conditional Gaussian, Gamma and Poisson distributions.

Gaussian auto-model: $E = \mathbb{R}$

Gaussian distributions $X \sim \mathcal{N}_n(\mu, \Sigma)$ on $S = \{1, 2, \ldots n\}$ with invertible covariance Σ and precision matrix $\Sigma^{-1} = Q$ are Gibbs energy models with singleton and pair potentials:

$$U(x) = -(1/2)^t(x-\mu)Q(x-\mu),$$

$$\Phi_i(x_i) = -\frac{1}{2}q_{ii}x_i^2 + \alpha_i x_i,$$

where $\alpha_i = \sum_j q_{ij}\mu_j$ and $\Phi_{ij}(x_i, x_j) = -q_{ij}x_i x_j$. This specification is admissible iff Q is p.d. The conditional distribution of $X_i|x^i$, with conditional energy $U_i(x_i|x^i) = \Phi_i(x_i) + \sum_{j \neq i} \Phi_{ij}(x_i, x_j)$ is normal with variance q_{ii}^{-1} and mean

$$\mu_i(x^i) = -q_{ii}^{-1} \sum_{j \neq i} q_{ij}(x_j - \mu_j).$$

Such models are also Gaussian CARs.

Auto-model with covariates

Without additional constraints, non-stationary models have parametric dimensions which are too large to be useful in practice. In each of the previous auto-models, we can reduce the dimension by modeling the parameter $\theta = \{(\alpha_i, \beta_{ij}),\ i \neq j, i, j \in S\}$ using the observable covariates $z = (z_i,\ i \in S)$ and/or the known weights $\{(a_i), (w_{ij})\}$, with w a symmetric matrix: for example, choosing $\beta_{ij} = \delta w_{ij}$ and $\alpha_i = \sum_{j=1}^{p} \gamma_j a_i z_{ij}$, where $z_i = {}^T(z_{i1}, \ldots, z_{ip}) \in \mathbb{R}^p$ is an observable covariate, leads to a model with $p+1$ parameters.

Mixed-state auto-model

Theorem (2.2) was generalized by (109) to exponential families (2.11) with multidimensional parameter $A_i(\cdot) \in \mathbb{R}^p$. In this case, the product $A_i(\cdot)B_i(x_i)$ is replaced by the scalar product $\langle A_i(\cdot), B_i(x_i)\rangle$. This generalization allows us to define mixed-state auto-models in the following way.

Suppose that X is a random variable taking values in the *mixed-state space* $E = \{0\} \cup]0, \infty[$ with mass p at 0 and density g_φ over $]0, \infty[$ if $X > 0$. For example, X could be the absolute speed of an object, equal to 0 if the object is at rest and > 0 otherwise; equally, X could be the amount of rainfall on a given day at a weather station, equal to 0 if it did not rain and > 0 otherwise. The parameter $(p, \varphi) \in \mathbb{R}^{1+k}$ is multidimensional whenever $\varphi \in \mathbb{R}^k$, $k \geq 1$. Denote δ_0 the Dirac delta function at 0 and suppose that g_φ belongs to the exponential family:

$$g_\varphi(x) = g_\varphi(0) \exp {}^t\varphi t(x).$$

A few simple calculations show that, relative to the reference measure $\lambda(dx) = \delta_0(dx) + \nu(dx)$ on the mixed-state space, X has probability density

$$f_\theta(x) = (1-p)g_\varphi(0)\exp {}^t\theta B(x),$$

where $B(x) = ({}^t\delta_0(x), t(x))$ and $\theta = -\log \dfrac{(1-p)g_\varphi(0)}{p}\varphi$.

To construct mixed-state auto-models, we thus work with conditional distributions $\pi_i(x_i|x_{\partial i})$ of density $f_\theta(\cdot|x_{\partial i})$ whose compatibility is guaranteed by the generalization of Theorem (2.2) to the multidimensional parameter case. These models are used by (31) in motion analysis, taking the absolute value of a Gaussian auto-model as the random field of conditional velocity, allowing positive probability for zero velocities.

Example 2.2. Cancer mortality in the Valence region (Spain)

Following Ferrándiz et al. (78), we present a model that allows us to analyze epidemiological cancer data (bladder, colon, prostate and stomach) in the Valence region of Spain. The goal is to see if the nitrate concentration in drinking water has an effect on cancer incidence, a relevant question as this is a region of intense agricultural activity using large quantities of fertilizer. The Valence region is made up of 263 districts, i being the main town of the district and X_i the aggregated number of deaths from a given type of cancer in that district over the years 1975-1980. Two covariates were kept: z_1, the percentage of the population over 40 years old and z_2, the concentration of nitrates in drinking water.

Usually, a statistical analysis of the $X = \{X_i, i \in S\}$ would use a log-linear Poisson regression incorporating the appropriate covariates $z_i = {}^T(z_{i1}, \ldots, z_{ip})$ (here, $p = 2$):

$$(X_i|x^i, z_i) \sim \mathscr{P}(\lambda_i),$$

where $\log(\lambda_i) = \alpha_i + \sum_{k=1}^p \beta_k z_{ik}$. This model assumes independence of counts, not necessarily true in the present example.

Poisson auto-models allow us to remove this hypothesis: the X_i are still characterized using covariates z but they are allowed to depend on the values of neighboring variables $x_{\partial i}$. Such models allow the detection of potential spatial dependency between the $\{X_i\}$ and/or detection of other covariates and risk factors, containing spatial dependencies, that may be missing in the model. The Poisson auto-model we select is the following:

$$(X_i|x^i, z_i) \sim \mathscr{P}(\lambda_i), \log(\lambda_i) = \alpha_i + \sum_{k=1}^p \beta_k z_{ik} + \sum_{j:\langle i,j\rangle} \gamma_{i,j} x_j. \qquad (2.15)$$

The parameters (α_i) represent disposition to disease at each specific site $i \in S$, (β_k) the influence of covariates (z_k) and $(\gamma_{i,j}, j \in \partial i, i \in S)$ the influence of the $x_{\partial i}$ at neighboring sites of X_i.

There are two possible interpretations of parameters γ:

1. They can measure a direct, real influence between neighboring variables, a natural interpretation when studying outbreaks (not the case in Ferrándiz et al. (78)).
2. They can measure other risk factors with spatial structure that are unaccounted for in the model: at site i, these hidden effects are supposed to be taken into

account inside the observable (auto)covariates $x_{\partial i}$. Testing whether $\gamma \neq 0$ means trying to discover if these factors are significant. In general, there is confusion between these two interpretations.

In order for the model to be admissible, the model parameters γ need to be ≤ 0, though positive values can be accepted on condition that the response variables x_i be truncated, which is reasonable here as x_i cannot be larger than district i's population.

Under (2.15), the model specification is useless because there are more parameters than observations. Moreover, we have to characterize the spatial neighbor relation between districts.

As for γ and neighbor relations, Ferrándiz et al. (78) suggest the following model: let u_i be the population of district i and d_{ij} the distance between the main centers of districts i and j. The neighbor relation $\langle i,j\rangle$ and parameters γ_{ij} are then deduced from the following proximity indices (a_{ij}):

$$\langle i,j\rangle \text{ if } a_{ij} = \frac{\sqrt{u_i u_j}}{d_{ij}} > a \text{ and } \gamma_{ij} = \gamma a_{ij}.$$

$a > 0$ is either a chosen constant or left as a model parameter.

The parameters α_i are modeled using a single parameter α:

$$\alpha_i = \alpha + \log(u_i).$$

The coefficient 1 in front of $\log(u_i)$ can be interpreted based on the supposition that the mean number of deaths λ_i is proportional to the size of the population u_i of district i. We end up with a $p+2$ parameter model (here 4 parameters), $\theta = (\alpha, (\beta_k), \gamma)$:

$$(X_i|x^i, z_i) \sim \mathscr{P}(\lambda_i), \qquad \log(\lambda_i) = \alpha + \log(u_i) + \sum_{k=1}^{p} \beta_k x_{ik} + \gamma \sum_{j:\langle i,j\rangle} a_{i,j} x_j. \quad (2.16)$$

If $\gamma = 0$, we have none other than a log-linear Poisson regression with intercept α, covariates x and individual effects $\log(u)$.

2.5 Markov random field dynamics

This section gives an illustration of the use of Gibbs random fields in modeling spatial dynamics. The models we present are semi-causal and well adapted to simulation and estimation: Markov chains (that may or may not be homogeneous) of (conditional) spatial Markov random fields (98). Another family is the STARMA (for Spatio-Temporal ARMA) linear models of Pfeifer and Deutsch (76; 174), easy to manipulate and often used in spatial econometrics (also cf. Cressie, (48, §6.8)).

2.5.1 Markov chain Markov random field dynamics

Let $X = \{X(t), t = 1,2,\ldots\}$, $X(t) = (X_i(t), i \in S)$ with $S = \{1,2,\ldots,n\}$ be a homogeneous Markov chain on $\Omega = E^S$. Denoting $y = x(t-1)$ and $x = x(t)$ two successive states of the chain, the transition $y \mapsto x$ can be written:

$$P(y,x) = Z^{-1}(y) \exp U(x|y),$$

where $Z(y) = \int_\Omega \exp U(z|y)dz < \infty$ if the conditional energy $U(\cdot|y)$ is almost surely admissible at y. If furthermore, conditional on y, $X(t)$ is a Markov random field on S, we can model $U(\cdot|y)$ with respect to potentials Φ_A and Φ_{BA}:

$$U(x|y) = \sum_{A\in\mathscr{C}} \Phi_A(x) + \sum_{B\in\mathscr{C}^-, A\in\mathscr{C}} \Phi_{B,A}(y,x), \tag{2.17}$$

where $\mathscr{C}$ and $\mathscr{C}^-$ are families of subsets of S that characterize two types of interactions associated with two graphs $\mathscr{G}$ and $\mathscr{G}^-$:

1. *Instantaneous interaction* potentials $\{\Phi_A, A \in \mathscr{C}\}$; $\mathscr{C}$ defines the *undirected* graph of instantaneous neighbors $\mathscr{G}(\mathscr{C})$.
2. *Temporal interaction* potentials $\{\Phi_{B,A}(y,x), B \in \mathscr{C}^-, A \in \mathscr{C}\}$; $\mathscr{C}^-$ defines a *directed graph* $\mathscr{G}^-$: $\langle j,i\rangle^-$ for $j \in B$ and $i \in A$ means that site j at time $(t-1)$ has an influence on site i at time t. In general, $\langle j,i\rangle^-$ does not imply $\langle i,j\rangle^-$.

Site $i \in S$ therefore has instantaneous neighbors $\partial i = \{j \in S : \langle i,j\rangle\}$ and neighbors from the past $\partial i^- = \{j \in S : \langle j,i\rangle^-\}$. Arrows representing dependencies are: $(j,t) \longleftrightarrow (i,t)$ if $\langle i,j\rangle$ and $(j,t-1) \longrightarrow (i,t)$ if $\langle j,i\rangle^-$. Model (2.17) is semi-causal, that is, a Markov chain with respect to time and a conditional Markov random field with respect to space. The distribution of $X_\Lambda(t)$, $\Lambda \subset S$, conditional on $(y = x(t-1), x^\Lambda = x^\Lambda(t))$ has energy:

$$U_\Lambda(x_\Lambda|y,x^\Lambda) = \sum_{A\in\mathscr{C}: A\cap\Lambda\neq\emptyset} \{\Phi^A(x) + \sum_{B\in\mathscr{C}^-} \Phi_{B,A}(y,x)\}.$$

Extending these models to inhomogeneous temporal contexts and/or cases with larger memory is not difficult.

2.5.2 Examples of dynamics

Auto-exponential dynamics

The conditional energy:

$$U(x|y) = -\sum_{i\in S}(\delta_i + \alpha_i(y))x_i - \sum_{\{i,j\}\in\mathscr{C}} \beta_{ij}(y)x_i x_j$$

defines an admissible auto-exponential dynamic if, for all $i, j \in S$, $\delta_i > 0$, $\alpha_i(\cdot) \geq 0$ and $\beta_{ij}(\cdot) = \beta_{ji}(\cdot) \geq 0$. It is possible to specify the functions $\alpha_i(\cdot)$ and $\beta_{ij}(\cdot)$, for example: $\alpha_i(y) = \sum_{j \in \partial i^-} \alpha_{ji} y_j$ and $\beta_{ij}(y) = \beta_{ij}$.

Auto-logistic dynamics

These dynamics, shown graphically in Figure 2.6-a have conditional transitions:

$$P(y,x) = Z^{-1}(y) \exp\{\sum_{i \in S} (\delta + \alpha \sum_{j \in S: \langle j,i \rangle^-} y_j) x_i + \beta \sum_{i,j \in S: \langle i,j \rangle} x_i x_j\}. \tag{2.18}$$

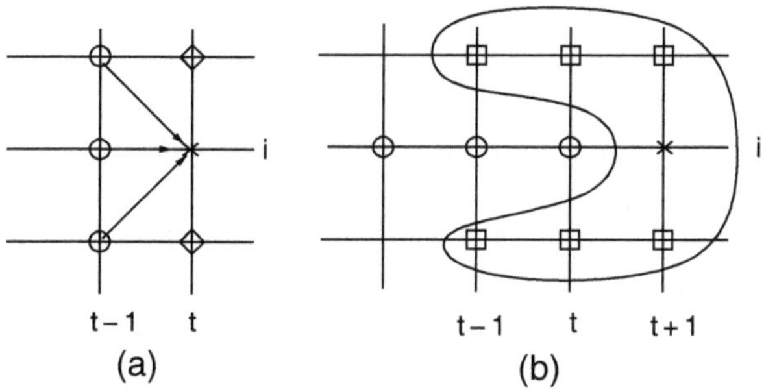

Fig. 2.6 (a) Graph showing auto-logistic dynamics: ○, neighbors of (i,t) from the past; ◇, present neighbors of (i,t). (b) Graph of contamination dynamics of rubber tree roots: □, neighbors of state $\times = (i,t+1)$, given state ○ $= (i,t)$ is healthy.

Example 2.3. Contamination dynamics of rubber tree roots

A simplified form of the model proposed by Chadoeuf et al. (41) to study contamination dynamics of rubber tree roots is the following: suppose that spatial localization is unidimensional, $i \in S = \{1,2,\ldots,n\} \subset \mathbb{Z}$ and that the binary-state process $\{Z_i(t),\, i \in S \text{ and } t = 0,1,2,\ldots\}$ satisfies:

1. $Z_i(t) = 0$ if the rubber tree root at (i,t) is healthy, otherwise $Z_i(t) = 1$.
2. Contamination is definitive, the root cannot be healed: if $Z_i(t) = 0$, then $Z_i(t') = 0$ for $t' < t$ and if $Z_i(t) = 1$, then $Z_i(t') = 1$ for $t' > t$.
3. Contamination is a Markov random field with respect to nearest neighbors, with temporal memory 2. Figure 2.6-b shows the set of neighboring sites (j,t') of site $(i,t+1)$.

Consider the vector $\tilde{X}_i(t) = (Z_i(t-2), Z_i(t-1), Z_i(t))$. We can associate $\tilde{X}_i(t)$ with $X_i(t)$, a Markov chain with 4 states $\{0,1,2,3\}$ defined by: (i) $X_i(t) = 0$ if $Z_i(t) = 0$; (ii) $X_i(t) = 1$ if $Z_i(t) = 1$ and $Z_i(t-1) = 0$; (iii) $X_i(t) = 2$ if $Z_i(t-1) = 1$ and $Z_i(t-2) = 0$; (iv) $X_i(t) = 3$ if $Z_i(t-2) = 1$.

It remains to model the transition $P(y,x)$ from $y = x(t)$ to $x = x(t+1)$. A possible choice of energy is:

$$U(y,x) = \sum_{i:\, y_i=0} \alpha(x_i) + \sum_{\langle i,j\rangle:\, y_i y_j=0} \beta(x_i,x_j),$$

with imposed identifiability constraints $\alpha(0)=0$ and, for all a,b, $\beta(a,0)=\beta(0,b)=0$.

In the definition of U, it is sufficient to limit the sum to singleton potentials $\alpha(y_i,x_i)$ at sites i for which $y_i(t)=y_i=0$ (we note $\alpha(x_i)=\alpha(0,x_i)$): in effect, if $y_i(t)\geq 1$, the tree at i is sick and will remain so. Similarly, the sum over pair potentials is limited to neighboring sites $\langle i,j\rangle$ such that $y_i(t)\times y_j(t)=0$: in effect, if this is not the case, the trees at i and j are sick at time t and will remain sick. The parametric dimension of this model is 12.

Exercises

2.1. Constraints on compatibility of conditional distributions.
Suppose $S=\{1,2,\ldots,n\}$, $E=\{0,1,2,\ldots,K-1\}$ and $\mathscr{F}=\{\nu_i(x_i|x^i),\, i\in S\}$ is an *unconstrained* family of conditional distributions with states in E.

1. What is the parametric dimension of $\mathscr{F}$? What number of constraints must we impose on $\mathscr{F}$ so that $\mathscr{F}$ is equivalent to the conditional distributions of a joint distribution π defined on E^S?
2. Consider the 2-NN kernel on $\mathbb{Z}$ for a process with states in $E=\{0,1,\ldots,K-1\}$:

$$Q(y|x,z) = P(X_0=y|X_{-1}=x, X_{+1}=z).$$

What is the parametric dimension of Q? Under what conditions are these conditional distributions those of a 2-NN Markov random field?

2.2. Compatibility of a bivariate distribution (8).

1. Using characterization (2.1), show that the two families of conditional densities,

$$f_{X|Y}(x|y) = (y+2)e^{-(y+2)x}\mathbf{1}(x>0),$$

$$f_{Y|X}(y|x) = (x+3)e^{-(x+3)y}\mathbf{1}(y>0),$$

are compatible as a joint distribution (X,Y). Identify the marginal and joint densities.
2. Do the following two densities have a compatible joint distribution?

$$f_{X|Y}(x|y) = y\mathbf{1}(0<x<y^{-1})\mathbf{1}(y>0),$$

$$f_{Y|X}(y|x) = x\mathbf{1}(0<y<x^{-1})\mathbf{1}(x>0).$$

2.3. Parametric dimension of a complete model.
Let $f: E^S \to \mathbb{R}$ be a real-valued function defined on $E=\{0,1,\ldots,K-1\}$, $S=\{1,2,\ldots,n\}$ such that $f(\mathbf{0})=0$. f is therefore associated with potentials $\Phi =$

$\{\Phi_A, A \subset S \text{ and } A \neq \emptyset\}$ by the Moëbius inversion formula. These potentials are identifiable and satisfy (2.3). Show that the $K^n - 1$ parameters of f can be found among those of Φ.

2.4. Compatibility of conditional Gaussian distributions.
Consider the energy U on $\mathbb{R}^2$ defined by:

$$U(x,y) = -\{a_1x + a_2y + b_1x^2 + b_2y^2 + cxy + d_1xy^2 + d_2x^2y + ex^2y^2\}.$$

1. Show that if $e > 0$, $b_1 > 0$ and $d_2 < 4eb_1$, the conditional energy $U(x|y)$ is admissible and is that of a Gaussian distribution. Find the parameters of this distribution. Same question for $U(y|x)$.
2. Define $\delta_1 = \inf_x\{b_2 + d_1x + (e/2)x^2\}$, $\delta_2 = \inf_y\{b_1 + d_2y + (e/2)y^2\}$ and $\delta = \inf\{\delta_1, \delta_2\}$. Show that $U(x,y)$ is admissible under the conditions:

$$b_1, b_2, e > 0, \ d_1 < 2b_2e, \ d_2 < 2b_1e, \ |c| < \delta.$$

   Deduce an example of a distribution on $\mathbb{R}^2$ which is non-Gaussian but which has conditional Gaussian distributions.

2.5. Gibbs models on the planar triangular network.

1. Characterize the 6-NN stationary model with 3 states $E = \{a,b,c\}$ over the planar triangular network: find the cliques, potentials, parametric dimension, conditional distributions at individual sites and conditional distributions over subsets of sites. Characterize the model if, instead, $E \subset \mathbb{R}$.
2. Same question, but: (i) with an isotropic model; (ii) with the model of at least pair potentials; (iii) when the potentials Φ_A are permutation-invariant on A.
3. Same questions for a two-state model.

2.6. Markov random fields for texture segmentation.
Let (X, Λ) be a field over $S = \{1,2,\ldots,n\}^2$ taking values in $E = \mathbb{R} \times \{1,2,\ldots,K\}$, defined hierarchically as follows: the "texture" Λ is the result of an 8-NN Potts model (2.6). Conditionally on λ, we choose one of the following models $(X|\lambda)$:

1. $(X_i|\lambda_i = k) \sim \mathcal{N}(0, \sigma_k^2)$, a Gaussian texture measuring roughness.
2. $(X_i|\lambda_i = k) \sim \mathcal{N}(\mu_k, \sigma^2)$, a grayscale texture.
3. On constant-labeled zones with label k, X is a covariation texture modeled by a 4-NN isotropic Gaussian CAR model with parameter (α_k, σ_e^2), $0 \leq \alpha_k < 1/4$.

1. Give details of each of the above models (cliques, potentials).
2. Suppose we observe X. Find the distributions $(\Lambda|X)$ and $(\Lambda_i|\Lambda^i, X)$. Use Gibbs sampling (cf. §4.2) to simulate $(\Lambda|X)$.

2.7. Restoration of a Gaussian signal.

1. $Y = h * X + \varepsilon$ is an observation resulting from the h-convolution of a Gaussian signal $X = \{X_i, i \in S\}$ where ε is an additive Gaussian WN with variance σ^2 independent of X. Characterize the distributions: (X,Y), $(X|y)$ and $(X_i|x^i, y)$, $i \in S$ in the two following situations:

a. $S = \{1,2,\dots,n\}$, X the 2-NN stationary CAR model and $(h*x)_i = a(X_{i-1} + X_{i+1}) + bX_i$.
b. $S = \{1,2,\dots,n\}^2$, X the 4-NN stationary CAR model and $h*X = X$.

2. Suppose we observe $Y = y$. Simulate $(X|y)$ using Gibbs sampling (cf. §4.2).

2.8. A distribution not satisfying positivity condition (2.9).
Consider the temporal process $\{X(t), t \geq 0\}$ on $S = \{1,2,3\}$ that takes the values $\{0,1\}^3$. $X_i(t)$ gives the state of site $i \in S$ at time $t \geq 0$ with: $X_i(t) = 0$ if the state is healthy and $X_i(t) = 1$ otherwise. Suppose that an infected state remains infected. If $x(0) = (1,0,0)$ and if contamination from t to $(t+1)$ happens to nearest neighbors independently with probability δ, find the distribution $\pi(2)$ of $X(2)$. Show that $\pi(2)$ does not satisfy the positivity condition, i.e., show that there exists $x = (x_1, x_2, x_3)$ such that $\pi_i(2)(x_i) > 0$ for $i = 1,2,3$, yet $\pi(2)(x) = 0$.

2.9. Causal models and corresponding bilateral model representations.

1. Let Y be a Markov chain on $E = \{-1,+1\}$ with transitions $p = P(Y_i = 1|Y_{i-1} = 1)$ and $q = P(Y_i = -1|Y_{i-1} = -1)$.

   a. Show that Y is a 2-NN bilateral Markov random field.
   b. Deduce that the bilateral conditional kernel can be written

   $$Q(y|x,z) = Z^{-1}(x,z)\exp\{x(\alpha + \beta(y+z)\},$$

   where $\alpha = (1/2)\log(p/q)$ and $\beta = (1/4)\log\{pq/[(1-p)(1-q)]\}$.
   c. Interpret the cases: $\alpha = 0$; $\beta = 0$.

2. For the state space $E = \{0,1\}$ on $\mathbb{Z}$, give the bilateral Markov representation of the homogeneous second-order Markov chain with transition

   $$P(Y_i = 1|Y_{i-2} = a, Y_{i-1} = b) = p(a,b).$$

   Can it be shown in general that a 4-NN Markov random field on $\mathbb{Z}$ is a second-order Markov chain?
3. Consider, on $S = \{1,2,\dots,n\}$ the energy model:

   $$U(y) = a\sum_{i=1}^{n} y_i + b\sum_{i=1}^{n-1} y_i y_{i+1} + c\sum_{i=1}^{n-2} y_i y_{i+2}.$$

   Partition S as $I \cup P$, where I is the set of odd indices and P the even ones. Show that (Y_I/y_P) is a Markov chain.
4. Consider on $\mathbb{Z}^2$ a causal binary random field Y with respect to the lexicographic order, whose conditional distribution at (i,j) depends only on the past sites $(i-1,j)$ and $(i,j-1)$. Give the bilateral model for Y.

2.10. Sampled Markov chains and random fields.

1. The distribution of a Markov chain $Y = (Y_1, Y_2, \dots, Y_n)$ with initial distribution $Y_1 \sim \nu$ and transitions $\{q_i, i = 1, \dots, n-1\}$ is:

$$P_\nu(y) = \nu(y_1) \prod_{i=1}^{n-1} q_i(y_i, y_{i+1}).$$

Show that Y observed only at every second time instant is still a Markov chain. Give its transitions and potentials for the state space $E = \{0,1\}$ if the transition is homogeneous.
2. Suppose a 4-NN Markov random field Y on the torus $S = \{1,2,\dots,n\}^2$ is observed at $S^+ = \{(i,j) \in S : i+j \text{ even}\}$. Show that Y_{S^+} is a Markov random field with maximal cliques $\{C_{ij} = \{(i,j), (i+2,j), (i+1,j+1), (i+1,j-1)\}, (i,j) \in S\}$. Find the distribution of Y_{S^+} if Y is stationary and $E = \{0,1\}$.
3. Show that the Markov property in (2) is lost: (i) if Y is an 8-NN Markov random field; (ii) if we sample Y at $S_2 = \{(i,j) \in S : i \text{ and } j \text{ even}\}$.

2.11. Noisy Markov random field models.
Let Y be a 2-NN binary Markov random field on $S = \{1,2,\dots,n\}$. Suppose that Y_i is transmitted with noise, that is, we observe Z_i with probability $1-\varepsilon$:

$$P(Y_i = Z_i) = 1 - \varepsilon = 1 - P(Y_i \neq Z_i).$$

Give the joint distribution of (Y,Z). Is the observed signal Z a Markov random field? Show that $(Y|Z)$ is a conditional Markov random field. Give the conditional distributions $(Y_i|Y^i, Z)$.

2.12. Restoration of a color image.
On $S = \{1,2,\dots,n\}^2$, given an observation y, we would like to reconstruct an image $x = \{x_i, i \in S\}$ that has four states $x_i \in E = \{a_1, a_2, a_3, a_4\}$. By passing through noisy channels, the signal x has been degraded by an i.i.d. noise at some rate $p < 1/4$:

$$P(Y_i = X_i) = 1 - 3p \text{ and } P(Y_i = a_k | X_i = a_l) = p, \qquad \forall i \in S, k \neq l.$$

1. Show that, with $n(y,x) = \sum_{i \in S} \mathbf{1}(x_i = y_i)$,

$$\pi(y|x) = c(p) \exp\left\{ n(y,x) \log \frac{1-3p}{p} \right\}.$$

2. Choose the four-state Potts model (2.6) with parameter β for the prior distribution on x. Show that:

   a. $\pi(x|y) = c(y,p) \exp\left\{ -\beta n(x) + n(y,x) \log \dfrac{1-3p}{p} \right\}$.
   b. $\pi_i(x_i|y, x^i) = c_i(y,p,x^i) \exp U(x_i|y, x^i)$, with

$$U(x_i|y,x^i) = -\beta \left\{ \sum_{j\in S\,:\,\langle i,j\rangle} \mathbf{1}(x_i = x_j) + \log \frac{1-3p}{p} \times \mathbf{1}(y_i = x_i) \right\}.$$

3. Reconstruction of x by the *Marginal Posterior Mode* method (MPM, (150); also cf. (96; 224)). This reconstruction method involves keeping, at each site $i \in S$, the marginal mode of $(X_i|y)$:

$$\widehat{x}_i = \underset{x_i\in E}{\operatorname{argmax}}\, \pi_i(x_i|y).$$

As the marginal distribution $\pi_i(x_i|y)$ of the Gibbs distribution $(X|y)$ can be analytically evaluated, we calculate this mode with Monte Carlo methods. Using Gibbs sampling (cf. Ch. 4) to simulate $(X|y)$ from its conditional distributions $\pi_i(x_i|y,x^i)$, describe an algorithm to restore x with the MPM method.
Application: For some choice of S and partition $S = A_1 \cup A_2 \cup A_3 \cup A_4$ defining x, introduce the channel noise $x \mapsto y$ with $p = 1/5$. Restore $\widehat{x}_\beta(y)$ using MPM for various choices of regularization parameter β.

Chapter 3
Spatial point processes

Whereas in Chapters 1 and 2 observation sites $S \subseteq \mathbb{R}^d$ were placed on a predetermined non-random network, here it is the *random spatial pattern* $x = \{x_1, x_2, \ldots\}$ of these site which is of interest. We say that x is the output of a point process (PP) X with state space defined over locally finite subsets of S. For what are known as marked PPs, a mark m_i is also attached to each observation site x_i.

PPs are used in a variety of situations (Diggle, (62)), in ecology and forestry (spatial distribution of plant species; (154)), spatial epidemiology (pointwise location of sick individuals; (141)), materials science (porosity models; (197)), seismology and geophysics (earthquake epicenters and intensities) and astrophysics (locations of stars in nebulae; (163)).

Figure 3.1 gives three examples of PPs: (a) a "random" distribution of ants' nests in a forest, where no spatial structure appears to exist; (b) a more regular distribution, showing the centers of cells in a histological section with each center surrounded by an empty space; (c) the distribution of pine saplings in a Finnish forest, where trees tend to aggregate around their parent tree.

The probabilistic theory of point processes is quite technical and we will not go into every last detail here. Our goal is to give a description of the most often used PP models as well as their most important statistics. We will heuristically present notions such as distributions of PPs, the Palm measure of PPs, Papangélou's conditional intensity and the Markov nearest neighbor property, all of which require deeper theoretical justifications, found for example in the books of Daley and Veres-Jones (56), Stoyan, Kendall and Mecke (204), van Lieshout (217) and Møller and Waagepetersen (160). Our approach is partly inspired by the review article by Møller and Waagepetersen (161) which gives a modern, concise and non-technical description of the main PP models and their statistics.

C. Gaetan, X. Guyon, *Spatial Statistics and Modeling*, Springer Series in Statistics,
DOI 10.1007/978-0-387-92257-7_3,

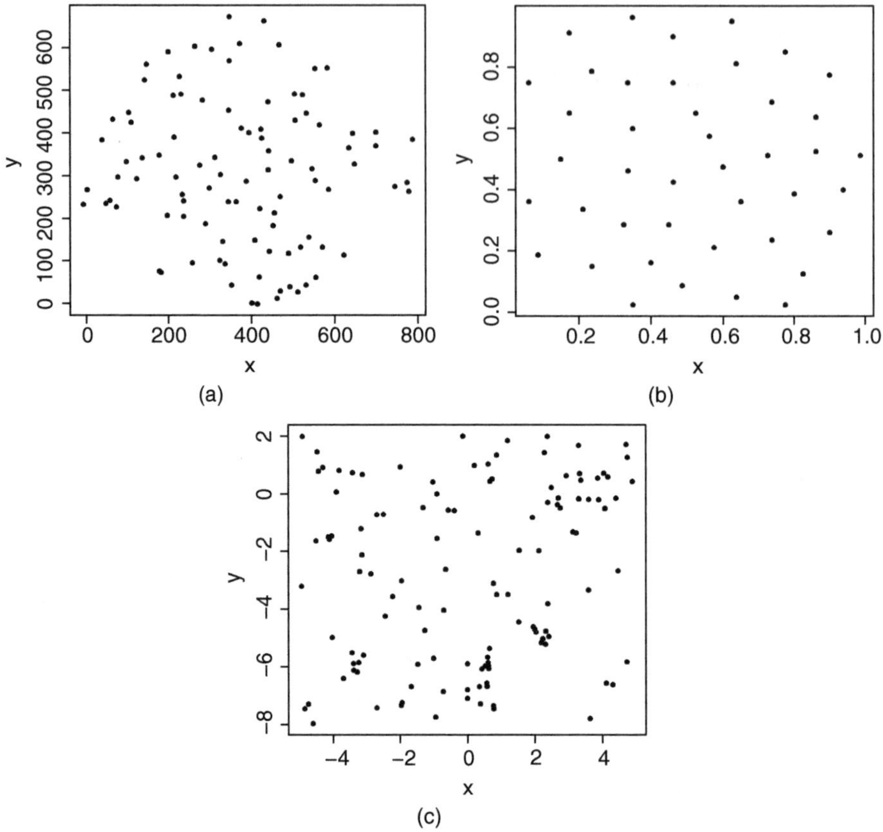

Fig. 3.1 Examples of point distributions: (a) 97 ants' nests (`ants` data in the `spatstat` package); (b) 42 cell centers of a histological section seen under a microscope (`cells` data in `spatstat`); (c) 126 pine saplings in a Finnish forest (`finpines` data in `spatstat`).

3.1 Definitions and notation

Let S be a closed subset of $\mathbb{R}^d$, $\mathscr{B}$ (resp. $\mathscr{B}(S)$, $\mathscr{B}_b(S)$) the set of all Borel sets of $\mathbb{R}^d$ (resp. Borel sets of S, bounded Borel sets of S) and ν the Lebesgue measure on $\mathscr{B}$. The output x of a point process X on S is a *locally finite* set of points of S,

$$x = \{x_1, x_2, \ldots\}, \qquad x_i \in S,$$

i.e., a subset $x \subset S$ such that $x \cap B$ is finite for any bounded Borel set B. Denote $\mathscr{N}_S$ the set of locally finite configurations, $x, y, \ldots$ configurations on S and x_i, y_i, ξ, η points of these configurations. Following Daley and Vere-Jones (56) (cf. also (217)), we have:

Definition 3.1. A point process on S is a mapping X from a probability space $(\Omega, \mathscr{A}, \mathbb{P})$ to the set $\mathscr{N}_S$ of locally finite configurations such that for each bounded

Borel set A, the number of points $N(A) = N_X(A)$ of X falling in A is a random variable.

For example, if $S = [0,1]^2$ and if U and $V : (\Omega, \mathscr{A}, \mathbb{P}) \to [0,1]$ are independent uniform random variables, $X = \{(U,V)\}$ and $X = \{(U,U),(U,V),(U^2,V)\}$ are PPs on S but $X = \{(U^n,V^m),\, n,m \in \mathbb{N}\}$ is not since 0 is a.s. a limit point of X.

In this definition, S can be replaced by a general complete metric space. Note that the output of a point process is at most countably infinite and without limit points. If S is bounded, $N_X(S)$ is almost surely finite and the PP is said to be finite. Here we will only consider *simple* PPs that do not allow multiple points; in such cases, the output x of the PP is merely a subset of S.

3.1.1 Exponential spaces

When S is bounded, the space E of configurations of a PP X on S is equivalent to the union of spaces E_n of configurations of n points of S, $n \geq 0$. $E = \bigcup_{n \geq 0} E_n$ is called the exponential space and is associated with the σ-field $\mathscr{E}$ for which all count variables $N(A) : E \longrightarrow \mathbb{N}$, $N(A) = \sharp(x \cap A)$ (where $A \in \mathscr{B}_b(S)$), are measurable. The σ-field $\mathscr{E}_n$ on E_n is the trace of $\mathscr{E}$ on E_n. Examples of events of $\mathscr{E}$ include: "there are at most 50 points in configuration x," "points of x are all at least a distance r apart, for some given $r > 0$," "0 is a point of x" and "there is no point in $A \subset S$."

The distribution of a point process X is the induced probability P on $(E, \mathscr{E})$ of $\mathbb{P}$. This distribution is characterized, on the sub σ-field of $\mathscr{A}$ that induces measurability of all count variables $N(A)$, $A \in \mathscr{B}_b(S)$, by the finite-dimensional joint distributions of these variables.

Definition 3.2. The finite-dimensional distribution of a point process X is defined by the choice of, for each $m \geq 1$ and m-tuple $(A_1, A_2, \ldots, A_m)$ of $\mathscr{B}_b(S)$, the distributions $(N(A_1), N(A_2), \ldots, N(A_m))$ on $\mathbb{N}^m$.

If S is not a compact space, the distribution of a PP can still be defined in a similar way since configurations x (potentially infinite) are finite on every bounded Borel set.

Stationary and isotropic point processes

We say that a PP X on $\mathbb{R}^d$ is *stationary* if for each $\xi \in \mathbb{R}^d$, the distribution of the translated PP $X_\xi = \{X_i + \xi\}$ is the same as that of X. We say X is *isotropic* if the distribution of ρX, obtained by rotating X by any ρ, has the same distribution as X. Isotropy implies stationarity.

Marked point process

Let K be a metric space (usually $K \subseteq \mathbb{R}^m$). A *marked point process* (MPP) (X,M) on $S \times K$ is a PP on $S \times K$ such that X is a PP on S: $(x,m) = \{(x_1,m_1),(x_2,m_2),\ldots\}$, where $m_i \in K$ is the mark associated with site x_i. Examples of mark spaces include: $K = \{m_1,m_2,\ldots,m_K\}$ (K types of points, cf. Fig. 3.2-a), $K = \mathbb{R}^+$ (the mark is a measure $r \geq 0$ associated with each point, cf. Fig. 3.2-b) and $K = [0,2\pi[\times\mathbb{R}^+$ (the mark is a segment centered at x with orientation $\theta \in [0,2\pi[$ and length $l \geq 0$).

Denote $B(x,r) \subset S$ the ball in $\mathbb{R}^2$ with center x and radius $r > 0$. An example of a marked PP is one that generates the set of centers $X = \{x_i\} \subset \mathbb{R}^2$, with marks the closed balls $B(x_i,r_i)$ centered at x_i with i.i.d. radii independent of X. In mathematical morphology (197; 204), the closed random set $\mathscr{X} = \cup_{x_i \in X} B(x_i,r_i)$ is called a *Boolean process*.

Fibre marked point processes (204) are associated with curvilinear marks m_i attached to x_i, for example, segments centered at x_i of length $l_i \sim \mathscr{E}xp(l^{-1})$ and at angle θ_i uniformly generated on $[0,2\pi[$ and independent of l_i, with (θ_i,l_i) i.i.d. and independent of X. Such MPPs are used in earth science to model the spatial distribution of root networks within a volume $S \subset \mathbb{R}^3$.

Multivariate PPs $X = (X(1),X(2),\ldots,X(M))$ can be seen as MPPs with a finite number M of marks: for each $m = 1,\ldots,M$, $X(m)$ is the subset of S giving the positions of species m. X can thus be equated with the MPP $\widetilde{X} = \cup_{m=1}^{M} X(m)$, a superposition of the $X(m)$. Figure 3.2-a shows a two-state (healthy and sick) MPP giving the geographic distribution of cases of child leukemia (cf. (53) and Exercise 5.14).

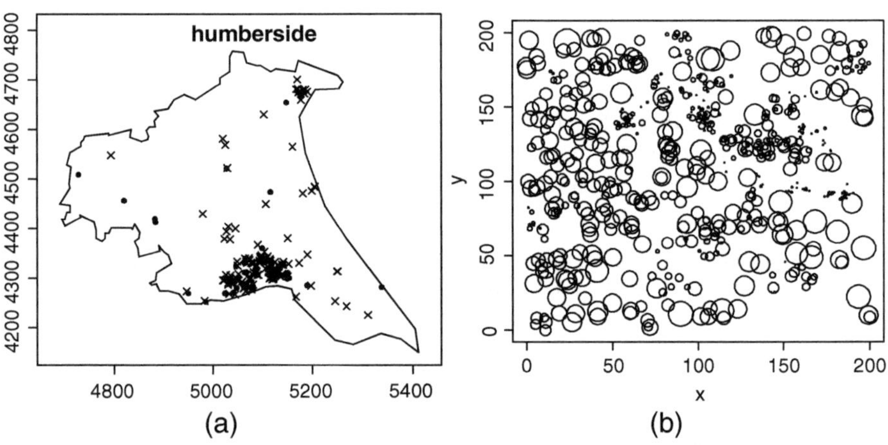

Fig. 3.2 (a) Example of a PP with 2 marks (`humberside` data from the `spatstat` package): location of 62 cases (•) of child leukemia (North-Humberside district, Great Britain, from 1974-82) and 141 houses (×) of healthy children randomly drawn from the birth register. (b) Example of a PP with continuous marks (`longleaf` data from `spatstat`): positions and sizes of 584 pine trees in a forest, with the size of pine trees proportional to the radii of the given circles.

3.1.2 Moments of a point process

In the same way that first-order (expectation) and second-order (variance) moments are fundamental quantities for studying real-valued processes, here the notions relevant to PPs are *moment measures of order p*, with $p \geq 1$. The moment of order p of a PP is the *measure* on $(S, \mathscr{B}(S))^p$ defined for products $B_1 \times \ldots \times B_p$ by:

$$\mu_p(B_1 \times \ldots \times B_p) = E(N(B_1) \ldots N(B_p)).$$

Moment of order 1 and intensity of a point process

The *intensity measure* λ of X is the moment measure of order 1:

$$\lambda(B) = \mu_1(B) = E(N(B)) = E\{\sum_{\xi \in X} \mathbf{1}(\xi \in B)\}.$$

In general, $\lambda(d\xi)$ can be interpreted as the probability that there is a point of X in the infinitesimal volume $d\xi$ around ξ and is modeled using an *intensity density* $\rho(\xi)$, $\lambda(d\xi) = \rho(\xi)d\xi$. If X is stationary, λ is translation-invariant: $\lambda(B) = \tau\nu(B)$, where τ, the constant intensity of X is the mean number of points of X per unit volume.

Factorial moment and intensity of order 2

The covariance between count variables is expressed with respect to the moment measure of order 2, $\mu_2(B_1 \times B_2) = E(N(B_1)N(B_2))$. However, denoting $\sum_{\xi,\eta \in X}^{\neq}$ the sum extended over distinct sites $\xi \neq \eta$ of X, the decomposition

$$\begin{aligned}
\mu_2(B_1 \times B_2) &= E\{\sum_{\xi \in X} \mathbf{1}(\xi \in B_1) \times \sum_{\eta \in X} \mathbf{1}(\eta \in B_2)\} \\
&= E\{\sum_{\xi,\eta \in X} \mathbf{1}((\xi,\eta) \in (B_1, B_2))\} \\
&= E\{\sum_{\xi \in X} \mathbf{1}(\xi \in B_1 \cap B_2)\} + \mathbb{E}\{\sum_{\xi,\eta \in X}^{\neq} \mathbf{1}((\xi,\eta) \in (B_1, B_2))\}
\end{aligned}$$

shows that μ_2 has a measure component on $\mathscr{B}(S)$ and another on the product $\mathscr{B}(S) \times \mathscr{B}(S)$. This situation disappears if we consider the *factorial moment measure* α_2 of order 2 defined on events $B_1 \times B_2$ by:

$$\begin{aligned}
\alpha_2(B_1 \times B_2) &= E\{\sum_{\xi,\eta \in X}^{\neq} \mathbf{1}((\xi,\eta) \in (B_1, B_2))\} \\
&= \mu_2(B_1 \times B_2) - \lambda(B_1 \cap B_2).
\end{aligned}$$

α_2 is the same as μ_2 over products of disjoint events as (Λ, α_2) gives the same information on X as (Λ, μ_2).

For two measurable mappings $h_1 : \mathbb{R}^d \to [0, \infty)$ and $h_2 : \mathbb{R}^d \times \mathbb{R}^d \to [0, \infty)$ a standard result from measure theory (160, Appendix C) leads to the results:

$$E\{\sum_{\xi \in X} h(\xi)\} = \int_{\mathbb{R}^d} h_1(\xi)\lambda(d\xi),$$

and

$$E\{\sum_{\xi,\eta \in X}^{\neq} h_2(\xi, \eta)\} = \int_{\mathbb{R}^d}\int_{\mathbb{R}^d} h(\xi, \eta)\alpha_2(d\xi, d\eta).$$

Furthermore, if B_1 and B_2 are measurable bounded subsets of S, it is easy to see that:

$$Cov(N(B_1), N(B_2)) = \alpha_2(B_1 \times B_2) + \mu_1(B_1 \cap B_2) - \mu_1(B_1)\mu_1(B_2).$$

When $\xi \neq \eta$, $\alpha_2(d\xi \times d\eta)$ gives the probability that X outputs a point in the infinitesimal volume $d\xi$ around ξ and a point in the infinitesimal volume $d\eta$ around η. If α_2 is absolutely continuous with respect to the Lebesgue measure on $(S, \mathscr{B}(S))^2$, its density $\rho_2(\xi, \eta)$ is the *intensity density of order two* of X. If X is stationary (resp. isotropic), $\rho_2(\xi, \eta)$ depends only on $\xi - \eta$ (resp. $\|\xi - \eta\|$).

Reweighted pair correlation

One of the main issues in the study of spatial point patterns is to know whether points tend to be attracted or repelled by each other, or neither. This third possibility, representing a spatial independence hypothesis denoted *Complete Spatial Randomness* (CSR), says that points are distributed independently of each other in S, though not necessarily uniformly. A Poisson PP exactly corresponds to this class of processes and a stationary Poisson PP to the case where the density ρ is constant.

Without supposing stationarity, Baddeley, Møller and Waagepetersen (13) define a function $g(\xi, \eta)$ known as the *reweighted pair correlation* function which is a good summary of second-order spatial dependency: if $\rho(\xi)$ and $\rho(\eta) > 0$,

$$g(\xi, \eta) = \frac{\rho_2(\xi, \eta)}{\rho(\xi)\rho(\eta)}. \tag{3.1}$$

It is easy to see that

$$g(\xi, \eta) = 1 + \frac{Cov(N(d\xi), N(d\eta))}{\rho(\xi)\rho(v)d\xi d\eta},$$

leading to the following interpretation:

1. $g(\xi, \eta) = 1$ if point locations are independent (CSR hypothesis).
2. $g(\xi, \eta) > 1$ represents attraction between points (positive pair covariance).
3. $g(\xi, \eta) < 1$ represents repulsion between points (negative pair covariance).

We say that PP X is *second-order stationary with respect to reweighted correlations* if:

$$\forall \xi, \eta \in S : g(\xi, \eta) = g(\xi - \eta). \tag{3.2}$$

Further on we will see examples of PPs (e.g., log-Gaussian Cox PPs) that can be second-order stationary with respect to reweighted correlations without being first-order stationary. The correlation g allows us to construct spatial independence tests without supposing stationarity of the intensity of the PP.

As for real-valued processes, first and second-order moment measures of two PPs can be the same even if their spatial configurations are quite different: in effect, distributions of PPs also depend on higher-order moment measures.

3.1.3 Examples of point processes

Model defined by its densities conditional on $n(\mathbf{x}) = n \geq 0$

When S is bounded, the distribution of X can be characterized by defining:

1. The probabilities $p_n = P(N(S) = n)$ that a configuration has n points, $n \geq 0$.
2. The densities g_n of x on E_n, the set of configurations with n points, $n \geq 1$.

Each density g_n over E_n is in a one-to-one correspondence with a density f_n (relative to the Lebesgue measure) on S^n that is invariant with respect to coordinate permutation:

$$f_n(x_1, x_2, \ldots, x_n) = \frac{1}{n!} g_n(\{x_1, x_2, \ldots, x_n\}).$$

Denoting $B_n^* = \{(x_1, x_2, \ldots, x_n) \in S^n \text{ s.t. } \{x_1, x_2, \ldots, x_n\} \in B_n\}$ the event associated with $B_n \in \mathscr{E}_n$, the probability of $B = \bigcup_{n \geq 0} B_n$ is given by:

$$P(X \in B) = \sum_{n \geq 0} p_n \int_{B_n^*} f_n(x_1, x_2, \ldots, x_n) dx_1 dx_2 \ldots dx_n.$$

The disadvantage of this approach is that the probabilities p_n must be given, which is unrealistic in practice. Later we will see that when X is defined by an *unconditional* joint density f (cf. §3.4), the p_n are implicitly given by f.

Independent points: binomial processes with n points

Let S be a bounded subset of $\mathbb{R}^d$ with volume $\nu(S) > 0$. A *binomial PP* on S with n points is made up of n uniformly generated i.i.d. points on S. If $\{A_1, A_2, \ldots, A_k\}$ is a partition of Borel subsets of S, then the random variable $(N(A_1), N(A_2), \ldots, N(A_k))$ is a multinomial random variable $\mathscr{M}(n; q_1, q_2, \ldots, q_k)$ with parameters $q_i = \nu(A_i)/\nu(S)$, $i = 1, \ldots, k$ (cf. Fig. 3.3-a). This model can be

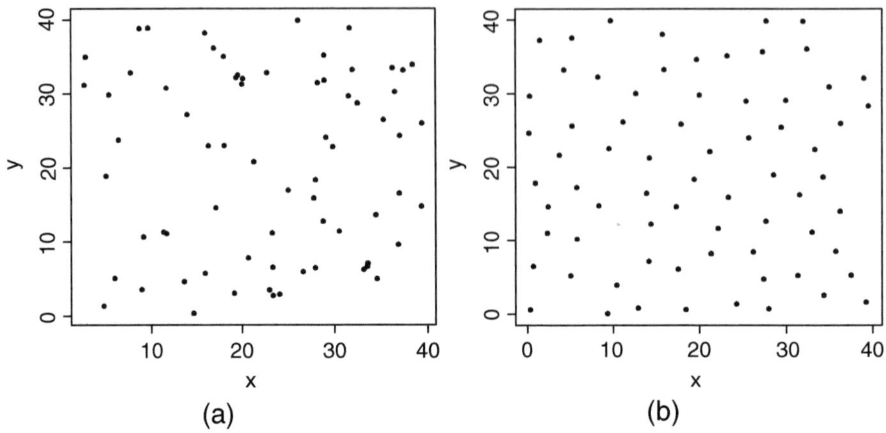

Fig. 3.3 Comparison of two spatial patterns with $n = 70$ points on $S = [0,40]^2$. (a) Random: data distributed according to a binomial PP; (b) More regular: data generated by a hard-core PP that disallows pairs of points closer than 3.5 units.

extended to i.i.d. distributions of n points on S with not-necessarily uniform density ρ. Such conditional distributions of $n(x) = n$ points correspond to Poisson PPs with n points and intensity ρ (cf. §3.2).

The hard-core model

One way to "regularize" spatial configurations is to disallow close points (cf. Fig. 3.3-b). This modeling strategy is well adapted to situations in which individual i located at x_i requires its own unique space, such as impenetrable sphere models in physics, tree distributions in forests, animal distributions in land zones, cell centers of cellular tissue and distribution of shops in towns. These models are special cases of Strauss models, that is, Gibbs models defined via their unconditional densities (cf. §3.4.2).

Aggregated patterns: the Neyman-Scott point process

Consider the following population dynamics context:

1. The positions of parents X is a Poisson PP.
2. Each parent $x_i \in X$ generates K_{x_i} descendants Y_{x_i} at positions Y_{x_i} around x_i, where the (K_{x_i}, Y_{x_i}) are i.i.d. random variables that are independent of X.

Neyman-Scott processes (163) are thus the superposition $D = \cup_{x_i \in X} Y_{x_i}$ of first generation descendants (cf. Fig. 3.4). These models give aggregation around parents if the spatial configuration of descendants is concentrated close to the parents.

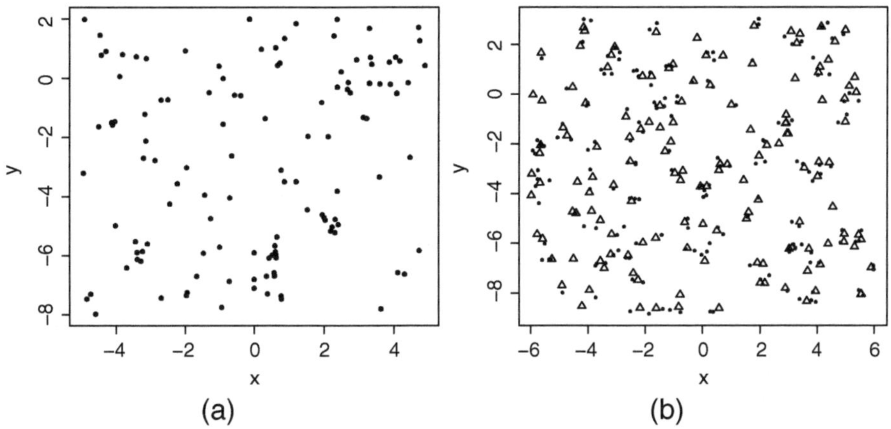

Fig. 3.4 (a) Real data: location of 126 pine saplings in a Finnish forest (`finpines` data from `spatstat`); (b) simulation of a Neyman-Scott process fitted to the data in (a) where: the number of descendants $K \sim \mathscr{P}(\mu)$; positions of descendants (•) around parents (△) follow a $\mathscr{N}_2(0, \sigma^2 I_2)$. Estimated parameters using the method described in §5.5.3 are $\widehat{\mu} = 0.672$, $\widehat{\lambda} = 1.875$ and $\widehat{\sigma}^2 = 0.00944$.

Many generalizations of these models are possible: different choices of spatial configuration of parents X, interdependence of descendants (competition), nonidentical distributions for descendants (variable parental fertility), etc. We will see further on that these models belong to the class of Cox PPs (cf. §3.3).

Overall, we distinguish three main classes of point process:

1. Poisson PPs model "random" spatial distributions (CSR, cf. Fig. 3.3-a). These are characterized by their not necessarily homogeneous intensity $\rho(\cdot)$.
2. Cox PPs are Poisson PPs that are conditional on some random environment. These are used to model *less regular* spatial distributions representing, for example, aggregation as in Neyman-Scott PPs (cf. Fig. 3.4).
3. Gibbs PPs are defined with respect to some conditional specification. These are useful for modeling *more regular* spatial patterns than Poisson PPs, for example the configuration of a hard-core model where each point retains a surrounding clear space (cf. Fig. 3.3-b and §3.4).

3.2 Poisson point process

Let λ be a positive measure on $(S, \mathscr{B}(S))$ with density ρ such that λ is finite on bounded Borel sets. A *Poisson point process (Poisson PP) with intensity measure* $\lambda(\cdot) > 0$ and intensity $\rho(\cdot)$ (we write PPP(λ) or PPP(ρ)) is characterized by:

1. For any $A \in \mathscr{B}_b(S)$ with measure $0 < \lambda(A) < \infty$, $N(A)$ has a Poisson distribution with parameter $\lambda(A)$.

2. Conditional on $N(A)$, the points of $x \cap A$ are i.i.d. with density proportional to $\rho(\xi)$, $\xi \in A$:

$$p_n = P(N(A) = n) = e^{-\lambda(A)} \frac{(\lambda(A))^n}{n!}$$

and

$$g_n(\{x_1, x_2, \dots, x_n\}) \propto \rho(x_1)\rho(x_2)\dots\rho(x_n).$$

This characterization implies that, if $A_i, i = 1, \dots, p$ are p disjoint Borel sets, the random variables $N(A_i), i = 1, \dots, p$ are independent. The Poisson PP is said to be *homogeneous* with intensity ρ if $\lambda(\cdot) = \rho\nu(\cdot)$; thus uniformity of the spatial distribution is added to independence in the spatial distribution.

A standard choice for modeling $\rho(\cdot) \geq 0$ is to use a log-linear model that depends on covariates $z(\xi), \xi \in S$:

$$\log \rho(\xi) = {}^t z(\xi)\beta, \qquad \beta \in \mathbb{R}^p.$$

Simulation of a Poisson point process.

A PPP(ρ) on some bounded subset S (cf. Fig. 3.5) can be simulated using an independent thinning of X (cf. Appendix A), also known as a *rejection sampling technique*: if the density $\rho(\cdot)$ is bounded above by $c < \infty$ on S, the following algorithm simulates a PPP(ρ) on S (cf. Ex. 3.8):

1. Simulate $X_h = \{x_i\}$, a homogeneous Poisson PP on S with intensity c.
2. Remove each x_i with probability $(1 - \rho(x_i)/c)$.

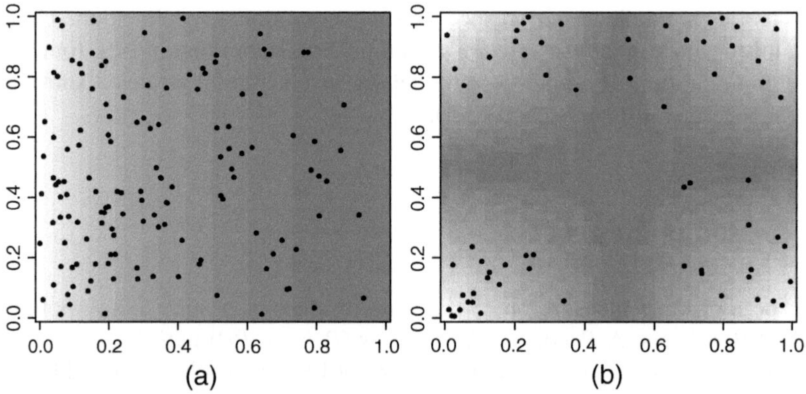

Fig. 3.5 Data generated from an inhomogeneous Poisson PP on $[0,1]^2$ with intensity: (a) $\lambda(x,y) = 400e^{-3x}$ $(E(N(S)) = 126.70)$; (b) $\lambda(x,y) = 800|0.5 - x|\,|0.5 - y|$ $(E(N(S)) = 50)$.

3.3 Cox point process

Suppose $\Lambda = (\Lambda(\xi))_{\xi \in S}$ is a locally integrable ≥ 0 process on S. Almost surely, for any bounded Borel set B, we have $\int_B \Lambda(\xi)d\xi < \infty$. A Cox point process X driven by $\Lambda = (\Lambda(\xi))_{\xi \in S}$ is a Poisson PP with random density Λ, where Λ models some random environment. If the density Λ is stationary, X is too. The simplest example of a Cox process is the mixed Poisson PP with $\Lambda(\xi) = \xi$ a positive random variable that is constant on S. Cox processes appear naturally in Bayesian contexts where the intensity $\lambda_\theta(\cdot)$ depends on some parameter θ following a prior distribution π.

3.3.1 log-Gaussian Cox process

Introduced by Møller, Syversveen and Waagepetersen (159), these models involve log-linear intensities with random effects:

$$\log \Lambda(\xi) = {}^t z(\xi)\beta + \Psi(\xi), \tag{3.3}$$

where $\Psi = (\Psi(\xi))_{\xi \in S}$ is a centered Gaussian random field with covariance $c(\xi,\eta) = Cov(\Psi(\xi),\Psi(\eta))$ ensuring local integrability of Λ. Fig. 3.6 gives examples of such models for two choices of covariance for the underlying field. Moment measures of these processes are easy to derive. In particular, we have (159):

$$\log \rho(\xi) = {}^t z(\xi)\beta + c(\xi,\xi)/2, \qquad g(\xi,\eta) = \exp(c(\xi,\eta)),$$

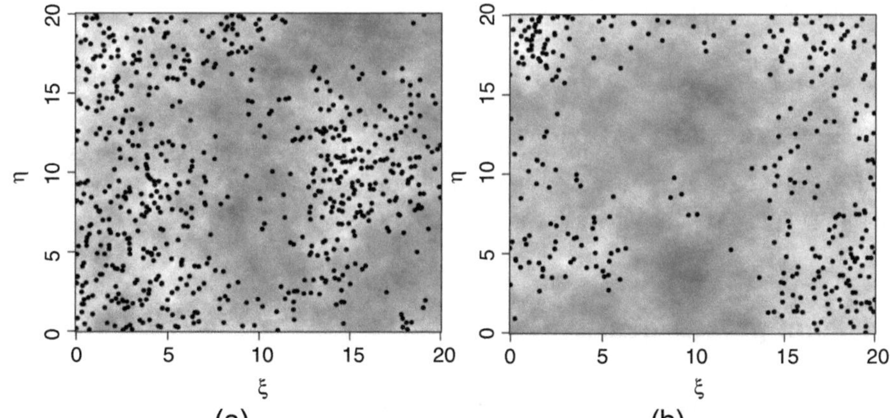

Fig. 3.6 Two spatial patterns of log-Gaussian Cox PPs with ${}^t z(\xi) = 1$ and $\beta = 1$. The grayscale background gives the intensity of a simulated underlying Gaussian random field with covariance (a) $c(\xi,\eta) = 3\exp\{-\|\xi-\eta\|/10\}$; (b) $c(\xi,\eta) = 3\exp\{-\|\xi-\eta\|^2/10\}$.

where g is the reweighted pair correlation (3.1), with a bijection between c and g. When Ψ is stationary (resp. isotropic), X is second-order stationary (resp. isotropic) with respect to the reweighted correlation g.

3.3.2 Doubly stochastic Poisson point process

Also called a *shot-noise process*, this is a Cox process X with intensity

$$\Lambda(\xi) = \sum_{(c,\gamma)\in\varphi} \gamma k(c,\xi), \tag{3.4}$$

where φ is the output of a Poisson PP on $S \times \mathbb{R}^+$ and $k(c,\cdot)$ a density on S centered at c. Thus, X is the superposition of independent Poisson PPs with intensity $\gamma k(c,\cdot)$ for the configuration (c,γ) issued from a Poisson PP. These doubly Poisson PPs model configurations with aggregation, such as for the spatial pattern of plant species due to a certain seed sowing process.

For Neyman-Scott processes ((163) and §3.1.3), centers c come from a stationary Poisson PP of intensity τ with constant γ corresponding to the mean number of descendants of each parent located at c. Neyman-Scott processes are stationary with intensity $\tau\gamma$. Thomas PPs (213) correspond to Gaussian dispersion distributions $k(c,\cdot) \sim \mathcal{N}_d(c,\sigma^2 Id)$ around each center c. Thomas processes are isotropic with reweighted correlations on $\mathbb{R}^2$ (160):

$$g(\xi,\eta) = g(\|\xi,\eta\|), \qquad g(r) = 1 + \frac{1}{4\pi\kappa\sigma^2}\exp\left\{-\frac{r^2}{4\sigma^2}\right\}.$$

We can also consider inhomogeneous models for which the measure Λ driving X depends on covariates $z(\xi), \xi \in S$ (220):

$$\Lambda(\xi) = \exp({}^t z(\xi)\beta) \sum_{(c,\gamma)\in\varphi} \gamma k(c,\xi).$$

This model, which has the same reweighted pair correlation g as (3.4), allows us to study second-order dependency properties while simultaneously allowing first-order inhomogeneity (cf. Example 5.15).

3.4 Point process density

In this section we suppose that S is a bounded subset of $\mathbb{R}^d$. One way to model a PP is to define its probability density f relative to that of a homogeneous Poisson PP with density 1. This approach allows us in particular to define Gibbs PPs.

In most cases, the density f of a PP is known only up to a multiplicative constant, $f(x) = cg(x)$, where g is analytically known. This is inconsequential when simulating PPs by MCMC methods as they only require knowledge of g (cf. §4.4). However, for maximum likelihood estimation of the model $f_\theta(x) = c(\theta)g_\theta(x)$, $c(\theta)$ is an analytically intractable constant and can be estimated by Monte Carlo Markov chain methods (cf. §5.5.6).

3.4.1 Definition

Let $S \subset \mathbb{R}^d$ be a bounded Borel set and Y_ρ a Poisson PP with intensity measure $\lambda > 0$, where λ induces a density ρ. The following Poisson representation gives the probability of each event $(Y_\rho \in F)$:

$$P(Y_\rho \in F) = \sum_{n=0}^{\infty} \frac{e^{-\lambda(S)}}{n!} \int_{S^n} \mathbf{1}\{x \in F\}\rho(x_1)\rho(x_2)\ldots\rho(x_n)dx_1dx_2\ldots dx_n,$$

where $x = \{x_1, x_2, \ldots, x_n\}$. This formula lets us define the density of a PP with respect to the distribution of Y_1, a PPP(1). In the following we denote by ν the Lebesgue measure on S.

Definition 3.3. We say that X has density f with respect to Y_1, the Poisson PP with intensity 1, if for each event $F \in \mathscr{E}$, we have:

$$\begin{aligned} P(X \in F) &= E[\mathbf{1}\{Y_1 \in F\}f(Y_1)] \\ &= \sum_{n=0}^{\infty} \frac{e^{-\nu(S)}}{n!} \int_{S^n} \mathbf{1}\{x \in F\}f(x)dx_1dx_2\ldots dx_n. \end{aligned}$$

The probability of a configuration of n points is thus:

$$p_n = P(n(S) = n) = \frac{e^{-\nu(S)}}{n!} \int_{S^n} f(x)dx_1dx_2\ldots dx_n.$$

Conditional on $n(x) = n$, the n points of X have a symmetric joint density $f_n(x_1, x_2, \ldots, x_n) \propto f(\{x_1, x_2, \ldots, x_n\})$: f_n is only known up to a multiplicative constant and p_n is analytically intractable due to the complexity of calculating the multiple integral. A very special case that can be completely characterized is that of the PPP(ρ):

$$f(x) = e^{\nu(S)-\lambda(S)} \prod_{i=1}^{n} \rho(x_i), \tag{3.5}$$

with $x = \{x_1, x_2, \ldots, x_n\}$. In general, we write $f(x) = cg(x)$, where $g(x)$ is known and c is an unknown normalization constant.

Papangélou conditional intensity

We say that a density f is *hereditary* if, for any finite configuration of points $x \subset S$,

$$f(x) > 0 \text{ and } y \subset x \Longrightarrow f(y) > 0. \tag{3.6}$$

This condition, satisfied for the most frequently encountered densities means that every subconfiguration of a configuration with positive density is itself a positive density.

For hereditary f, the Papangélou intensity (168) for $\xi \notin x$ conditional on x is defined by:

$$\lambda(\xi,x) = \frac{f(x \cup \{\xi\})}{f(x)} \text{ if } f(x) > 0, \qquad \lambda(\xi,x) = 0 \text{ otherwise.} \tag{3.7}$$

If $f(x) = cg(x)$, $\lambda(\xi,x)$ does not depend on the normalization constant c. Furthermore, (3.7) shows that there is a bijection between f and λ; a PP can thus be modeled using its conditional intensity and this intensity is used both in the Monte Carlo Markov chain simulation procedure and the parametric estimation by conditional pseudo-likelihood.

$\lambda(\xi,x)$ can be interpreted as the probability density of there being a point of X at ξ, conditional on the fact that the rest of the points of X are located at x, where the expectation with respect to X of this conditional probability is the density $\rho(\xi)$ of X at ξ (160),

$$E(\lambda(\xi,X)) = \rho(\xi).$$

For a PPP(ρ), $\lambda(\xi,x) = \rho$ does not depend on x; for Markov PPs, the conditional intensity $\lambda(\xi,x) = \lambda(\xi,x \cap \partial\xi)$ only depends on the configuration x in a neighborhood $\partial\xi$ of ξ (cf. §3.6.1).

3.4.2 Gibbs point process

Suppose that X has density $f > 0$:

$$f(x) = \exp\{-U(x)\}/Z, \tag{3.8}$$

where the energy $U(x)$ is *admissible*, i.e., satisfies for each $n \geq 0$:

$$q_n = \int_{S^n} \exp(-U(x))dx_1dx_2\ldots dx_n < \infty, \qquad \sum_{n=0}^{\infty} \frac{e^{-\nu(S)}}{n!} q_n < \infty.$$

To be admissible, U must be admissible conditional on $n(x) = n$, for all n. A sufficient condition is that $n(x)$ be bounded by some $n_0 < \infty$ and that U be conditionally admissible for all $n \leq n_0$; for example, hard-core models, which exclude configurations x with pairs of sites closer than r, $r > 0$ always have a bounded number of points $n(x)$: in effect, in $\mathbb{R}^2$, if $n(x) = n$, the surface of the union of n balls centered at $x_i \in x$ with radius r can not be larger than that of S^r, the set enlarged by dilation by a circle of radius r: $n(x) \times \pi r^2 \leq \nu(S^r)$. A different condition that ensures admissibility of U is that U be bounded.

Strauss and hard-core point processes

A standard example of Gibbs energy is associated with *singletons* and *pair potentials* with energy:

$$U(x) = \sum_{i=1}^{n} \varphi(x_i) + \sum_{i=1}^{n} \sum_{j>i}^{n} \psi(\|x_i - x_j\|).$$

For fixed radius $r > 0$, *Strauss* PPs (206) correspond to the choice $\varphi(x_i) = a$ and $\psi_{\{x_i,x_j\}}(x) = b\mathbf{1}(\|x_i - x_j\| \le r)$, with density $f_\theta(x) = c(\theta)\exp({}^t\theta\, T(x))$, $\theta = (a,b) \in \mathbb{R}^2$, where:

$$T_1(x) = n(x), \qquad T_2(x) = s(x) = \sum_{i<j} \mathbf{1}(\|x_i - x_j\| \le r).$$

T_2 counts the number of "r-neighbor" pairs of points. Denoting $\beta = e^a$ and $\gamma = e^b$, the density f can also be written:

$$f_\theta(x) = c(\theta)\beta^{n(x)}\gamma^{s(x)}, \qquad \theta = {}^t(\beta,\gamma). \tag{3.9}$$

Homogeneous Poisson PPs correspond to $\gamma = 1$ while $\gamma = 0$ defines a *hard-core model* with density:

$$f_{\beta,r}(x) = c\beta^{n(x)}\mathbf{1}\{\forall i \ne j, \|x_i - x_j\| > r\}.$$

The indicator function in the hard-core density excludes configurations with pairs of points at distance $\le r$ (impenetrable sphere models); as S is bounded, the number of points $n(x)$ of the configuration is necessarily bounded.

We now describe the conditional (cf. Fig. 3.7-a-b) and unconditional (cf. Fig. 3.7-c-d) distributions of Strauss PPs.

Conditional on $n(x) = n$, $f_{\theta,n}(x) \propto \gamma^{s(x)}$:

(i) If $\gamma < 1$, the process X is more regular than a binomial PP, increasingly so as γ gets closer to 0.
(ii) $\gamma = 1$ corresponds to a binomial process with n points.
(iii) If $\gamma > 1$, X is less regular than binomial PPs, with regularity decreasing as γ increases. In this case, aggregates appear.

Unconditionally, f_θ is admissible only if $\gamma \le 1$. In effect:

(i) If $\gamma \le 1$ and if $n(x) = n$, $f_\theta(x) \le c(\theta)\beta^n$ and:

$$\sum_{n\ge 0} \frac{1}{n!} \int_{S^n} f_\theta(x)dx_1 \ldots dx_n \le c(\theta) \sum_{n\ge 0} \frac{1}{n!}\beta^n(\nu(S))^n = e^{\beta\nu(S)} < +\infty.$$

(ii) If $\gamma > 1$, f_θ is not admissible: in effect, denoting E_n the expectation of the uniform distribution on S^n, Jensen's inequality gives:

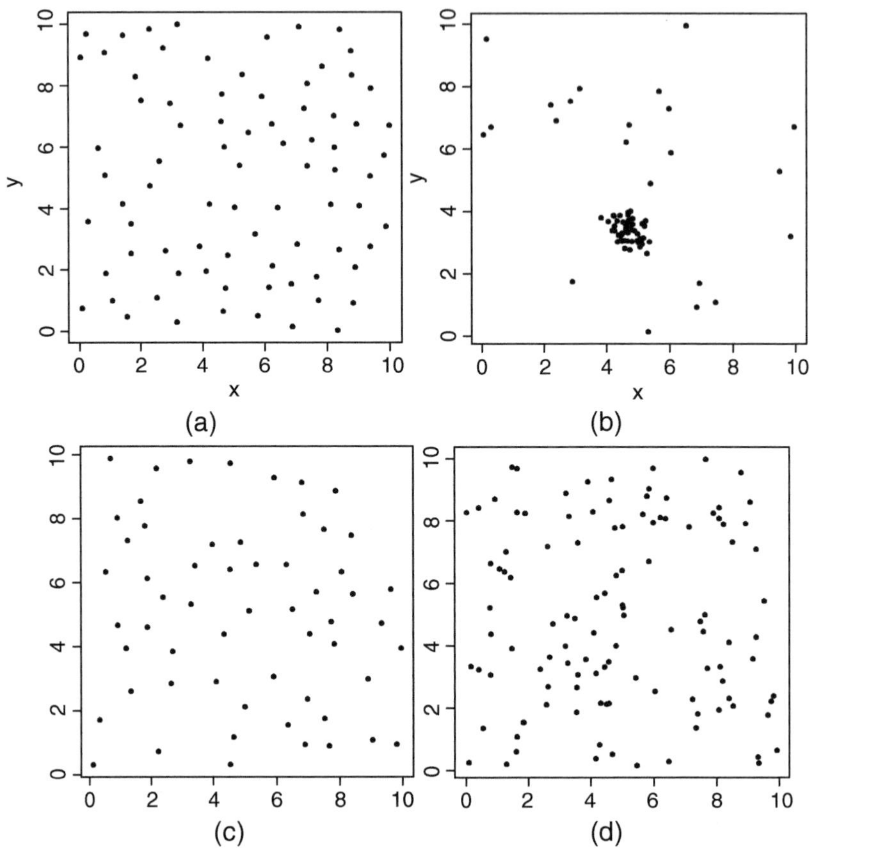

Fig. 3.7 Results of a Strauss PP on $S = [0,10]^2$, $r = 0.7$: conditional on $n = 80$, (a) $\gamma = 0$, (b) $\gamma = 1.5$; unconditional with $\beta = 2$, (c) $\gamma = 0$ (hard-core), (d) $\gamma = 0.8$.

$$\frac{1}{\nu(S)^n}\int_{S^n}\gamma^{s(x)}dx_1\ldots dx_n = E_n(\gamma^{s(X)})$$

$$\geq \exp E_n(bs(X)) = \exp\frac{n(n-1)}{2}bp(S),$$

where $p(S) = P(\|X_1 - X_2\| < r)$ if X_1 and X_2 are uniform i.i.d. on S. It is easy to see that $p(S) > 0$ and that $\sum u_n$ diverges, where

$$u_n = \frac{(\nu(S)\beta)^n}{n!}\gamma^{n(n-1)bp(S)/2}.$$

Several generalizations of Strauss PPs are possible:

1. By not allowing pairs of distinct points to be closer than some r_0, $0 < r_0 < r$ (*hard-core Strauss process*):

$$s(x) = \sum_{i<j} \mathbf{1}(r_0 \le \|x_i - x_j\| \le r).$$

As the number of points generated in any event x is bounded, the density is always admissible.

2. By "saturating" the statistic $n_{x_i}(x) = \sum_{\xi \in x:\, \xi \neq x_i} \mathbf{1}_{\|x_i - \xi\| \le r}$ at some fixed value $0 < \delta < \infty$, i.e., putting an upper bound on the influence of any single point. If $g(x) = \sum_{x_i \in x} \min\{\delta, n_{x_i}(x)\}$, the density:

$$f_\theta(x) = c(\theta)\beta^{n(x)}\gamma^{g(x)}$$

is always admissible (*Geyer's saturation process*; (87)).

3. By modeling the pair *potential* by a *step function* with k steps:

$$\phi(x_i, x_j) = \gamma_k \text{ if } d(x_i, x_j) \in (r_{k-1}, r_k], \qquad \phi(x_i, x_j) = 0 \text{ otherwise, } k = 1, \ldots, p,$$

for some predetermined choice of thresholds $r_0 = 0 < r_1 < r_2 < \ldots < r_p = r$, $p \ge 1$. Letting $s_k(x)$ be the number of distinct pairs of points of x separated by a distance found in $(r_{k-1}, r_k]$, $k = 1, \ldots, p$, the density belongs to the exponential family:

$$f_\theta(x; \beta, \gamma_1, \ldots, \gamma_p) = c(\theta)\beta^{n(x)} \prod_{k=1}^{p} \gamma_k^{s_k(x)}.$$

f_θ is admissible iff $\gamma_1 \le 1$.

4. By including *triplets* of points of x that are pairwise closer than r. If $t(x)$ is the count of the number of triplets, the density

$$f_\theta(x) = c(\theta)\beta^{n(x)}\gamma^{s(x)}\delta^{t(x)}$$

is admissible iff $\{\gamma \text{ and } \delta \le 1\}$.

Area interaction and connected component Gibbs point process

Gibbs PPs with *area interaction* or *connected component interaction* are further alternatives which allow us to model, without constraints on the parameters, distributions of varying regularity (17; 217). Their density is of the form:

$$f_\theta(x) = c\beta^{n(x)}\gamma^{h(x)},$$

with h left to be defined. For $r > 0$, denote $B(x) = \bigcup_{x_i \in x} B(x_i, r/2) \cap S$ the union (restricted to S) of balls with centers $x_i \in x$ and radius r, $a(x)$ the area of $B(x)$ and $c(x)$ its number of connected components (cf. Fig. 3.8). An *area interaction* PP corresponds to the choice $h(x) = a(x)$, whereas a *connected component interaction* PP corresponds to the choice $h(x) = c(x)$.

As S is bounded and $r > 0$, the functions a and c are bounded and both densities are admissible without constraints on the parameters (β, γ). For each of the two

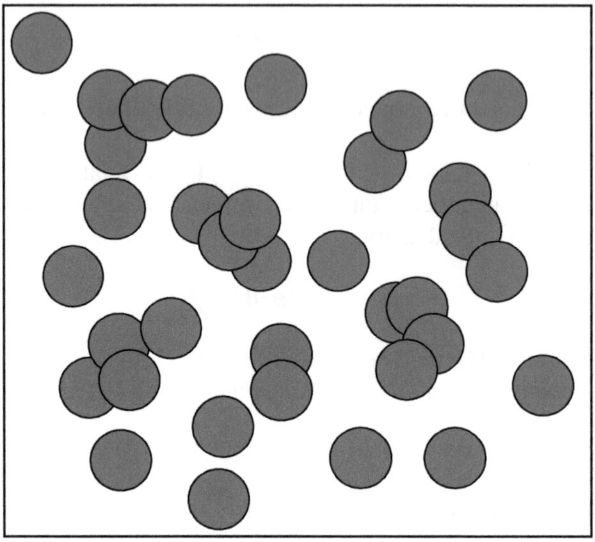

Fig. 3.8 A random generation of the set $B(x)$ that allows us to define the area $a(x)$ and the number of connected components $c(x)$ (here $c(x) = 19$).

models, the spatial distribution will be more (resp. less) regular if $\gamma < 1$ (resp. $\gamma > 1$), with $\gamma = 1$ corresponding to a homogeneous Poisson PP of intensity β. One difficulty in the use of these models is the numerical calculation of $a(x)$ and $c(x)$, requiring appropriate discretization techniques.

3.5 Nearest neighbor distances for point processes

Nearest neighbor distances (NN) are useful statistics for testing the CSR hypothesis for independence of points in spatial PPs. Their distributions are linked to the notion of Palm measures for PPs, which we now present heuristically.

3.5.1 Palm measure

Suppose X is a PP on S with distribution P. The Palm probability P_ξ of X at point ξ is the distribution of X conditional on the presence of a point of X at ξ:

$$\forall F \in \mathscr{E} : P_\xi(F) = P(F|\{\xi \in X\}).$$

Palm measures P_ξ allow us to define statistics which are conditional on the presence of a point of X at ξ: for example, distance to the nearest neighbor of $\xi \in X$,

$d(\xi,X) = \inf\{\eta \in X \text{ and } \eta \neq \xi | \xi \in X\}$, or perhaps the number of points of the configuration found in the ball $B(\xi,r)$, given that $\xi \in X$.

The difficulty in defining P_ξ is due to the fact that the conditioning event $\{\xi \in X\}$ has zero probability; the heuristic approach involves conditioning on the event $\{X \cap B(\xi,\varepsilon) \neq \emptyset\}$, with probability > 0 if $\varepsilon > 0$ and the first-order density of X is > 0, then to see if for an event F "outside" ξ, the conditional probability

$$P_{\xi,\varepsilon}(F) = \frac{P(F \cap \{X \cap B(\xi,\varepsilon) \neq \emptyset\})}{P(\{X \cap B(\xi,\varepsilon) \neq \emptyset\})}$$

converges as $\varepsilon \to 0$. If such a limit exists, it is what we mean by the Palm measure P_ξ; for example, for X a Poisson PP, the two events seen in the numerator are independent when ε is small and thus $P_{\xi,\varepsilon}$ equals P; this can be interpreted as saying that for a Poisson PP, conditioning (or not) on the presence of a point of X at ξ has no consequence on probabilities of events away from ξ.

3.5.2 Two nearest neighbor distances for X

We distinguish between two different NN distances for X: distance from a point belonging to X to its nearest neighbor and distance from a given point in S to its NN in X (cf. Fig. 3.9).

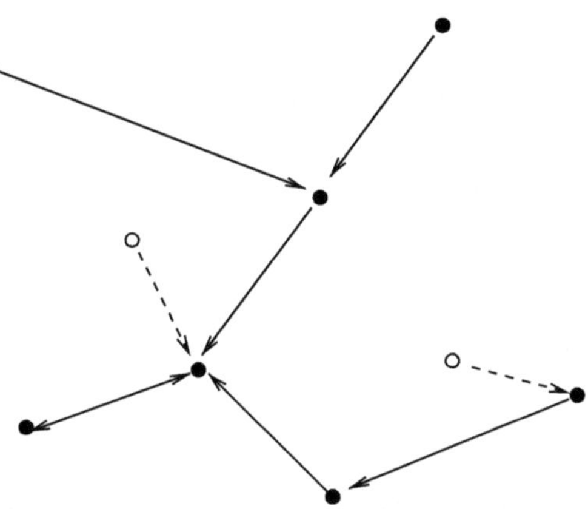

Fig. 3.9 The two NN distances. (•) are observations from the point process (here $n(x) = 7$), (∘) are the selected points (here, 2 sites). The solid line shows the distance from each event to the closest other event. The dashed line gives the distance from a selected point to the closest event.

Distance from a point $\xi \in X$ to its nearest neighbor in X

The distance $d(\xi,X\backslash\{\xi\})$ of a point $\xi \in X$ to its nearest neighbor in X has the cumulative distribution function G_ξ:

$$G_\xi(r) = P_\xi(d(\xi,X\backslash\{\xi\}) \leq r), \qquad \xi \in X, r \geq 0.$$

If X is stationary, $G_\xi = G$ is independent of ξ. If X is a homogeneous PPP(λ) on $\mathbb{R}^2$ and $N(B(\xi,r))$ a Poisson distribution with parameter $\lambda\pi r^2$, $G(r) = P(N(B(\xi,r)) = 0) = 1 - \exp\{-\lambda\pi r^2\}$. G has expectation $(2\sqrt{\lambda})^{-1}$ and variance $\lambda^{-1}(\pi^{-1} - 0.25)$.

Distance from a point u to its nearest neighbor in X

The distance $d(\xi, X\backslash\{\xi\})$ from a point $\xi \in X$ to its NN in X must be distinguished from the distance $d(u,X)$ of a point $u \in \mathbb{R}^d$ (not necessarily in X) to its NN in X. The distribution of this second distance is:

$$F_u(r) = P(d(u,X) \leq r), \qquad u \in S, r \geq 0. \tag{3.10}$$

If X is stationary, $F_u = F$. If furthermore X is a homogeneous Poisson PP, $G(r) = F(r)$. A *summary statistic* giving an indication of the closeness of X to a Poisson PP is given by the statistic J:

$$J(r) = \frac{1 - G(r)}{1 - F(r)}.$$

If X is a stationary PP, $J > 1$, $J = 1$ and $J < 1$ indicate respectively that X is more, equally or less regular than a Poisson PP. An estimation of J provides a test statistic for the CSR hypothesis.

3.5.3 Second-order reduced moments

Ripley's K function

Suppose X is an *isotropic* PP on $\mathbb{R}^d$ with intensity ρ. An alternative second-order indicator of the spatial distribution of points of X is Ripley's K function (183) or second-order reduced moment:

$$K(h) = \frac{1}{\rho} E_\xi[N(B(\xi,h)\backslash\{\xi\})], \qquad h \geq 0,$$

where E_ξ is the expectation of Palm's distribution P_ξ. The function K can be interpreted in two ways:

1. $\rho K(h)$ is proportional to the mean number of points of X in the ball $B(\xi,h)$, not including ξ and conditional on $\xi \in X$.
2. $\rho^2 K(h)/2$ is the mean number of (unordered) pairs of distinct points at distance $\leq h$, provided that one point belongs to a subset A of unit surface.

This second-order reduced mean is invariant by uniform random thinning of points. If points of the PP X with moment K are thinned following an i.i.d. Binomial, the resulting process still has the second-order reduced moment K.

Denoting $\rho_2(\xi,\eta) = \rho_2(\|\xi - \eta\|)$ the second-order density of X and b_d the volume of the unit sphere in $\mathbb{R}^d$, we have:

$$K(h) = \frac{d \times b_d}{\rho} \int_0^h \xi^{d-1} \rho_2(\xi) d\xi.$$

In particular, in dimension $d = 2$,

$$\rho^2 K(h) = 2\pi \int_0^h u\rho_2(u)du, \qquad \rho_2(h) = \frac{\rho^2}{2\pi h} K'(h). \tag{3.11}$$

More generally, Baddeley, Møller and Waagepetersen (13) have extended the moment K to the second-order reduced factorial moment measure $\mathcal{K}$ for stationary PPs X with intensity ρ by the formula:

$$\rho^2 \mathcal{K}(B) = \frac{1}{\nu(A)} E \sum_{\xi,\eta \in X} \mathbf{1}[\xi \in A, \eta - \xi \in B], \qquad B \subseteq \mathbb{R}^d. \tag{3.12}$$

$\mathcal{K}(B)$ does not depend on $A \subseteq \mathbb{R}^d$ if $0 < \nu(A) < \infty$. $\mathcal{K}$ is linked to the second-order factorial moment measure by the formula:

$$\alpha_2(B_1 \times B_2) = \rho^2 \int_{B_1} \mathcal{K}(B_2 - \xi) d\xi, \qquad B_1, B_2 \subseteq \mathbb{R}^d. \tag{3.13}$$

If ρ_2, like K is related to the distribution of the distance between pairs of points, equation (3.12) shows that it can be considered a natural nonparametric estimator of K (where A is an observation window; cf. §5.5.3).

If X is a homogeneous Poisson PP, the graph of $h \mapsto L(h) = h$ is a straight line; a concave function L indicates patterns with aggregates; convex L indicates more regular patterns than that of a Poisson PP. Estimation of L provides another statistic for testing the CSR hypothesis.

K can also be used to set up parametric estimation methods (for example, least squares, cf. §5.5.4) provided we have its analytic representation.

For example, if X is a homogeneous Poisson PP on $\mathbb{R}^d$, $K(h) = b_d \times h^d$; if X is a Neyman-Scott PP on $\mathbb{R}^d$ for which the position of parents follows a homogeneous PPP(ρ) with each parent generating a random number N of offspring, with dispersion distributions around each parent having isotropic density $k(\xi) = g(\|\xi\|)$, then, setting $G(h) = \int_0^h g(u)du$ (48):

$$K(h) = b_d h^d + \frac{E(N(N-1))G(h)}{\rho [E(N)]^2}.$$

A further *summary statistic* useful for checking the CSR hypothesis is the function:

$$L(h) = \left\{ \frac{K(h)}{b_d} \right\}^{1/d}.$$

The Baddeley-Møller-Waagepetersen K_{BMW} function

Suppose that X is a second-order stationary PP for the reweighted correlation g (cf. (3.1) and (3.2)): $g(\xi,\eta) = g(\xi-\eta)$. Baddeley, Møller and Waagepetersen (13) extended Ripley's second-order reduced moment function to:

$$K_{BMW}(h) = \int_{\mathbb{R}^d} \mathbf{1}\{\|\xi\| \leq h\} g(\xi) d\xi = \frac{1}{\nu(B)} E\left[\sum_{\xi,\eta \in X \cap B}^{\neq} \frac{\mathbf{1}\{\|\xi-\eta\| \leq h\}}{\rho(\xi)\rho(\eta)} \right]. \tag{3.14}$$

The equality on the left can be compared with equations (3.11) (in dimension $d = 2$) and (3.13) (in general). The equality on the right suggests a natural nonparametric estimator of $K_{BMW}(h)$ (cf. (5.5.3)).

3.6 Markov point process

The notion of Markov PPs was introduced by Ripley and Kelly (186). Its generalization to nearest neighbor Markov properties was given by Baddeley and Møller (12).

3.6.1 The Ripley-Kelly Markov property

Let X be a PP with density f with respect to some PPP(λ), where λ is a positive density measure with finite mass on the set of bounded Borel sets. Let $\xi \sim \eta$ be a symmetric neighbor relation on S: for example, for some fixed $r > 0$, $\xi \sim_r \eta$ if $\|\xi - \eta\| \leq r$. The neighborhood of $A \subset S$ is

$$\partial A = \{\eta \in S \text{ and } \eta \notin A : \exists \xi \in A \text{ s.t. } \xi \sim \eta\}.$$

We denote $\partial\{\xi\} = \partial\xi$ if $\xi \in S$.

Definition 3.4. A process X with hereditary (3.6) density f is Markov for the relation $\sim$ if, for each configuration x with density > 0, the Papangélou conditional intensity (3.7) $\lambda(\xi,x) = f(x \cup \{\xi\})/f(x)$ depends only on ξ and $\partial\xi \cap x$:

$$\lambda(\xi,x) = \frac{f(x \cup \{\xi\})}{f(x)} = \lambda(\xi; x \cap \partial\xi).$$

The Markov property translates that this conditional intensity depends only on neighboring points of ξ belonging to x.

Examples of Markov point processes

1. The Poisson PP with intensity ρ and conditional intensity $\lambda(u,x) \equiv \rho(\xi)$ is Markov for any neighbor relation on S.
2. The Strauss PP (3.9) with conditional intensity

$$\lambda(\xi,x) = \beta \exp\{\log\gamma \sum_{x_i \in x} \mathbf{1}\{\|x_i - \xi\| \le r\}$$

is $\sim_r$-Markov; its generalizations to hard-core Strauss processes and/or those with step potential functions or saturated pair interaction potentials (cf. §3.4.2) are also $\sim_r$-Markov.
3. The hard-core process with conditional intensity $\lambda(\xi,x) = \beta \mathbf{1}\{\partial\xi \cap x = \emptyset\}$ is $\sim_r$-Markov.

However, as we will see in §3.6.2, for all $r > 0$, the PP with connectivity-interaction (cf. §3.4.2) is not $\sim_r$-Markov: in effect, as two points of S can be connected in $B(x) = \bigcup_{x_i \in x} B(x_i, r/2) \cap S$ whilst being arbitrarily far apart, the conditional intensity $\lambda(\xi,x)$ can depend on points of x arbitrarily far from ξ.

The Markov property (3.4), local at ξ, can be extended to any Borel set A of S. If X is Markov, the distribution of $X \cap A$ conditional on $X \cap A^c$ is dependent only on $X \cap \partial A \cap A^c$, the configuration of X on $\partial A \cap A^c$.

Like Markov random fields on networks (cf. §2.3.2), we can give a Hammersley-Clifford theorem characterizing Markov PP densities in terms of potentials defined on graph cliques. For the neighbor relation $\sim$, a clique is some configuration $x = \{x_1, x_2, \dots, x_n\}$ such that for each $i \ne j$, $x_i \sim x_j$, with the convention that singletons are also considered cliques. We note $\mathscr{C}$ the family of cliques of $(S, \sim)$.

Proposition 3.1. *(186; 217) A PP with density f is Markov for the relation $\sim$ if and only if there exists a measurable function $\Phi : E \to (\mathbb{R}^+)^*$ such that:*

$$f(x) = \prod_{y \subset x,\, y \in \mathscr{C}} \Phi(y) = \exp \sum_{y \subset x,\, y \in \mathscr{C}} \phi(y).$$

$\phi = \log \Phi$ is the Gibbs interaction potential and the Papangélou conditional intensity is:

$$\lambda(\xi,x) = \prod_{y \subset x,\, y \in \mathscr{C}} \Phi(y \cup \{\xi\}), \qquad \xi \in S \backslash \{x\}. \tag{3.15}$$

An example of a Markov PP density with pair interactions is:

$$f(x) = \alpha \prod_{x_i \in x} \beta(x_i) \prod_{x_i \sim x_j\,,\, i<j} \gamma(x_i, x_j).$$

The Strauss process corresponds to the potentials

$$\beta(x_i) = \beta \quad \text{and} \quad \gamma(x_i, x_j) = \gamma^{\mathbf{1}\{\|x_i - x_j\| \le r\}}.$$

Markov property for a marked point process

The definition of the Markov property and the Hammersley-Clifford theorem remain unchanged if $Y = (X,M)$ is a marked PP on $S \times K$ with symmetric neighbor relation $\sim$ on $S \times K$. If (X,M) has independent marks and X is Markov for some neighbor relation $\sim$ on S, (X,M) is a Markov marked PP for the relation $(x,m) \sim (y,o) \Longleftrightarrow x \sim y$ on S. An example of an isotropic model with pair interactions and a finite number of marks $M = \{1,2,\ldots,K\}$ is given by the density in $y = \{(x_i,m_i)\}$:

$$f(y) = \alpha \prod_i \beta_{m_i} \prod_{i<j} \gamma_{m_i,m_j}(\|x_i - x_j\|).$$

If for fixed real $r_{kl} > 0$, $k,l \in K$ and $k \neq l$, $\gamma_{kl}(d) \equiv 1$ when $d > r_{k,l}$, then conditions (1) and (2) are satisfied for the neighbor relation:

$$(\xi,m) \sim (\xi',m') \Longleftrightarrow \|\xi - \xi'\| \leq r_{m,m'}.$$

Y is a Markov marked PP.

Further examples of Markov point models are presented in (160; 11).

Example 3.1. Canopy interaction in forestry

Suppose that the zone of influence of a tree centered at x_i is defined by a circle $B(x_i;m_i)$ centered at x_i with radius $m_i > 0$ bounded by some $m < \infty$. A pair interaction representing competition between trees i and j can be modeled by pair potentials:

$$\Phi_2((x_i;m_i),(x_j;m_j)) = b \times \nu(B(x_i;m_i) \cap B(x_j;m_j)),$$

with singleton potentials for K predetermined values $0 = r_0 < r_1 < r_2 < \ldots < r_{K-1} < r_K = m < \infty$ of:

$$\Phi_i(x_i;m_i) = \alpha(m_i) = a_k \quad \text{if} \quad r_{k-1} < m_i \leq r_k,\ k = 1,\ldots,K.$$

The associated energy is admissible if $b < 0$ (competition between trees) and defines a Markov marked PP on $\mathbb{R}^2 \times \mathbb{R}^+$ with conditional intensity

$$\lambda((u,h);(x,m)) = \exp\{\alpha(h) + b \sum_{j:\|x_j - u\| \leq 2m} \nu(B(u;h) \cap B(x_j;m_j))\}$$

for the neighbor relation $(x,m) \sim (x',m') \Longleftrightarrow \|x - x'\| \leq 2m$.

3.6.2 Markov nearest neighbor property

A more general Markov property, known as the *Markov nearest neighbor property* was developed by Baddeley and Møller (12). We are going to briefly present this for the special case of PPs with connectivity-interactions (cf. §3.4.2), not Markov in the Ripley-Kelly sense but instead for a new neighbor relation that depends on the configuration x, the x-plus nearest neighbor relation $\sim_x$.

Let x be some configuration on S, $r > 0$ fixed and $B(x) = \bigcup_{x_i \in x} B(x_i, r/2) \cap S$. We say that two points ξ and η of x are connected for x if ξ and η are in the same connected component of $B(x)$. We note $\xi \sim_x \eta$ this neighbor relation. Thus, $\sim_x$ is a relation between points of x that depends on x. We remark that two points of S can be neighbors for $\sim_x$ whilst being arbitrarily far apart with respect to the euclidean distance: in effect, connectivity by connected components joins pairs of points if there exists a chain of r-balls centered on points of x that joins each point to the next.

Let $c(x)$ be the number of connected components of $B(x)$. If S is bounded, the density of the PP with connectivity-interactions is $f(x) = c\alpha^{n(x)}\beta^{c(x)}$, α and $\beta > 0$. The Papangélou conditional intensity is:

$$\lambda(\xi, x) = \alpha\beta^{c(x\cup\{\xi\})-c(x)}.$$

To see that $\lambda(\xi, x)$ can depend on points $\eta \in x$ arbitrarily far from u in S, we can choose a configuration z such that for $x = z \cup \{\eta\}$, $c(z) = 2$, $c(x) = 1$, $c(x) = c(x \cup \{\xi\}) = 1$. Thus, for any $R > 0$, X is not Markov in the Ripley-Kelly sense for the usual R-neighbor relation.

However, if $\eta \in x$ is not connected to ξ in $B(x \cup \{\xi\})$, η does not contribute to the difference $c(x \cup \{\xi\}) - c(x)$ and $\lambda(\xi, x)$ is independent of η. Then, X is Markov for the connected components NN relation.

Markov nearest neighbor property

Let $\sim_x$ be a family of relations between points of x, $x \in E$, with the neighborhood of $z \subset x$ given by:

$$\partial(z|x) = \{\xi \in x : \exists \eta \in z \text{ s.t. } \xi \sim_x \eta\}.$$

A process with density f is called *Markov nearest neighbor* if f is hereditary and if the conditional density $\lambda(\xi, x) = f(x \cup \{\xi\})/f(x)$ only depends on ξ and $(\partial_{x\cup\{\xi\}}\{\xi\}) \cap x$. The density of a Markov NN process is characterized by a Hammersley-Clifford theorem.

As the neighbor relation $\sim_x$ depends on x, it makes it more difficult to prove the Markov nearest neighbor property: for example, consider the following density associated with singleton and pair potentials:

$$f(x) = c(\alpha, \beta)\alpha^{n(x)} \prod_{x_i \sim_x x_j} \beta(x_i, x_j). \tag{3.16}$$

In general, further constraints must be imposed on the relation $\sim_x$ so that f is Markov nearest neighbor (12). These constraints are implicitly satisfied when, as we have just seen, $\sim_x$ is the connected component neighbor relation, but also if

$\{\xi \sim_x \eta\} \equiv \{\xi \sim \eta\}$, where $\sim$ is a symmetric relation on S and independent of x.

We now give another example of Markov NN PPs.

Example 3.2. Nearest neighbor relations for Delaunay triangulations

Let x be a locally finite configuration of points of $S \subset \mathbb{R}^2$. To each site x_i associate a *zone of influence* $\mathscr{P}_i(x)$ defined as the following subset of $\mathbb{R}^2$:

$$\xi \in \mathscr{P}_i(x) \Longleftrightarrow \forall j \neq i,\ \|\xi - x_i\| \leq \|\xi - x_j\|.$$

For $x = \{x_1, x_2, \ldots, x_n\}$, the decomposition $S = \cup_{i=1}^{n} \mathscr{P}_i(x)$ is called the *Voronoi diagram.* Except possibly on the boundary of the observation window S, $\mathscr{P}_i(x)$ is a convex polygon. What we call the *Delaunay triangulation* of x can be associated with this decomposition: two distinct points x_i and x_j are defined to be neighbors if $\mathscr{P}_i(x)$ and $\mathscr{P}_j(x)$ share a common edge. This defines a neighbor relation $\sim_{t(x)}$, i.e., the NN relation for the Delauney triangulation $t(x)$ of x (cf. Fig. 3.10).

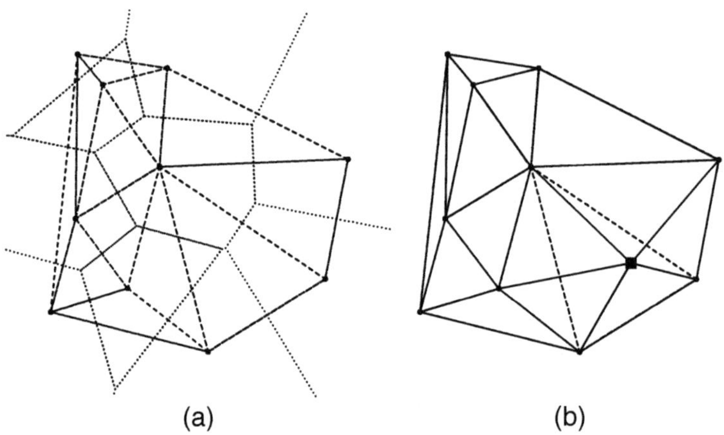

Fig. 3.10 (a) Example showing a Voronoi diagram (dotted line) and a Delaunay triangulation (dashed line) associated with a configuration x of $n = 10$ points; (b) modification of the triangulation when we add the point ■.

For ϕ and ψ bounded singleton and pair potentials, a PP model for this NN relation is defined by the Gibbs density

$$f(x) = c \exp\{\sum_{i=1}^{n} \phi(x_i) + \sum_{x_i \sim_{t(x)} x_j} \psi(x_i, x_j)\}.$$

It can be shown that f is a Markov NN density with respect to $\sim_{t(x)}$.

3.6.3 Gibbs point process on $\mathbb{R}^d$

The study of Gibbs PPs is important in the effort to characterize asymptotic properties of estimators of PP models observed over all $\mathbb{R}^d$. Given a symmetric relation $\sim$ on $\mathbb{R}^d$ and a locally finite configuration x, the definition of the Papangélou conditional intensity can be extended by setting

$$\lambda(\xi,x) = \exp\left\{-\sum_{y\subset x\cup\{\xi\},\, y\in\mathscr{C}} \phi(y\cup\{\xi\})\right\}, \qquad \xi\in\mathbb{R}^d,$$

where $\mathscr{C}$ is the family of cliques with respect to $\sim$ and ϕ a potential on $\mathscr{N}_{\mathbb{R}^d}$. There are several ways to define stationary Gibbs PPs on $\mathbb{R}^d$ (84; 165; 160):

1. *Starting from local specifications (Georgii, (84)):* for each bounded Borel set A of $\mathbb{R}^d$, the distribution of $X\cap A$ conditional on $X\cap A^c$ is independent of $X\cap\partial A\cap A^c$, the configuration of X on $\partial A\cap A^c$ with density with respect to a PPP(1) of:

$$\pi(x_A|x_{\partial A}) = Z^{-1}(x_{\partial A})\exp\{-\sum_{y\subset x_A\cup x_{\partial A},\, y\cap x_A\neq\emptyset} \phi(y)\}.$$

2. *Starting from integral representations (Nguyen and Zessin (165))*: for any function $h:\mathbb{R}^d\times\mathscr{N}_{\mathbb{R}^d}\to[0,\infty)$,

$$E[\sum_{\xi\in X} h(\xi,X\backslash\{\xi\})] = \int_{\mathbb{R}^d} E\left[h(\eta,X)\lambda(\eta,X)\right]d\eta. \tag{3.17}$$

This integral representation is the basis of the definition of PP "residuals" (cf. §5.5.7).

Specifying PPs using local representations introduces the same problems we met for Gibbs random fields on networks: existence and/or uniqueness, stationarity, ergodicity and weak dependency. There are two strategies for dealing with these issues. Denote $\mathscr{G}(\phi)$ the set of distributions of Gibbs PPs with potential ϕ.

1. For ϕ a translation-invariant singleton and pair interaction potential: $\phi(x_i)=\alpha$, $\phi(\{x_i,x_j\}) = g(\|x_i-x_j\|)$, Ruelle (190) and Preston (176) showed that $\mathscr{G}(\phi)$ is non-empty for a large class of functions g. It suffices for example that there exists two positive and decreasing functions g_1 and g_2, g_1 on $[0,a_1[$ and g_2 on $[a_2,\infty[$ with $0<a_1<a_2<\infty$, such that $g(t)\geq g_1(t)$ for $t\leq a_1$ and $g(t)\geq -g_2(t)$ for $t\geq a_2$ that satisfy:

$$\int_0^{a_1} g_1(t)dt=\infty \quad\text{and}\quad \int_{a_2}^{\infty} g_2(t)<\infty.$$

2. A second approach (Klein, (131)) uniquely associates with X a lattice process X^* with potential ϕ^* (cf. §5.5.5.2) and uses results on Gibbs lattice random fields (Georgii, (85)) ensuring $\mathscr{G}(\phi^*)\neq\emptyset$.

Exercises

3.1. Uniform simulation in balls.

1. Let U_1 and U_2 be uniform i.i.d. on $[0,1]$. Show that the polar coordinate $(R,\Theta) = (\sqrt{U_1}, 2\pi U_2)$ is uniformly distributed over the disk $B_2(0,1) \subset \mathbb{R}^2$. How much is gained by this circular procedure compared to the square method consisting of retaining (U_1,U_2) only if $U_1^2 + U_2^2 \leq 1$?
2. Suggest a "spherical" simulation strategy for uniformly generating points in the sphere $B_3(0,1) \subset \mathbb{R}^3$. How much is gained compared to the cube-based method?

3.2. Uniform simulation on a Borel set.

Show that the following algorithm simulates a uniform distribution on a bounded Borel set $W \subset \mathbb{R}^2$: start by including $W \subset R = \cup_{k=1}^K R_k$ in a finite union of rectangles R_k with pairwise measure zero intersections, then:

1. `Choose` R_k `with probability` $\nu(R_k)/\nu(R)$.
2. `Generate a point` x `uniformly in` R_k.

   `If` $x \in W$, `keep it. Otherwise, repeat from step 1.`

3.3. Cox processes.

Let X be a Cox process on S.

1. Show that X is overdispersed, i.e., show that $var(N_X(B)) \geq E(N_X(B))$.
2. Calculate the distribution of $N_X(B)$ when $B \subset S$ if X is driven by $Z \equiv \xi$, where ξ follows a $\Gamma(\theta_1, \theta_2)$.

3.4. Simulation of a Poisson point process on $\mathbb{R}$.

Let $X = \{X_i\}$ be a homogeneous PPP(λ) on $\mathbb{R}^1$ such that $X_0 = \inf\{X_i, X_i \geq 0\}$ and for all $i \in \mathbb{Z}$, $X_i \leq X_{i+1}$.

1. Show that variables $X_i - X_{i-1}$ are i.i.d. $\sim \mathscr{E}xp(\lambda)$.
2. Find the distributions of the following r.v.: (i) X_0 and X_1; (ii) $D_1 = \inf\{X_1 - X_0, X_0 - X_{-1}\}$; (iii) $D_1' = \inf\{X_0, -X_{-1}\}$; (iv) $D_2 =$ distance to the 2nd-nearest neighbor of X_0.
3. Suggest possible methods for simulating X on $[a,b]$.

3.5. Second-nearest neighbor distance.

Let X be a homogeneous Poisson PP on $\mathbb{R}^2$ with intensity ρ.

1. If $x_0 \in X$ and if $x_1 \in X$ is the closest point to x_0, find the distribution of $D_2 = \inf\{d(x_0,x), x \in X \backslash \{x_0,x_1\}\}$.
2. Same question if x_0 is some point in $\mathbb{R}^2$.

3.6. The Matérn-I model.

Matérn-I models (154; 48, p. 669) are "hard-core" models obtained in the following way. Let X_0 be a homogeneous PPP(ρ_0) on $\mathbb{R}^2$ and $r > 0$. Looking at all pairs of points of X_0, we remove all points of X_0 appearing in (at least) one pair of points $\leq r$ apart. The resulting process X_1 is called a Matérn-I model, it is more regular than the initial Poisson PP and has no points closer than r to each other.

1. Show that X_1 is a homogeneous PP with intensity $\rho_1 = \rho_0 \exp\{-\pi\rho_0 r^2\}$.
2. Show that the second-order intensity of X_1 is $\alpha_2(h) = \rho_0^2 k(\|h\|)$, where $k(\|s-u\|) = \exp\{-\rho_0 V_r(\|s-u\|)\}$ if $\|s-u\| \geq r$, $k(\|s-u\|) = 0$ otherwise, with $V_r(z)$ the surface of the union of two spheres of radius r a distance of z apart.
3. Deduce K, Ripley's second-order reduced moment for X_1.

3.7. The Bernoulli lattice process and the Poisson point process.

1. Let $S_\delta = (\delta\mathbb{Z})^d$ be the d-dimensional δ-skeleton in $\mathbb{R}^d$. A Bernoulli process $Y(\delta) = \{Y_i(\delta), i \in S_\delta\}$ with parameter p is a set of i.i.d. Bernoulli variables with parameter p. Show that if $p = p_\delta = \lambda\delta^d$, the Bernoulli process converges to a PPP(λ) as $\delta \to 0$.
2. Let X be a homogeneous PPP(ρ) on $\mathbb{R}^d$ and $Y_i = \mathbf{1}((X \cap [i, i+1[) \neq \emptyset)$, where for $i \in \mathbb{Z}^d$, $[i, i+1[$ is the cube with sides of length 1 and base i. Show that Y is a Bernoulli process.
3. Let X be a homogeneous PPP(ρ) on $\mathbb{R}^2$. Define $Z = \{Z_i,\ i \in \mathbb{Z}^2\}$, where $Z_i = N(i+A)$ for $A =]-1,1]^2$. Calculate $E(Z_i)$ and $Cov(Z_i, Z_j)$.

3.8. Simulation of point processes.
Simulate the following point processes on $S = [0,1]^2$:

1. An inhomogeneous Poisson PP with intensity $\rho(x,y) = 1 + 4xy$.
2. 20 points of a hard-core process with $r = 0.05$.
3. The following Neyman-Scott process: (i) parents come from a $PPP(20)$; (ii) the distribution of the number of descendants is given by $\mathscr{B}in(10, 1/2)$; (iii) dispersion around x_i uniformly on $B_2(x_i, 0.1)$; also, suppose independence of all distributions.
4. A marked PP $\{(x_i, m_i)\}$ where $X = (x_i)$ is a PPP(20) with mark:
   (a) A ball $B_2(x_i, r_i)$ with radius $r_i \sim \mathscr{U}([0, 0.1])$.
   (b) A segment centered at x_i of length $\mathscr{E}xp(10)$ and uniformly oriented.

3.9. Bivariate point process models (217).
Suppose $Y = (X_1, X_2)$ is a bivariate PP such that X_1 and X_2 are independent PPP(α) on $S \subset \mathbb{R}^2$, conditional on $d(X_1, X_2) \geq r$ for some fixed $r > 0$.

1. Show that Y has, with respect to the product of two independent PPP(1) the density: $f(x_1, x_2) = c\alpha^{n(x_1)+n(x_2)}\mathbf{1}\{d(x_1, x_2) \geq r\}$.
2. Show that the Papangélou conditional intensity for Y when adding a point ξ_1 to x_1 is: $\lambda(\xi_1, (x_1, x_2)) = \alpha\mathbf{1}\{d(\xi_1, x_2) \geq r\}$. Deduce that Y is a Markov process with respect to the relation: $(\xi, i) \sim (\eta, j) \Longleftrightarrow i \neq j$ and $\|\xi - \eta\| < r$: Y is Markov with range r.
3. Show that X_1 is the area interaction PP: $f_1(x_1) = c_1\alpha^{n(x_1)}e^{\alpha a(x_1)}$.
4. Show that $X = X_1 \cup X_2$ has density $g(x) = c'\sum^* f(x_1, x_2) = c''\alpha^{n(x)}k(x)$ where $\sum^*$ (resp. $k(x)$) is the set (resp. the number) of partitions of x into $x_1 \cup x_2$ so that the distance from x_1 to x_2 is $\geq r$.
5. Let $B(x) = \cup_{x_i \in x} B(x, r/2)$ and let $c(x)$ be the number of connected components of $B(x)$. Show that $k(x) = 2^{c(x)}$. X is thus a connectivity-interaction PP with parameter $\beta = 2$.
6. Show that these properties hold true when X_1 and X_2 have different parameters.

Chapter 4
Simulation of spatial models

Being able to simulate probability distributions and random variables is useful whenever we lack an analytic solution to a problem, be it combinatorial (number of ways to put 32 dominoes on an 8×8 grid), a search for maxima (Bayesian image reconstruction, cf. §2.2.2) or calculating integrals. For example, calculating the expectation of real statistics $g(X)$ of the random variable X with distribution π:

$$\pi(g) = \int_{\Omega} g(x)\pi(dx),$$

becomes impractical if X is high-dimensional. Monte Carlo simulation involves estimating $\pi(g)$ with

$$\overline{g}_n = \frac{1}{n}\sum_{i=1}^{n} g(X_i)$$

using n values $\{X_i, i = 1, \dots, n\}$ sampled from the distribution π. The Strong Law of Large Numbers (SLLN) ensures that $\overline{g}_n \to \pi(g)$ if $n \to \infty$ and $g(X)$ is integrable. Furthermore, if $Var\{g(X)\} < \infty$, the Central Limit Theorem (CLT) shows that convergence occurs at a rate $n^{-1/2}$.

A further example that uses Monte Carlo simulation is the empirical calculation of the distribution of a statistic (e.g., real-valued) $T(X)$. If the distribution of T is unknown (difficulty of calculation, lack of knowledge of asymptotic results), we can estimate its cumulative distribution function F from its empirical distribution over n samples from T:

$$F_n(t) = \frac{1}{n}\sum_{i=1}^{n} \mathbf{1}(T(X_i) \le t).$$

In particular, the quantile $t(\alpha)$ of T, defined by $P(T \le t(\alpha)) = \alpha$ can be estimated by $T_{([n\alpha])}$ as n increases, where $T_{(1)} \le T_{(2)} \le \dots \le T_{(n)}$ is the order statistic for the $(T_i = T(X_i), i = 1, \dots, n)$ and $[r]$ the integer part of the real-valued r. This method is therefore based on simulating n times from X. For large n, $T_{([n\alpha])}$ is a good approximation of $t(\alpha)$. We can thus calculate statistical tables or approximate confidence intervals for the statistic T that are asymptotically exact.

C. Gaetan, X. Guyon, *Spatial Statistics and Modeling*, Springer Series in Statistics,
DOI 10.1007/978-0-387-92257-7_4,

When X can be simulated using a standard method (inverting the cumulative distribution function or using the acceptance-rejection method (cf. (59; 188) and Appendix A), it is easy to approximately calculate the desired quantile.

However, if π is "complex", these standard methods are impractical. The following are two examples in the case of spatial random fields:

1. For an Ising model X on $S = \{1,2,\dots,10\}^2$ (cf. §2.2.2), inversion of the cumulative distribution function, classically used for simulating r.v. with a finite number of states (here $\Omega = \{-1,+1\}^S$ is finite) can not be implemented because Ω is too big ($\sharp\Omega = 2^{100} \simeq 1.27 \times 10^{30}$). First, the distribution $\pi(x) = c^{-1} \exp U(x)$, $c = \sum_{y\in\Omega} \exp U(y)$ cannot be evaluated; second, dividing $[0,1]$ into $\sharp\Omega$ subintervals, necessary for inverting the cumulative distribution function, is entirely impractical.
2. Suppose we want to simulate a hard-core point process X (cf. §3.4.2) on $S = [0,1]^2$ with the hard-core radius $r = 0.1$, conditional on $n = 42$ generated points. To do this, we propose the following acceptance-rejection method: simulate n uniform i.i.d. points on S and keep the configuration if all pairs of points are at least a distance r apart. As the probability that a pair of points be $\leq r$ apart is πr^2, the mean total number of pairs $\leq r$ apart is of the order $n^2\pi r^2$. When n is not too large, this number approximately follows a Poisson distribution (33) and the probability that a generated configuration is an r-hard-core is approximately $\exp(-n^2\pi r^2)$, that is, 1.48×10^{-22} for our particular case ($n = 42$). This value becomes 5.2×10^{-13} when $n = 30$, 4×10^{-6} when $n = 20$ and 0.04 when $n = 10$. That is, the reject method is useless if $n \geq 20$.

To get around these problems, algorithms using *Markov chain dynamics* have been introduced. If a Markov chain (X_n) "converges" in distribution to π, X_n gives an asymptotic simulation from the distribution π for large n. Such methods are called *Monte Carlo Markov Chains* (MCMC). We are going to present here the two principal algorithms, *Gibbs sampling* for sampling distributions on *product spaces* E^S and the *Metropolis-Hastings* algorithm (MH) for *general spaces*.

Let us begin by recalling some basic definitions and properties of Markov chains that are useful for constructing MCMC algorithms and controlling their convergence. For additional details, we suggest taking a look at (157; 72; 103; 229).

4.1 Convergence of Markov chains

Let π be a probability distribution that is absolutely continuous with respect to a reference measure μ defined on $(\Omega, \mathscr{E})$. With a slight abuse of notation, we also denote π the density of this distribution. In order to simulate π we construct a Markov chain $X = (X_k, k \geq 0)$ on Ω with transition P such that the distribution of X_n converges to π. Thus for large k, X_k is close to π. We now recall some useful details on Markov chains.

Let $X = (X_n, n \geq 0)$ be a process with values in $(\Omega, \mathscr{E})$ and distribution $\mathbb{P}$. We say that X is a Markov chain if, for all $n \geq 1$, each event $A \in \mathscr{E}$ and sequence $x_0, x_1, x_2, \ldots$ from Ω,

$$\begin{aligned}\mathbb{P}(X_{n+1} &\in A | X_n = x_n, \ldots, X_0 = x_0) \\ &= \mathbb{P}(X_{n+1} \in A | X_n = x_n) = P_n(x_n, A).\end{aligned}$$

The conditional probability P_n represents the *chain transition* at the nth step. we say the chain is homogeneous if $P_n \equiv P$ for all n. P will simply be called the transition (probability) from now on. Roughly speaking, X is a Markov chain if its temporal memory is 1.

Definitions and examples

The *transition* P of a homogeneous Markov chain on $(\Omega, \mathscr{E})$ is a mapping $P : (\Omega, \mathscr{E}) \to [0,1]$ such that:

1. For all $A \in \mathscr{E}$, $P(\cdot, A)$ is measurable.
2. For all $x \in \Omega$, $P(x, \cdot)$ is a probability on $(\Omega, \mathscr{E})$.

If $X_0 \sim \nu_0$, the distribution of $(X_0, X_1, \ldots, X_n)$ for a homogeneous chain is characterized by:

$$P_n(dx_0, dx_1, \ldots, dx_n) = \nu(dx_0) \prod_{i=1}^{n} P(x_{i-1}, dx_i).$$

For $k \geq 1$, the marginal distribution ν_k of X_k is

$$\nu_k(dx_k) = \int_\Omega \nu_0(dx) P^k(x, dx_k),$$

where $P^k(x, \cdot) = \int_{\Omega^{k-1}} \prod_{i=1}^{k} P(x_{i-1}, dx_i)$ is the *transition after k steps* of the chain. In particular, the distribution of X_1 if $X_0 \sim \nu$ is

$$(\nu P)(dy) = \int_\Omega \nu(dx) P(x, dy). \tag{4.1}$$

For finite state spaces $\Omega = \{1, 2, \ldots, m\}$, distributions ν on Ω are represented by vectors in $\mathbb{R}^m$ and Markov chain transitions by the $m \times m$ matrix of probabilities $P_{i,j} = \Pr(X_{n+1} = j | X_n = i)$. The ith row represents the initial state and the jth column the final state. We say that P is a stochastic matrix in the sense that each row of P is a discrete probability distribution on Ω: for each $i \in \{1, 2, \ldots m\} : \sum_{j=1}^{m} P(i, j) = 1$. Under this notation, the transition P^k for k steps of a discrete, homogeneous chain can simply be found by taking the kth power of P. The distribution of X_k is thus νP^k. This formulation (and these properties) can be easily extended to discrete countable spaces $\Omega = \{1, 2, 3, \ldots\}$.

The following processes are Markov chains:

1. Real-valued random walks defined by the recurrence relation: $X_{n+1} = X_n + e_n$, $n \geq 0$ where $e = (e_n)$ is an i.i.d. sequence in $\mathbb{R}$.
2. $AR(1)$ processes with i.i.d. noise.
3. If X is an $AR(p)$, the sequence of vectors $X_n^{(p)} = (X_n, X_{n+1}, \ldots, X_{n+p-1})$ is a Markov chain on $\mathbb{R}^p$. X has memory p and $X^{(p)}$ memory 1.
4. $X_{n+1} = \Phi(X_n, U_n)$, where $\Phi : \Omega \times [0,1] \longrightarrow \Omega$ is measurable and (U_n) a uniformly distributed i.i.d. sequence of random variables on $[0,1]$. If Ω is discrete, any Markov chain can be written in this way (cf. Appendix A2).
5. *Random walks on graphs*: Let $S = \{1, 2, \ldots, m\}$ be a finite state space endowed with a symmetric graph $\mathcal{G}$ in which there exists a path between all pairs of points. Then, denote $d_i > 0$ the number of neighbors of point i and $\langle i, j \rangle$ the fact that i and j are neighbors. The following transition defines a random Markov walk on S:
$$P(i,j) = \frac{1}{d_i} \text{ if } \langle i,j \rangle \qquad \text{and} \qquad P(i,j) = 0 \text{ otherwise.} \tag{4.2}$$
6. *Birth and death processes:* Suppose S is finite and associated with a symmetric graph $\mathcal{G}$ without loops and $x \in \Omega = \{0,1\}^S$ some configuration of $\{0,1\}$ on S. Consider the dynamics $X_n = x \mapsto X_{n+1} = y$ defined by: choose uniformly at random a site $s \in S$.
   a. If $X_n(s) = 1$, retain $X_{n+1}(s) = 1$ with probability $\alpha > 0$; otherwise, $X_{n+1}(s) = 0$.
   b. If $X_n(s) = 0$ and if a neighboring site has the value 1, then let $X_{n+1}(s) = 1$ with probability $\beta > 0$; otherwise $X_{n+1}(s) = 0$.
   c. Lastly, if $X_n(s) = 0$ and all neighbors of i have the value 0, let $X_{n+1}(s) = 1$ with probability $\gamma > 0$; otherwise, $X_{n+1}(s) = 0$.

Irreducibility, aperiodicity and P-invariance of a Markov chain

Suppose π is a distribution on $(\Omega, \mathcal{E})$. Convergence of Markov chains is linked to the following three properties:

1. P is said to be *π-irreducible* if for each $x \in \Omega$ and $A \in \mathcal{E}$ such that $\pi(A) > 0$, there exists $k = k(x,A)$ such that $P^k(x,A) > 0$.
2. π is said to be *P-invariant* if $\pi P = \pi$, with πP defined as in (4.1).
3. P is *periodic* if there exists some $d \geq 2$ and some partition $\{\Omega_1, \Omega_2, \ldots, \Omega_d\}$ of Ω such that for all $i = 1, \ldots, d$, $P(X_1 \in \Omega_{i+1} | X_0 \in \Omega_i) = 1$, with the convention $d + 1 \equiv 1$. If this is not the case, we say the chain is *aperiodic*.

π-irreducibility means that any event with positive π-probability can be reached from any initial state in a finite number of steps with positive probability. When Ω is finite, we say that the transition (matrix) P is *primitive* if there exists some integer $k > 0$ such that, for each i, j, $P^k(i,j) > 0$. That is, we can pass from any state i to any state j in k steps. For finite Ω, P being primitive implies irreducibility.

If Ω is discrete, finitely or countably and if π is positive (noted $\pi > 0 : \forall i \in \Omega$, $\pi(i) > 0$), π-irreducibility is equivalent to there being communication between all states: we say that i communicates with j if there exists $k(i,j) > 0$ such that $P^{k(i,j)}(i,j) > 0$.

d-periodicity means that the transition P^d lives separately in d disjoint state subspaces Ω_l, $l = 1,\dots,d$. A sufficient condition ensuring aperiodicity of chains is that the transition (density) satisfies $P(x,x) > 0$ for some set of x's with μ-measure > 0.

The random walk (4.2) is irreducible. The same is true for the birth and death process (6) as long as $\alpha, \beta, \gamma \in]0,1[$. The random walk is aperiodic if $\mathscr{G}$ has at least one loop but is no longer so if there are $m = 2$ sites and no loops.

Stationary distribution of a Markov chain

If π is a P-invariant distribution, we say that π is a stationary distribution of X because if $X_0 \sim \pi$, $X_k \sim \pi$ for all k. In this case, the process $X = (X_k)$ is stationary.

Convergence of a Markov chain

It is easy to show that if (X_k) is a Markov chain with transition P such that:

$$\text{for all } x \in \Omega \text{ and } A \in \mathscr{E} : P^n(x,A) \to \pi(A) \text{ if } n \to \infty,$$

then π is P-invariant and P is π-irreducible and aperiodic. The remarkable property giving the foundation of MCMC algorithms is the converse of this result. Before describing this property, let us define the *total variation norm* (TV) between two probabilities ν_1 and ν_2 on $(\Omega, \mathscr{E})$:

$$\| \nu_1 - \nu_2 \|_{TV} = \sup_{A \in \mathscr{E}} |\nu_1(A) - \nu_2(A)| .$$

If the sequence $\nu_n \to \nu$ with respect to the TV norm, then $\nu_n \to \nu$ in distribution. When Ω is discrete, the TV norm is none other than the half-norm l_1 :

$$\| \nu_1 - \nu_2 \|_{TV} = \frac{1}{2} \|\nu_1 - \nu_2\|_1 = \frac{1}{2} \sum_{i \in \Omega} |\nu_1(i) - \nu_2(i)| .$$

Theorem 4.1. *Convergence of Markov chains (Tierney, (214))*

1. *Let P be a π-irreducible and π-invariant aperiodic transition. Then, π-a.s. for $x \in \Omega$, $\left\|P^k(x,\cdot) - \pi\right\|_{TV} \to 0$ when $k \to \infty$.*
2. *If furthermore for all $x \in \Omega$, $P(x,\cdot)$ is absolutely continuous with respect to π, this convergence happens for all $x \in \Omega$.*

Under conditions (1-2) we have for any initial distribution ν, $\left\|\nu P^k - \pi\right\|_{VT} \to 0$. If Ω is finite, we have, more precisely (129):

Theorem 4.2. *Convergence of a Markov chain with a finite state space*
Let P be a transition on Ω. The following statements are equivalent:

1. For all ν, $\nu P^k \to \pi$, and $\pi > 0$ is the unique invariant distribution of P.
2. P is irreducible and aperiodic.
3. P is primitive.

Furthermore, if for some $k > 0$, $\varepsilon = \inf_{i,j} P^k(i,j) > 0$, then, defining $m = \sharp\Omega$ and $[x]$ the integer part of x, we have:

$$\|\nu P^n - \pi\|_{TV} \leq (1 - m\varepsilon)^{[n/k]}. \tag{4.3}$$

To construct an MCMC algorithm for simulating π, it is thus necessary to propose a transition P that is "easy" to simulate, π-irreducible and aperiodic. Also, π must be P-invariant. π-irreducibility and aperiodicity must be verified on a case by case basis. Finding a P such that π is P-invariant is not easy in general. If for example Ω is finite, this question is linked to the search for a left eigenvector of P associated with the eigenvalue 1, which is difficult to find when Ω is large.

A simple strategy that ensures π is P-invariant is to propose a P that is *π-reversible*.

Definition 4.1. π-reversible chains
P is *π-reversible* if

$$\forall A, B \in \mathscr{E}, \int_A P(x,B)\pi(dx) = \int_B P(x,A)\pi(dx).$$

If π and P are distributions with densities, π-reversibility of P is written:

$$\forall x, y \in \Omega: \pi(x)p(x,y) = \pi(y)p(y,x).$$

If the transition P of a chain X is π-reversible, then the distributions of pairs (X_n, X_{n+1}) and (X_{n+1}, X_n) are identical if $X_n \sim \pi$. The chain's distribution remains unchanged when time is reversed. Furthermore, we have:

Proposition 4.1. *If P is π-reversible, then π is P-invariant.*

Proof. π-reversibility of P gives directly:

$$(\pi P)(A) = \int_\Omega \pi(dx)P(x,A) = \int_A \pi(dx)P(x,\Omega) = \pi(A).$$

□

Example 4.1. Simulating hard-core models on discrete networks

Let $S = \{1, 2, \dots, m\}$ be a finite set of sites associated with a symmetric neighbor graph $\mathscr{G}$ without loops. A hard-core configuration on $(S, \mathscr{G})$ is a configuration of $x_i \in \{0,1\}$ at each site $i \in S$ ($x_i = 1$ if i is busy, $x_i = 0$ if i is free) such that neighboring sites of S cannot be simultaneously busy. The space of possible configurations is thus:

$$\Omega_0 = \{x \in \{0,1\}^S \text{ such that } x_i x_j = 0 \text{ if } i \text{ and } j \text{ are neighbors}\}.$$

Hard-core models are used in 3-dimensional physics to analyze behavior of gases over networks where particles of non-negligible radius cannot physically overlap. For uniform π on Ω_0, we can for example investigate the mean number of occupied sites of particular hard-core configurations or confidence intervals for this number. As Ω_0 has a large cardinal number, simulation of this uniform distribution is performed using the following algorithm (Häggström, (103)):

1. `Choose uniformly at random a site` i `of` S.
2. `Flip an unbiased coin.`
3. `If it comes up "Heads" and if the neighboring sites of` i `are free, then` $x_i(n+1)=1$`; otherwise` $x_i(n+1)=0$.
4. `For` $j \neq i$, $x_{n+1}(j)=x_n(j)$. `Return to 1.`

We now show that P is *π-reversible*. As π is uniform, reversibility of P is equivalent to P being symmetric: $P(x,y)=P(y,x)$ when $x \neq y$. Let i be the site where the change $x_i \neq y_i$ occurs (no other changes occur elsewhere). There are two possibilities: (i) $x_i = 1 \mapsto y_i = 0$ occurs with probability $1/2$ because only flipping 'Tails' gives this result (if we had flipped 'Heads' and all neighboring sites of i were unoccupied, we would have had $y_i = 1$); (ii) $x_i = 0 \mapsto y_i = 1$ can only happen if we flip 'Heads' and if all neighboring sites are vacant. This transition thus occurs with probability $1/2$. P is therefore symmetric and hence π-reversible.

Furthermore, P is *irreducible*: let $x,y \in \Omega_0$; we show that we can move from any x to any y in Ω_0 in a finite number of steps; this is for example possible by moving from x to $\mathbf{0}$ (the configuration with 0 everywhere) by deleting one by one the points where $\mathbf{x}$ equals 1, then from $\mathbf{0}$ to y by adding one by one the points where y equals 1. If $n(z)$ is defined as the number of sites occupied by z, moving from x to $\mathbf{0}$ is possible in $n(x)$ steps, each with probability $\geq 1/2m$, moving from $\mathbf{0}$ to y is similarly possible in $n(y)$ steps, each with probability $\geq 1/2m$. We thus obtain

$$P^{n(x)+n(y)}(x,y) \geq (1/2m)^{n(x)+n(y)} > 0, \qquad \forall x,y \in \Omega_0.$$

To prove aperiodicity, it is sufficient to remark that for any configuration x, $P(x,x) \geq 1/2$. The Markov chain with transition P therefore gives us a way to (approximately) simulate from π.

To calculate the mean number M_0 of points in a hard-core configuration, we begin by generating the sequence $X_0, X_1, X_2, \ldots, X_M$ with respect to transition P (with for example the initial state $X_0 = 0$). Denoting $n(X)$ the number of points in trial X, M_0 can be estimated by $\widehat{M}_0 = M^{-1}\sum_{k=1}^{M} n(X_k)$. If we are interested in the variability of this distribution around M_0, we can use the estimation $\widehat{\sigma}^2 = M^{-1}\sum_{k=1}^{M} n(X_k)^2 - (\widehat{M}_0)^2$.

4.1.1 Strong law of large numbers and central limit theorem for a homogeneous Markov chain

The rate of convergence of $\overline{g}_n = n^{-1}\sum_{i=1}^{n} g(X_i) \to \pi(g)$ can be quantified using the following notion of *geometric ergodicity* (157; 71; 188):

$$\exists \rho < 1,\ \exists M \in L^1(\pi) \text{ s.t. } \forall x \in \Omega\, \forall k \geq 1:\ \|P^k(x,\cdot) - \pi(\cdot)\|_{TV} \leq M(x)\rho^k.$$

Theorem 4.3. *Strong law of large numbers and central limit theorem for Markov chains*

1. *SLLN: if X is π-irreducible with invariant distribution π and if $g \in L^1(\pi)$, then π-a.s.:*
$$\overline{g}_n = \frac{1}{n}\sum_{k=0}^{n-1} g(X_k) \to \pi g = \int g(x)\pi(dx).$$
We denote
$$\sigma^2(g) = Var(g(X_0)) + 2\sum_{k=1}^{\infty} Cov(g(X_0), g(X_k)),$$
provided that this quantity exists. In this formula, the covariances are evaluated under the stationary distribution.
2. *CLT: if P is geometrically ergodic, then $\sigma^2(g)$ exists. If furthermore $\sigma^2(g) > 0$, we have:*
$$\sqrt{n}\,(\overline{g}_n - \pi g) \xrightarrow{d} \mathcal{N}(0, \sigma^2(g))$$
if one of the following conditions is satisfied: (i) $g \in L^{2+\varepsilon}(\pi)$ for some $\varepsilon > 0$; (ii) P is π-reversible and $g \in L^2(\pi)$.

4.2 Two Markov chain simulation algorithms

Let π be a distribution with density $\pi(x)$ with respect to a measure μ on $(\Omega, \mathscr{E})$. If Ω is discrete, we choose μ to be the counting measure. Suppose positivity, i.e., for all $x \in \Omega$, $\pi(x) > 0$, if necessary shrinking the state space until this is true. To simulate π, we aim to construct a π-irreducible aperiodic chain with transition P for which π is P-invariant.

Gibbs sampling is only applicable when working with product spaces: $\Omega = E^S$. In contrast, the Metropolis algorithm is applicable for general state spaces.

4.2.1 Gibbs sampling on product spaces

Let $X = (X_i,\ i \in S)$ be a variable on $S = \{1, 2, \ldots, n\}$ with distribution π on $\Omega = \prod_{i \in S} E_i$. For each $x = (x_1, x_2, \ldots, x_n) \in \Omega$, suppose that the conditional distributions $\pi_i(\cdot|x^i)$, where $x^i = (x_j, j \neq i)$, can be easily simulated. We consider transitions

$$P_i(x,y) = \pi_i(y_i|x^i)1(x^i = y^i) \tag{4.4}$$

which allow a change at one site i only, this being $x_i \mapsto y_i$, with probability $\pi_i(y_i|x^i)$, x being unchanged elsewhere ($x^i = y^i$). We now take a look at two types of sampling.

Gibbs sampling with sequential sweep

Suppose that we *sequentially visit* S, for example in the order $1 \to 2 \to \ldots \to (n-1) \to n$. A sequence of visits covering all sites of S is called a *sweep* or *scan* of S. At each step, the value x_i at i is relaxed, i.e., drawn randomly with respect to π_i, conditional on the current state. The transition (density) from state $x = (x_1, x_2, \ldots, x_n)$ to state $y = (y_1, y_2, \ldots, y_n)$ after sweeping S is given by:

$$P_S(x,y) = \prod_{i=1}^{n} \pi_i(y_i | y_1, \ldots, y_{i-1}, x_{i+1}, x_{i+2}, \ldots, x_n).$$

At the ith sweep step, π_i is conditioned by the $(i-1)$ already simulated values y and the $(n-i)$ values of x than are yet to be simulated.

Gibbs sampling with random sweep

Let $p = (p_1, p_2, \ldots, p_n)$ be a positive probability distribution on S (for all i, $p_i > 0$). At each step, a site i is chosen at random following the distribution p and the value at this site is simulated following the distribution π_i conditional on the current state. The transition for one sweep is thus

$$P_R(x,y) = \sum_{i=1}^{n} p_i \pi_i(y_i | x^i) \mathbf{1}(x^i = y^i).$$

Theorem 4.4. *Convergence of Gibbs sampling.*

Suppose that for all $x \in \Omega$, $\pi(x) > 0$. Then, the transitions P_S of the sequential sampler and P_R of the random sampler are π-irreducible and aperiodic with invariant distribution π. Furthermore, for any initial distribution ν, $\nu P^k \to \pi$.

Proof. *Sequential sampler P_S*: Positivity of π implies positivity of the conditional densities:

$$\pi_i(x_i | x^i) > 0, \qquad \forall x = (x_i, x^i).$$

We deduce that for all x,y, $P_S(x,y) > 0$. P_S is thus π-irreducible and aperiodic. Next we show that π is P_S-invariant. As P_S is made up of transitions P_i (4.4), it suffices to show that each P_i is π-invariant. Remark that each P_i is π-reversible as, in effect,

$$\begin{aligned}
\pi(x)P_i(x,y) &= \pi(x_i, x^i)\pi_i(y_i | x^i)\mathbf{1}(x^i = y^i) \\
&= \frac{\pi(x_i, x^i)\pi_i(y_i, x^i)}{\pi^i(x^i)}\mathbf{1}(x^i = y^i) \\
&= \pi(y)P_i(y,x)
\end{aligned}$$

is symmetric for (x,y). π is thus P_S-invariant. As $P_S(x,\cdot)$ is absolutely continuous with respect to π, we deduce that for any initial distribution ν, $\nu P^k \to \pi$.

If Ω is finite, $\eta = \inf_{i,x} \pi_i(x_i|x^i) > 0$ and $\varepsilon = \inf_{x,y} P(x,y) \geq \eta^n > 0$. From (4.3), we have the inequality:

$$\forall \nu, \left\| \nu P^k - \pi \right\|_{VT} \leq (1 - m\varepsilon)^k, \qquad m = \sharp\Omega.$$

Random sampler P_R: n transitions P_R must be lined up to show π-irreducibility and aperiodicity of P_R: if for example we first choose site 1, then 2 and on to n, we obtain the lower bound

$$\forall x, y \in \Omega : P_R^n(x,y) \geq p_1 p_2 \ldots p_n P_S(x,y) > 0.$$

P_R is π-reversible since

$$\begin{aligned}
\pi(x) P_R(x,y) &= \sum_{i=1}^{n} p_i \pi(x_i, x^i) \pi_i(y_i|x^i) \mathbf{1}(x^i = y^i) \\
&= \sum_{i=1}^{n} p_i \frac{\pi(x_i, x^i)\pi_i(y_i, x^i)}{\pi^i(x^i)} \mathbf{1}(x^i = y^i) \\
&= \pi(y) P_R(y,x).
\end{aligned}$$

π is thus P_R-invariant and the absolute continuity of $P_R(x,\cdot)$ with respect to π ensures convergence of the random sampler. If Ω is finite, $\delta = \inf_{x,y} P^n(x,y) \geq \nu_1 \nu_2 \ldots \nu_n \eta^n > 0$ and

$$\left\| \nu P^k - \pi \right\| \leq 2(1 - m\delta)^{[k/n]}.$$

□

To construct P, all that is needed is to know π up to a multiplicative constant. In effect, if $\pi(x) = c\,e(x)$, the conditional distributions π_i are independent of c. This is an important remark as usually (for example for Gibbs random fields), π is only known up to a multiplicative constant.

Different ways to sweep S that possibly lead to repeated visits, periodicity etc. of sweeping, or simulation on subsets of S other than single site visiting can also lead to convergence of $\nu P^n \to \pi$. The only condition required to ensure convergence of a sampler is that the sweep sequence visits infinitely often every site in S (81).

4.2.2 The Metropolis-Hastings algorithm

Proposed by Metropolis in 1953, this algorithm was stated in its general form by Hastings (110; 188) in 1970. The property of optimality in terms of asymptotic variance of the Metropolis algorithm within the family of Metropolis-Hastings (MH) algorithms was shown by Peskun ((172) and cf. §4.2.2). An important difference with Gibbs sampling is that here the state space Ω is not necessarily a product space.

Description of the Metropolis-Hastings algorithm

The MH algorithm is founded on the idea of constructing a transition P that is π-reversible, making P thus π-invariant. This is done in two steps:

1. *Proposal transition*: we start by proposing a change $x \mapsto y$ according to a transition $Q(x,\cdot)$.
2. *Acceptance probability*: we accept the change with probability $a(x,y)$, where $a : \Omega \times \Omega \to]0,1]$.

The two choices needed in the algorithm are Q, the transition for the proposed change and a the probability of accepting the change. Denoting by $q(x,y)$ the density of $Q(x,\cdot)$, the MH transition P is written:

$$P(x,y) = a(x,y)q(x,y) + \mathbf{1}(x=y)\left[1 - \int_\Omega a(x,z)q(x,z)dz\right]. \tag{4.5}$$

The choice (Q,a) ensures π-reversibility of P if the following *detailed balance equation* is satisfied:

$$\forall x,y \in \Omega : \pi(x)q(x,y)a(x,y) = \pi(y)q(y,x)a(y,x). \tag{4.6}$$

By imposing reversibility, we get that q satisfies the weak symmetry condition $q(x,y) > 0 \Leftrightarrow q(y,x) > 0$. Q must be such that the change $x \mapsto y$ is allowed, it must also allow the change $y \mapsto x$. For such couples (x,y), we define the MH ratio as

$$r(x,y) = \frac{\pi(y)q(y,x)}{\pi(x)q(x,y)}.$$

If q is symmetric, $r(x,y) = \pi(y)/\pi(x)$.

For it to be true that P provides a simulation of π, it remains to show that P is irreducible and aperiodic. For P to be irreducible, Q must be too, but this is not sufficient: irreducibility, like aperiodicity of P, must be examined case by case. If Ω is finite, aperiodicity is ensure if one of the following conditions holds:

1. There exists x_0 s.t. $Q(x_0,x_0) > 0$ (for at least one state, we allow no change).
2. There exists (x_0,y_0) s.t. $r(x_0,y_0) < 1$: for the Metropolis algorithm (4.8), this signifies that a change can be refused.
3. Q is symmetric and π is not uniform.

For (1) and (2), it suffices to show that $P(x_0,x_0) > 0$. Let us examine (3): as Q is irreducible, all states communicate under the relation: $x \sim y \Leftrightarrow q(x,y) > 0$; as π is not uniform and all states communicate, there exists $x_0 \sim y_0$ s.t. $\pi(x_0) > \pi(y_0)$. Aperiodicity thus results from the lower bound:

$$P(x_0,x_0) \geq Q(x_0,y_0)\left\{1 - \frac{\pi(y_0)}{\pi(x_0)}\right\} > 0.$$

As is the case for Gibbs sampling, it suffices to know π up to a multiplicative constant in order to construct the MH transition P.

The detailed balance equation (4.6) that ensures π-reversibility of P is satisfied if the probability of accepting changes a is written $a(x,y) = F(r(x,y))$ for some function $F :]0,\infty[\to]0,1]$ satisfying:

$$\forall \xi > 0\colon F(\xi) = \xi F(\xi^{-1}). \tag{4.7}$$

In effect, under this condition,

$$a(x,y) = F(r(x,y)) = r(x,y)F(r(x,y)^{-1}) = r(x,y)a(y,x).$$

Two classical dynamics satisfy (4.7): *Barker dynamics* (19) and *Metropolis dynamics*. Barker dynamics are associated with the function $F(\xi) = \frac{\xi}{1+\xi}$. If q is symmetric,

$$F(r(x,y)) = F\left(\frac{\pi(y)}{\pi(x)}\right) = \frac{\pi(y)}{\pi(x)+\pi(y)}.$$

We accept the proposed change $x \mapsto y$ if y is more probable than x.

The Metropolis algorithm

Metropolis dynamics correspond to the choice $F(\xi) = \min\{1,\xi\}$. In this case,

$$a(x,y) = \min\left\{1, \frac{\pi(y)q(y,x)}{\pi(x)q(x,y)}\right\} \tag{4.8}$$

and

$$a(x,y) = \min\left\{1, \frac{\pi(y)}{\pi(x)}\right\}$$

if q is symmetric.

If so, the Metropolis algorithm is:

1. Let x be the initial state, choose y according to the distribution $Q(x,.)$.
2. If $\pi(y) \geq \pi(x)$, keep y. Return to 1.
3. If $\pi(y) < \pi(x)$, generate a uniform variable U on $[0,1]$:
   a. If $U \leq p = \pi(y)/\pi(x)$, keep y.
   b. If $U > p$, keep the initial value x.
4. Return to 1.

Gibbs sampling is a special case of the Metropolis algorithm, the one for which the proposed change consists of randomly drawing a site i, relaxing x_i to y_i with respect to the density $q_i(x,y_i)$, then accepting y_i with probability

$$a_i(x,y) = \min\left\{1, \frac{\pi(y)q_i(y,x_i)}{\pi(x)q_i(x,y_i)}\right\}.$$

The choices $q_i(x,y_i) = \pi_i(y_i|x^i)$ lead to $a(x,y) \equiv 1$ and thus reduce to Gibbs sampling.

Example 4.2. A max cut problem for graphs

Let $S = \{1,2,\ldots,n\}$ be a set of sites, $w = \{w_{i,j}, i,j \in S\}$ a family of real-valued symmetric weights on $S \times S$, $\Omega = \mathscr{P}(S)$ the set of subsets of S and $U : \Omega \to \mathbb{R}$ defined by:

$$U(A) = \sum_{i\in A, j\notin A} w_{i,j} \quad \text{if } A \neq S \text{ and } A \neq \emptyset, \qquad U(A) = +\infty \text{ otherwise.}$$

Consider the following combinatorial problem: "*Find the subset $\Omega_{\min}$ of subsets $A \in \Omega$ that gives the minimum for U,*"

$$\Omega_{\min} = \{A \in \Omega : U(A) = \min\{U(B) : B \subseteq S\}\}.$$

It is not possible to resolve this problem by listing the possible values of U as the cardinality of Ω is too large. One way to get around this problem is the following: simulate a variable whose distribution π_β on Ω is:

$$\pi_\beta(A) = c(\beta)\exp\{-\beta U(A)\}. \tag{4.9}$$

In effect, if $\beta > 0$ is quite small, the mode of π_β will be found at configurations $A \in \Omega_{\min}$. Thus, simulating π_β for small β is a way to get close to $\Omega_{\min}$. We remark that this algorithm mimics another algorithm used to resolve this kind of optimization problem called *Simulated Annealing* (SA): SA simulations are performed on a sequence of parameters $\beta_n \to 0_+$. Convergence of β_n to 0 must be slow in order to ensure convergence of X_n to the uniform π_0 on $\Omega_{\min}$ (82; 106; 1; 10; 40).

To simulate π_β, we propose the following Metropolis algorithm:

1. The only changes $A \mapsto B$ allowed are: (i) $B = A \cup \{s\}$ if $A \neq S$ and $s \notin A$; (ii) $B = A \setminus \{s\}$ if $A \neq \emptyset$ and $s \in A$.
2. The proposal transition Q is created by uniformly choosing s in S: $Q(A,B) = 1/n$ for (i) and (ii); otherwise, $Q(A,B) = 0$.
3. Evaluate
   - $\Delta U = U(B) - U(A) = \sum_{j\notin B} w_{s,j} - \sum_{i\in A} w_{i,s}$ if (i),
   - $\Delta U = \sum_{i\in B} w_{i,s} - \sum_{j\notin A} w_{s,j}$ if (ii).
4. If $\Delta U \leq 0$, keep B.
5. If $\Delta U > 0$, generate uniformly U on $[0,1]$. If $U \leq \exp\{-\beta\Delta U\}$, keep B; otherwise stay at A.
6. Return to 1.

Q is symmetric and P is irreducible: in effect, if E and F are two subsets of S with respectively $n(E)$ and $n(F)$ points, paths from E to F consist in deleting one by one the points of E to get to the empty set $\emptyset$, then adding one by one the points in F; this path involves $n_E + n_F$ steps and can be obtained with probability $P^{n_E+n_F}(E,F) > 0$.

The transition is aperiodic if U is not constant: in effect, in this case there exists two configurations A and B which are Q-neighbors such that $U(A) < U(B)$. The proposed change $A \mapsto B$ is refused with probability $p = [1 - \exp\{-\beta \Delta U\}] > 0$. $P(A,A) = p/n > 0$ and P is thus aperiodic. This version of the Metropolis algorithm therefore represents a simulation giving convergence to π_β.

Variance optimality of the Metropolis algorithm

For Ω a finite state space and H a transition defining a chain $(X_0, X_1, \ldots)$ converging to π, the limit

$$v(f,H) = \lim_{n \to \infty} \frac{1}{n} Var\{\sum_{i=1}^{n} f(X_i)\}$$

exists and is independent of the initial distribution of the chain (128). Variance optimality of the Metropolis kernel P_M within the family of all MH kernels represents the fact that for any f and any MH kernel P_{MH}, we have (172):

$$v(f, P_M) \le v(f, P_{MH}). \tag{4.10}$$

A heuristic explanation of this property is that the Metropolis kernel promotes to a greater extent state changes than the other MH kernels (P_M "mixes" more than P_{MH}). This optimality result remains true for general state spaces (215).

4.3 Simulating a Markov random field on a network

4.3.1 The two standard algorithms

Let X be a Markov random field on $S = \{1, 2, \ldots, n\}$ with states in $\Omega = \prod_{i \in S} E_i$ (for example $\Omega = E^S$) and density:

$$\pi(x) = Z^{-1} \exp U(x), \qquad U(x) = \sum_{A \in \mathscr{C}} \Phi_A(x), \tag{4.11}$$

where $\mathscr{C}$ is the family of cliques of the Markov graph (cf. Ch. 2, Definition 2.2) and $\Phi = (\Phi_A, A \in \mathscr{C})$ the Gibbs potential of X. We can simulate π in two ways:

1. Using Gibbs sampling for the conditional distributions:

$$\pi_i(x_i | x^i) = Z_i^{-1}(x_{\partial i}) \exp U_i(x_i | x^i),$$

where $U_i(x_i|x^i) = \sum_{A:A\ni i} \Phi_A(x)$ and $Z_i(x_{\partial i}) = \sum_{u_i \in E_i} \exp U_i(u_i|x^i)$. As the space E_i does not have the complexity of a product space, it is sufficient to know the conditional energies $U_i(x_i|x^i)$ in order to simulate the distribution X_i conditional on x^i.

2. Using Metropolis dynamics: if the proposal transition Q has a density denoted q and if q is symmetric, then defining $a^+ = \max\{a,0\}$, the Metropolis transition has density p given by:

$$p(x,y) = q(x,y) \exp -\{U(x) - U(y)\}^+ \text{ if } x \neq y.$$

For Metropolis dynamics, changes can occur site by site or in other ways.

Further MCMC algorithms exist: Exercise 4.11 studies the set of MCMC transitions able to simulate X when relaxation at each step occurs site by site; Gibbs sampling and the Metropolis algorithm are two particular cases of such transitions.

4.3.2 Examples

4-nearest neighbor isotropic Ising model

For $S = \{1,2,\dots,n\}^2$ and $E = \{-1,+1\}$, the joint distribution and conditional distributions are respectively, where $v_i = \sum_{j\in\partial i} x_j$ is the contribution of the 4-NN of i,

$$\pi(x) = Z^{-1} \exp\{\alpha \sum_i x_i + \beta \sum_{<i,j>} x_i x_j\}, \tag{4.12}$$

$$\pi_i(x_i|x^i) = \frac{\exp x_i(\alpha + \beta v_i)}{2ch(\alpha + \beta v_i)}. \tag{4.13}$$

Spin-flip Metropolis dynamics

The proposal transition Q for the change $x \mapsto y$ is the following: randomly choose two sites $i \neq j$, swap the spins of i and j and change nothing else: $y_i = x_j$, $y_j = x_i$ and $y^{\{i,j\}} = x^{\{i,j\}}$. The proposal transition is thus:

$$Q(x,y) = \begin{cases} \frac{2}{n^2(n^2-1)} & \text{for such swaps,} \\ 0 & \text{otherwise.} \end{cases}$$

We emphasize that if $x(0)$ is the initial configuration, the algorithm evolves in the subspace $\Omega_{x(0)} \subset \{-1,+1\}^S$ of configurations having the same number of spins $+1$ (and thus of spins -1) as $x(0)$. The simulation will therefore be of π, restricted to the subspace $\Omega_{x(0)}$.

We must calculate $\Delta U(x,y) = U(y) - U(x)$ in order to identify the Metropolis transition. This is a local calculation if X is Markov: for example, for the 4-NN isotropic Ising model (4.12), we have:

$$\Delta U(x,y) = \begin{cases} \beta(x_j - x_i)(v_i - v_j) & \text{if } \|i-j\|_1 > 1, \\ \beta(x_j - x_i)(v_i - v_j) - \beta(x_j - x_i)^2 & \text{otherwise.} \end{cases}$$

One iteration of the spin-flip Metropolis algorithm is thus:

1. `Draw from two independent uniform distributions on` $\{1,2,\ldots,n\}^2$, `thus selecting two sites` i `and` j.
2. `We make the move` $x \mapsto y$ `by swapping` x_i `and` x_j.
3. `Calculate` $\Delta U(x,y)$: `if` $\Delta U(x,y) > 0$, `accept` y.
4. `Otherwise, draw` $U \sim \mathscr{U}([0,1])$ `independently from 1:`

   a. `If` $U < \exp \Delta U(x,y)$, `keep` y.
   b. `If` $U \geq \exp \Delta U(x,y)$, `stay with` x.

Q is irreducible on Ω_0 because any two configurations of Ω_0 are related by a permutation and all permutations are finite products of transpositions. The Metropolis transition P is irreducible because at each elementary step, we accept the change with probability > 0.

If $\beta \neq 0$, U is not constant; if furthermore $x(0)$ is not constant, there exist two configurations x and y that are Q-neighbors such that $U(y) > U(x)$. Thus the change $x \mapsto y$ is refused with probability $P(x,x) \geq 2\{1 - \exp \Delta U(x,y)\}/n^2(n^2-1) > 0$. The chain is therefore aperiodic and the algorithm simulates π restricted to Ω_0.

Simulation of auto-models

For auto-models (cf. §2.4), as the product of the conditional distributions π_i is absolutely continuous with respect to π, Gibbs sampling converges to π for any initial distribution. Figure 4.1 shows this evolution for two auto-logistic binary textures π.

Simulating Gaussian vectors

We now consider Gibbs sampling of Gaussian vectors $X \sim \mathscr{N}_n(m,\Sigma)$, where Σ is invertible. Noting $Q = \Sigma^{-1}$, the transition associated with the sweeping $1 \mapsto 2 \mapsto \ldots \mapsto n$ is the product of the ith relaxations conditional on $z^i = (y_1,\ldots,y_{i-1},x_{i+1},\ldots,x_n)$, $i = 1,\ldots,n$, with respect to

$$\pi_i(x_i|z^i) \sim \mathscr{N}_1(-q_{ii}^{-1} \sum_{j:j\neq i} q_{ij} z_j, q_{ii}^{-1}).$$

If the random field is Markov, $\pi_i(x_i|x^i) = \pi_i(x_i|x_{\partial i})$.

A comparison of this simulation method with the standard method using Cholesky decomposition $\Sigma = T\,{}^tT$ of Σ does not lead to a conclusive advantage of one method over the other. For example, for $S = \{1,2,\ldots,100\}^2$, Σ is a matrix of dimension $10^4 \times 10^4$:

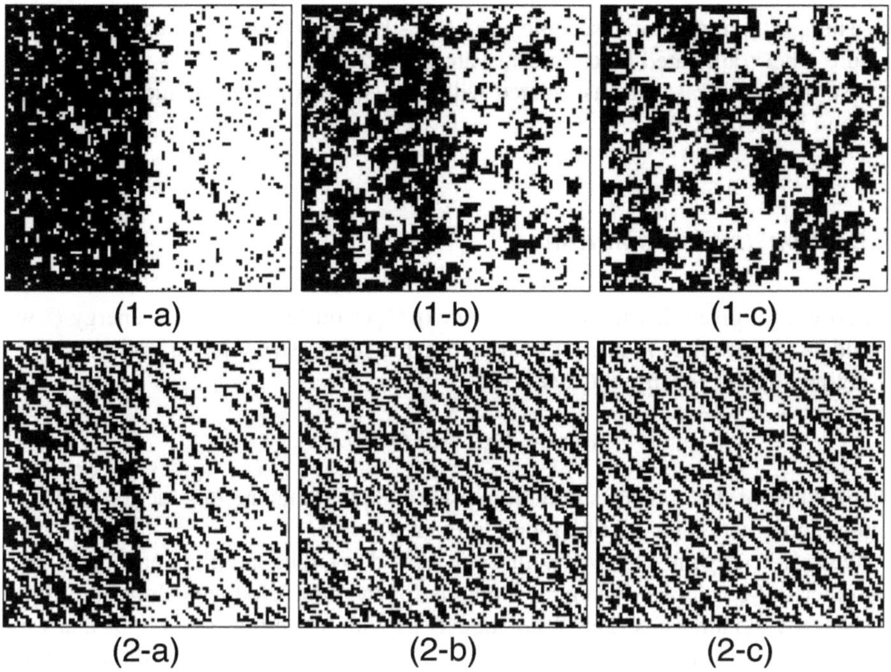

Fig. 4.1 Simulation using Gibbs sampling of two auto-logistic binary textures on the $\{1,2,\ldots,100\}^2$ grid. The initial configuration is 1 on the left half and 0 on the right. Simulations: (a) after one sweep, (b) after 10 sweeps, (c) after 1000 sweeps. The two models considered are: (1) 4-NN isotropic model with $\alpha = -3$, $\beta = 1.5$; (2) 8-NN model with $\alpha = 2$, $\beta_{0,1} = -1.2$, $\beta_{1,0} = -1.2$, $\beta_{1,1} = -0.8$ and $\beta_{1,-1} = 1.4$. Configurations 1 and 0 are equiprobable for both models.

1. *Via a Cholesky decomposition*: if ε is a sample of 10^4 Gaussian random variables with unit variance, $X = T\varepsilon \sim \mathcal{N}_{10^4}(0,\Sigma)$. Calculating T is costly but does not need to be repeated if we want to generate X again; furthermore, the simulation is exact.
2. *Via Gibbs sampling*: this method requires a large number N of sweeps so that (we judge that) the algorithm has entered its stationary phase (cf. §4.5.1 and (89)). If for example $N = 100$ sweeps is "sufficient," 100×10^4 Gaussian simulations are necessary in order to generate X. It is not necessary to know the Cholesky form of Σ, nor to be able to diagonalize Σ. However, in order to generate a new X, we must start from scratch and perform another 100×10^4 simulations. Furthermore, the obtained simulated result is an approximation (i.e., not exact).

Product state spaces

Gibbs sampling is well adapted to simulating models which have product state spaces, $E = \Lambda \times \mathbb{R}^n$. This type of space can be found in numerous applications:

for example, spatial remote sensing, $\lambda \in \Lambda = \{1, 2, \ldots, r\}$ representing a qualitative texture label (cultivated land, forest, water, arid zones, etc.) and $x \in \mathbb{R}^n$ the quantitative multispectral value associated with each site. Examples of such models are given in Exercise 2.6.

4.3.3 Constrained simulation

Suppose for a given distribution $\pi(x) = c \exp U(x)$ on Ω with known energy U we would like to simulate π_C, i.e., π restricted to the subset $\Omega_C = \{x \in \Omega : C(x) = 0\} \subset \Omega$ defined by the constraint $C : \Omega \to \mathbb{R}^+$,

$$\pi_C(x) = \mathbf{1}_{\Omega_C}(x) Z_C^{-1} \exp U(x).$$

An example of this type of simulation, used in constrained reconstruction of a system of geological faults can be found in Exercise 4.13.

We now briefly present two results, one related to Gibbs sampling on $\Omega = E^n$ and the other to Metropolis dynamics. Here, as for the simulated annealing algorithm, the Markov dynamics used are *inhomogeneous* with transitions $(P_k, k \geq 0)$ that vary with respect to time. Their convergence is a consequence of an ergodicity criteria for inhomogeneous chains (cf. Isaacson and Madsen, (120)).

The algorithm is as follows: let (λ_k) be a real-valued positive sequence, (π_k) the sequence of distributions with penalized energy U_k with respect to U defined by:

$$U_k(x) = U(x) - \lambda_k C(x),$$

and π_k the distribution with energy U_k. If $\lambda_k \to +\infty$, configurations x not satisfying $C(x) = 0$ will be progressively eliminated and it is easy to see that $\pi_k(x) \longrightarrow \pi_C(x)$ for all $x \in \Omega$. We have the results:

1. *Inhomogeneous Gibbs dynamics (81)*: let P_k be the Gibbs transition associated with π_k for the kth sweep of S. If $\lambda_k = \lambda_0 \log k$ for some small enough λ_0, the inhomogeneous chain (P_k) converges to π_C.
2. *Inhomogeneous Metropolis dynamics (228)*: let Q be a proposal transition that is homogeneous in time, irreducible and symmetric. Let P_k be the Metropolis kernel with energy U_k:

$$P_k(x, y) = q(x, y) \exp -\{U_k(x) - U_k(y)\}^+ \qquad \text{if } y \neq x.$$

   Then, if $\lambda_k = \lambda_0 \log k$ and if λ_0 is sufficiently small, the inhomogeneous chain (P_k) converges to π_C.

The necessary conditions for slow convergence of $\lambda_k \to \infty$ are analogous (with $\beta_k = T_k^{-1}$) to the slow convergence of $T_k \to 0$ for the sequence of temperatures T_k of the simulated annealing algorithm. These conditions ensure ergodicity of the inhomogeneous chain (P_k) (120; 82; 1).

4.3.4 Simulating Markov chain dynamics

Suppose that the dynamic $X = \{X(t), t = 1,2,\ldots\}$ of a Markov random field $X(t) = (X_i(t),\, i \in S)$, $S = \{1,2,\ldots,n\}$ is characterized as follows (cf. §2.5). We are given:

1. An initial distribution $X(0)$ with energy U_0.
2. Temporal transitions $P_t(x,y)$ from $x = x(t-1)$ to $y = x(t)$ with energy $U_t(x,y)$.

The simulation can be performed *recursively* with successive Gibbs sampling: we simulate $X(0)$, then $X(t)$, $t \geq 1$ conditional on $x(t-1)$, for $t = 1,2,\ldots$. For random fields that are conditionally Markov, conditional distributions at space-time "sites" (i,t) are easily expressed using conditional energies $U_{t,i}(\cdot|x, y_{\partial i})$: for example, for auto-logistic dynamics (2.18) with potentials:

$$\Phi_i(x,y) = y_i\{\alpha_i + \sum_{j \in \partial i^-} \alpha_{ij} x_j(t-1)\} \text{ and } \Phi_{ij}(x,y) = \beta_{ij} y_i y_j \text{ if } i \in S \text{ and } j \in \partial i,$$

$$U_{t,i}(y_{it}|x, y_{\partial i}) = y_{it}\{\alpha_i + \sum_{j \in \partial i^-} \alpha_{ij} x_j(t-1)\} + \sum_{j \in \partial i} \beta_{ij} y_j.$$

Gibbs sampling is easy to use whenever we are dealing with a model that is Markov in both time and space, allowing us to simulate either homogeneous or inhomogeneous Markov dynamics in time: for example, growing stain models, fire or spatial diffusion models and systems of particles (74; 73).

Example 4.3. Simulating growing stain models

Consider the following growing stain model in $\mathbb{Z}^2$: let the random field $X(t) = (X_s(t),\, s \in \mathbb{Z}^2)$ take states in $\{0,1\}^{\mathbb{Z}^2}$ where 0 represents healthy and 1 sick. Suppose that X is incurable, i.e., if $X_s(t) = 1$, then $X_s(t') = 1$ for $t' > t$. Next, let the state x represent the subset of $\mathbb{Z}^2$ where we have a 1. One possible model is the specification of an initial state (for example $X(0) = \{0\}$) along with a nearest neighbor spatio-temporal transition $X(t) = x \mapsto y = X(t+1)$ for only $y \subseteq x \cup \partial x$, with energy transition:

$$U(x,y) = \alpha \sum_{i \in \partial x} y_i + \beta \sum_{i \in \partial x,\, j \in x \text{ s.t. } \langle i,j \rangle} x_j y_i.$$

Gibbs sampling conditional on x thus sweeps the set of sites of ∂x with a simulation distribution at site $i \in \partial x$ of conditional energy

$$U_i(y_i|x, y^i \cap \partial x) = y_i\{\alpha + \beta v_i(x)\},$$

where $v_i(x)$ is the number of neighbors of i in x.

4.4 Simulation of a point process

Suppose X is a point process (PP) on a bounded Borel set S in $\mathbb{R}^d$. If X is a Cox PP or a shot-noise PP (cf. §3.3), simulation methods follow directly from those for Poisson PPs (cf. §4.4.3).

If X is defined by its unconditional density f (cf. §3.4), it can be simulated using MH dynamics (88; 160). An implementation of these dynamics can be found in the `R` package `spatstat`. Let us first take a look at simulations conditional on a fixed number of points $n(x) = n$.

4.4.1 Simulation conditional on a fixed number of points

Suppose we fix $n(x) = n$. Writing $(x \cup \xi) \backslash \eta$ to mean $(x \cup \{\xi\}) \backslash \{\eta\}$, the MH algorithm is written:

1. Change $x \to y = (x \cup \xi) \backslash \eta$, $\eta \in x$ and $\xi \in S$, with density $q(x,y) = q(x,\eta,\xi)$.
2. Accept y with probability $a(x,y) = a(x,\eta,\xi)$.

One possible acceptance probability $a(x,y)$ ensuring that the kernel is π-reversible is

$$a(x,y) = \min\{1, r(x,y)\}, \text{ where } r(x,y) = \frac{f(y)q(y,x)}{f(x)q(x,y)}.$$

If the change $x \mapsto y$ is obtained upon uniformly deleting a point in x and then uniformly generating an ξ in S, the algorithm gives irreducibility and aperiodicity and converges to π; aperiodicity results from positivity of the transition density, irreducibility from the fact that we can pass from x to y in n steps by changing, step by step, x_i (death) into y_i (birth) with positive probability density.

4.4.2 Unconditional simulation

Suppose that the density of f is *hereditary*, i.e., satisfying:

$$\text{if for } x \in \Omega \text{ and } \xi \in S,\ f(x \cup \xi) > 0, \text{ then } f(x) > 0.$$

In this context, as the algorithm allows one birth or death at each step, it visits different spaces E_n of configurations with n points. If $x \in E_n$, $n \geq 1$, one loop of the algorithm is:

1. With probability $\alpha_{n,n+1}(x)$, add a point $\xi \in S$ that has been chosen with respect to the density $b(x,\cdot)$ on S.
2. With probability $\alpha_{n,n-1}(x) = 1 - \alpha_{n,n+1}(x)$, delete a point η of x with probability $d(x,\eta)$.

When $n = 0$, we stay in the empty configuration with probability $\alpha_{0,0}(\emptyset) = 1 - \alpha_{0,1}(\emptyset)$. The MH ratio is:

$$r(x, x \cup \xi) = \frac{f(x \cup \xi)(1 - \alpha_{n+1,n}(x \cup \xi))d(x,\xi)}{f(x)\alpha_{n,n+1}(x)b(x,\xi)} \quad \text{and} \quad r(x \cup \xi, x) = r(x, x \cup \xi)^{-1}.$$

The general condition for reversibility of the MH algorithm is given by (4.6). For the Metropolis algorithm, the acceptance probability is:

- $a(x, x \cup \xi) = \min\{1, r(x, x \cup \xi)\}$ of accepting the birth: $x \to y = x \cup \xi$.
- $a(x, x \backslash \eta) = \min\{1, r(x, x \backslash \eta)\}$ of accepting the death: $x \to y = x \backslash \eta$.

Proposition 4.2. *Ergodicity of the MH algorithm*
Suppose that the two following conditions hold:

1. *For the birth and death distributions b and d: if* $n(x) = n$, $f(x \cup \xi) > 0$, $\alpha_{n+1,n}(x \cup \xi) > 0$ *and* $\alpha_{n,n+1}(x) > 0$, *then* $d(x, \xi) > 0$ *and* $b(x, \xi) > 0$.
2. *The probabilities of visiting different spaces* E_n *satisfy:* $\forall n \geq 0$ *and* $x \in E_n$, $0 < \alpha_{n+1,n}(x) < 1$, *and (4.6) is satisfied.*

Then, the previously described MH algorithm simulates a PP with density f.

The first condition is satisfied when $b(x, \cdot) = b(\cdot)$ is uniform on S and if $d(x, \cdot) = d(\cdot)$ is uniform on x.

Proof. The chain is π-invariant. It is also aperiodic since $\alpha_{0,1}(\emptyset)$ is > 0, implying a probability > 0 to stay in the empty configuration. Lastly, the algorithm is π-irreducible: in effect, we move from $x = \{x_1, x_2, \ldots, x_n\}$ to $y = \{y_1, y_2, \ldots, y_p\}$ with a positive probability density by first deleting the n points of x and then adding the p points of y. □

For the uniform choices $\alpha_{n+1,n} = 1/2$, $b(x, \cdot) = 1/\nu(S)$, $d(x, \eta) = 1/n$ if $x = \{x_1, x_2, \ldots, x_n\}$ and $\eta \in x$, we have:

$$r(x, x \cup \xi) = \frac{\nu(S) f(x \cup \xi)}{n(x) f(x)}.$$

If $n(x) > 0$, one loop of the Metropolis algorithm $x \to y$ is as follows:

1. With probability 1/2, a birth $\xi \in S$ is proposed uniformly; it is retained with probability $\inf\{1, r(x, x \cup \xi)\}$: $y = x \cup \{\xi\}$; otherwise, configuration x stays the same: $y = x$.
2. With probability 1/2, a death occurs at η, a uniformly chosen point of x; this death is retained with probability $\min\{1, r(x \backslash \eta, x)^{-1}\} : y = x \backslash \{\eta\}$; otherwise, configuration x stays as it was: $y = x$. If $n(x) = 0$, only the first step is performed with probability 1.

4.4.3 Simulation of a Cox point process

Simulation of Cox PPs (cf. §3.3) follows naturally from how we simulate Poisson PPs. Simulation of a PPP(ρ) with intensity ρ occurs as follows: denoting $\lambda(S) = \int_S \rho(u) du < \infty$:

1. `Generate a non-negative integer` n `from the Poisson distribution` $\mathscr{P}(\lambda(S))$.
2. `Draw` n `points` $x_1, x_2, \ldots, x_n$ `i.i.d. with density` ρ `on` S.

$x = \{x_1, x_2, \ldots, x_n\}$ is then the result of a PPP(ρ) on S.

Simulation of a Cox PP of random intensity $(\Lambda(u))_{u\in S}$ is achieved by simulating a conditional Poisson PP with intensity Λ. Simulation of the random field of density Λ, continuous on S, depends on the Cox process under consideration. For example, simulating the conditional log-intensity $\log \rho(u) = {}^t z(u)\beta + \Psi(u)$ of a log-Gaussian Cox PP (3.3) is performed using the observable covariates $(z(u))_{u\in S}$ and a simulated Gaussian process Ψ along with one of the methods proposed in §4.7.

Simulation of a doubly stochastic Poisson PP (3.4) is performed in two steps: first, determine the positions c and intensities γ of the "parent" Poisson PPs; second, simulate the *cluster* around each parent using spatial densities $\gamma k(c,\cdot)$. As parents c outside S can have descendants in S, we have to make sure that the simulation method takes into account this boundary effect (cf. (160)).

4.5 Performance and convergence of MCMC methods

The difficulty both in theory and practice with MCMC algorithms is to know at what instant n_0 (burn-in time) we can suppose that the associated chain $X = (X_n)$ has entered its stationary regime. Should we choose n_0 as a function of when the distribution of X_n, $n \geq n_0$ is close to the stationary distribution π of the chain? By bounding $\|X_n - \pi\|_{VT}$? We consider only briefly this question here and refer the reader to (89; 188) among others, as well as to the articles mentioned in the following paragraphs for more details.

4.5.1 Performance of MCMC methods

Two criteria are useful for evaluating the performance of MCMC methods:

1. The rate of convergence to 0 of the total variation norm $\|X_k - \pi\|_{VT} = \left\|\nu P^k - \pi\right\|_{VT}$, where $X_0 \sim \nu$.
2. The variance $v(f,P) = \lim_{n\to\infty} \frac{1}{n} Var(\sum_{i=1}^n f(X_i))$ of an empirical mean along the chain's trajectory.

Peskun (4.10) showed that for the variance criteria $v(f,P)$, the Metropolis algorithm is the best of all MH algorithms. The variance $v(f,P)$ can be characterized using the spectrum of P if Ω is finite with m points. In this case, as the chain is reversible, P is self-adjoint in $l^2(\pi)$, where the space $\mathbb{R}^m$ is endowed with the scalar product $\langle u,v\rangle_\pi = \sum_1^m u_k v_k \pi(k)$. P can thus be diagonalized in $\mathbb{R}$. If we note $\lambda_1 = 1 > \lambda_2 > \ldots > \lambda_m > -1$ its spectrum and $e_1 = 1, e_2, \ldots, e_m$ the associated basis chosen to be orthonormal of $l^2(\pi)$, we have (128):

$$v(f,P) = \sum_{k=2}^{m} \frac{1+\lambda_k}{1-\lambda_k} \langle f, e_k \rangle_\pi^2.$$

In order for this to be useful, the spectral decomposition of P needs to be known, which is rarely the case.

Quantifying the rate of convergence of $\|\nu P^k - \pi\|_{VT} \to 0$ is a fundamental question that in theory allows us to propose a stopping rule for the algorithm. While several results exist (if Ω is finite, the Perron-Frobénius Theorem gives $\|\nu P^k - \pi\|_{VT} \leq C(\nu)(\sup\{|\lambda_2|, |\lambda_m|\})^k$ for transitions that are reversible), even here effective evaluation of the rate of convergence of the algorithm is in general impossible because it is linked to having a precise description of the spectrum of P, possible only in rare cases (61; 191). Furthermore, when explicit bounds exist ((4.3), bounding the transition using Dobrushin's contraction coefficient (65; 120; 96, §6.1)), they are generally impractical, bounds of the type $\|\nu P^k - \pi\|_{VT} \leq (1 - m\varepsilon)^k$ (cf. Th. 4.2) being useful only if $m\varepsilon$ is not too small. However, to take an example, for sequential Gibbs sampling for 4-NN isotropic Ising models on $S = \{1,2,\ldots,10\}^2$ and with parameters $\alpha = 0$ and $\beta = 1$, $m = 2^{100}$, we have $\varepsilon = \{\inf_{i,x_i,x^i} \pi_i(x_i|x^i)\}^m$ and $m\varepsilon \sim (6.8 \times 10^{-4})^{100}$!

An alternative approach, presented in §4.6, consists of using an exact simulation algorithm (177).

In practice, when simulating using MCMC methods, several aspects need to be considered: choice of algorithm, ease of implementation, processing time, choice of initial state, use of a single chain or several independent chains, time n_0 to chain stationarity and subsampling at every K steps to guarantee near independence of subsequences $(X_{Kn})_n$. We limit ourselves here to giving responses to the question of choosing the time to stationarity n_0 and suggest (89) and (188) for a more thorough coverage.

4.5.2 Two methods for quantifying rates of convergence

A first approach consists in following the evolution in k of summary statistics $h(X_k)$ of the chain $X = (X_k)$. For example, if π belongs to an exponential family, we could choose h to be the exhaustive statistic of π. Figure 4.2 shows both evolution of $n(X_k)$ and $s(X_k)$ in an MH simulation of a Strauss process as well as empirical autocorrelations for each statistic. A choice of $n_0 = 300$ to mark the entry time into the stationary regime seems reasonable and a time lag of $K = 100$ approximately removes the correlation between $h(X_t)$ and $h(X_{t+K})$ (cf. Fig. 4.3).

A second approach (80) consists of generating m independent chains in parallel after having initialized the chains in various different states and then comparing their variability. We thus follow the m trajectories $\{h_k^{(i)} = h(X_k^{(i)}), k \geq 0\}$. When the chains reach their stationary regimes, their variance should be similar. We analyze this set of m variances by calculating B (*between*) and W (*within*) the inter- and intra-chain variability:

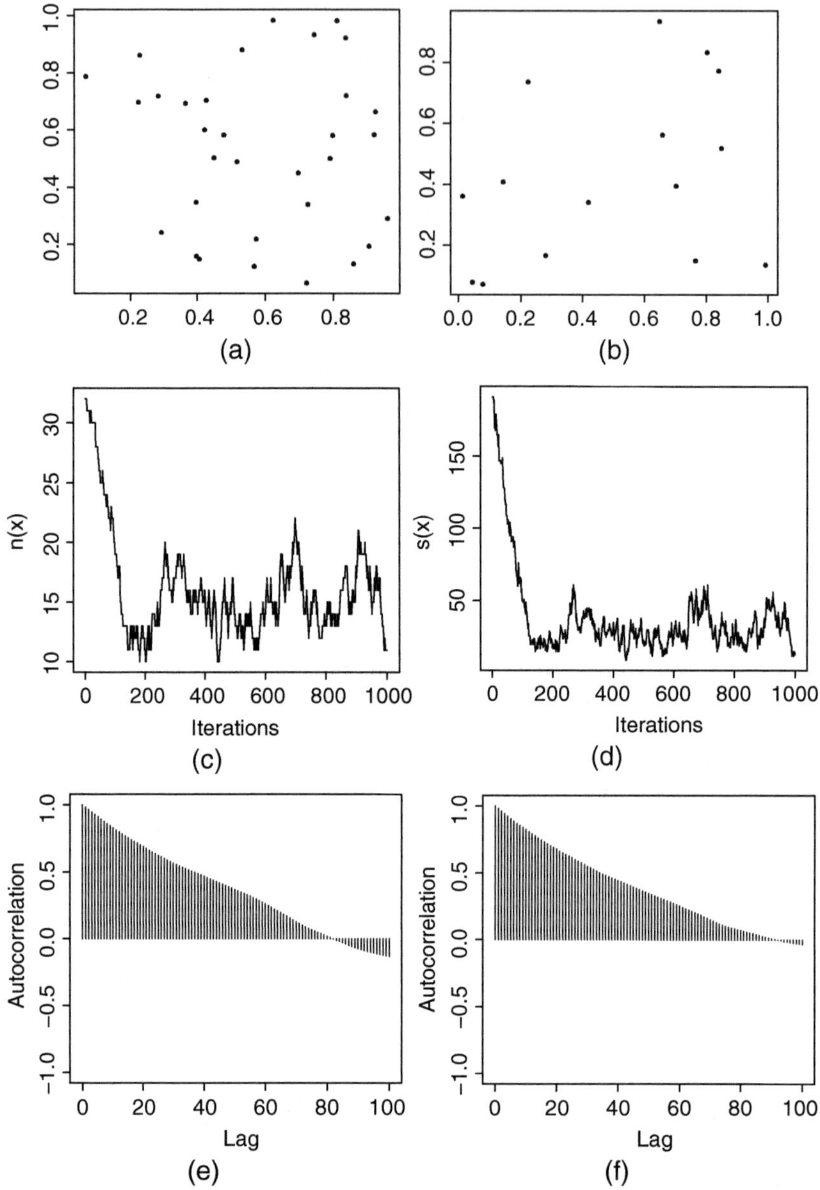

Fig. 4.2 Unconditional simulation of a Strauss process X with parameters $\beta = 40$, $\gamma = 0.8$, $r = 0.4$ on $S = [0,1]^2$ using the Metropolis algorithm with the uniform choices: (a) $X_0 \sim PPP(40)$, (b) X_{1000}, after 1000 iterations. Checking of the convergence of the algorithm is done using summary statistics $n_S(X_k)$ (c) and $s(X_k)$ (d) over time and their autocorrelation functions (e) and (f).

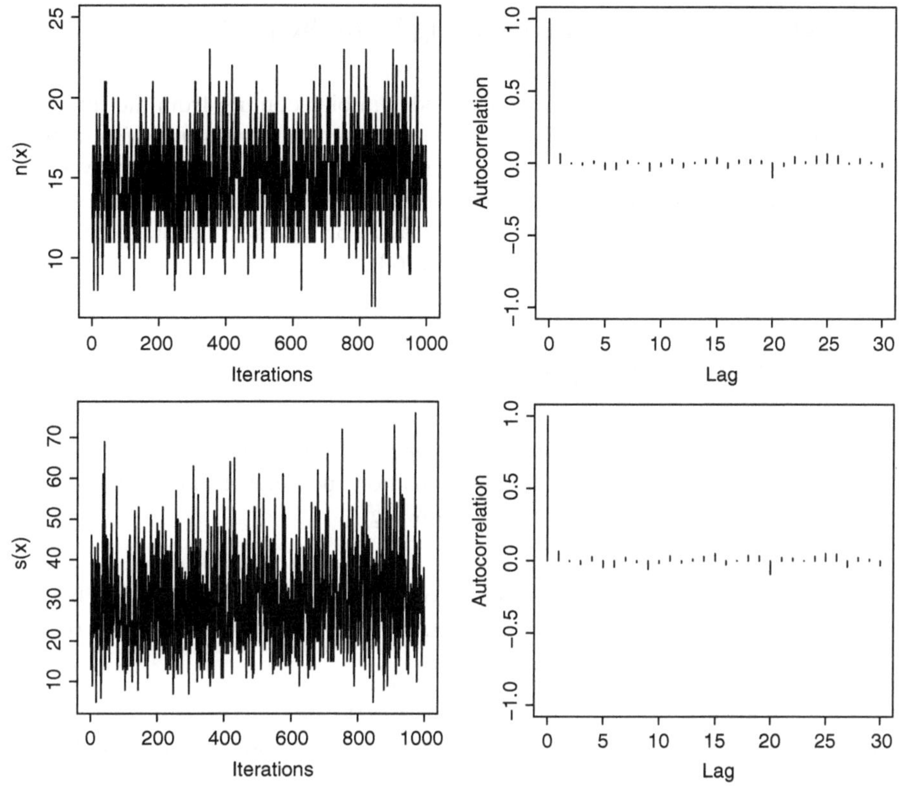

Fig. 4.3 Subsampling the simulation chain of a Strauss process (cf. Fig. 4.2): to the left, the statistics $n_S(X_{sk})$ and $s(X_{sk})$, $k = 0,1,2,\ldots$, with $s = 100$; to the right, the corresponding autocorrelations.

$$B = \frac{1}{m-1}\sum_{i=1}^{m}(\overline{h}^{(i)} - \overline{h})^2, \qquad W = \frac{1}{m-1}\sum_{i=1}^{m} S_i^2,$$

where $S_i^2 = (n-1)^{-1}\sum_{k=0}^{n-1}(h_k^{(i)} - \overline{h}^{(i)})^2$ and $\overline{h}^{(i)}$ is the empirical mean of h for the ith chain, $\overline{h}$ the mean across all chains. We can either estimate $Var_\pi(h)$ using $V = W + B$ or W. The choice of V gives an unbiased estimate of this variance when $X_0^{(i)} \sim \pi$ but otherwise it overestimates the variance. As for W, it underestimates the variance if n is finite as the chain has not reached all states of X. Nevertheless, both estimators converge to $Var_\pi(h)$. We can therefore quantify convergence of chains based on the statistic $R = V/W$: when the chain has entered the stationary regime, $R \cong 1$; if not, $R \gg 1$ and we should perform further iterations.

4.6 Exact simulation using coupling from the past

The difficulty in using MCMC methods resides in bias coming from the initiation step $\| \nu P^k - \pi \|_{VT}$. Propp and Wilson (177) proposed an *exact simulation* method based on coupling from the past (CFTP), a simulation technique that removes this problem. Their simple yet powerful idea revolutionized the field.

For a general overview, we suggest `http://dbwilson.com/exact/` as well as the article by Diaconis and Freedman (60).

4.6.1 The Propp-Wilson algorithm

We limit ourselves here to describing the Propp-Wilson algorithm for *finite spaces* $\Omega = \{1,2,\ldots,r\}$, where P is an ergodic transition such that π is P-invariant. The ingredients of the CFTP method are the following:

(a) *The simulator* $\mathscr{S} = (\mathscr{S}_t)_{t\geq 1}$ with $\mathscr{S}_t = \{f_{-t}(i), i \in \Omega\}$, where $f_{-t}(i)$ follows the distribution $P(i,\cdot)$. The generators $(\mathscr{S}_t)$ are i.i.d. for different t but the $\{f_{-t}(i), i \in \Omega\}$ are potentially dependent. An iteration from $-t$ to $-t+1$ ($t \geq 1$) is:

1. `For each` $i \in \Omega$, `simulate` $i \mapsto f_{-t}(i) \in \Omega$.
2. `Memorize the transitions` $\{i \mapsto f_{-t}(i), i \in \Omega\}$ `from` $-t$ `to` $-t+1$.

The simulation moves back in time, starting from $t = 0$.

(b) *The map* $F_{t_1}^{t_2} : \Omega \to \Omega$ from t_1 to t_2, $t_1 < t_2 \leq 0$, is the transformation

$$F_{t_1}^{t_2} = f_{t_2-1} \circ f_{t_2-2} \ldots \circ f_{t_1+1} \circ f_{t_1}.$$

$F_{t_1}^{t_2}(i)$ is the state at t_2 of the chain initialized at i at time t_1. F_t^0, $t < 0$ satisfies:

$$F_t^0 = F_{t+1}^0 \circ f_t \text{, with } F_0^0 = Id. \tag{4.14}$$

F_t^0 is obtained recursively using a stack of length r.

(c) *A coupling time for* $(\mathscr{S})$ is some $T < 0$ for which F_T^0 is constant:

$$\exists i_* \in \Omega \text{ s.t.: } \forall i \in \Omega, F_T^0(i) = i_*.$$

An example of running this algorithm is given in Figure 4.4. We start by choosing the random arrows $i \mapsto j$ that link states i at time $t = -1$ to states j at time $t = 0$ (j is chosen with distribution $P(i,\cdot)$), this for all $i \in \Omega$. If all arrows have the same final state $j = i_*$, i_* is the coupling state for the coupling time $T = -1$. If not, we repeat the same process between times $t = -2$ and $t = -1$ and draw all paths of length 2 leaving states i at time $t = -2$ and using any of the two families of previously drawn arrows. Again, either all the paths (of length 2) end at the same state i_* and hence i_*

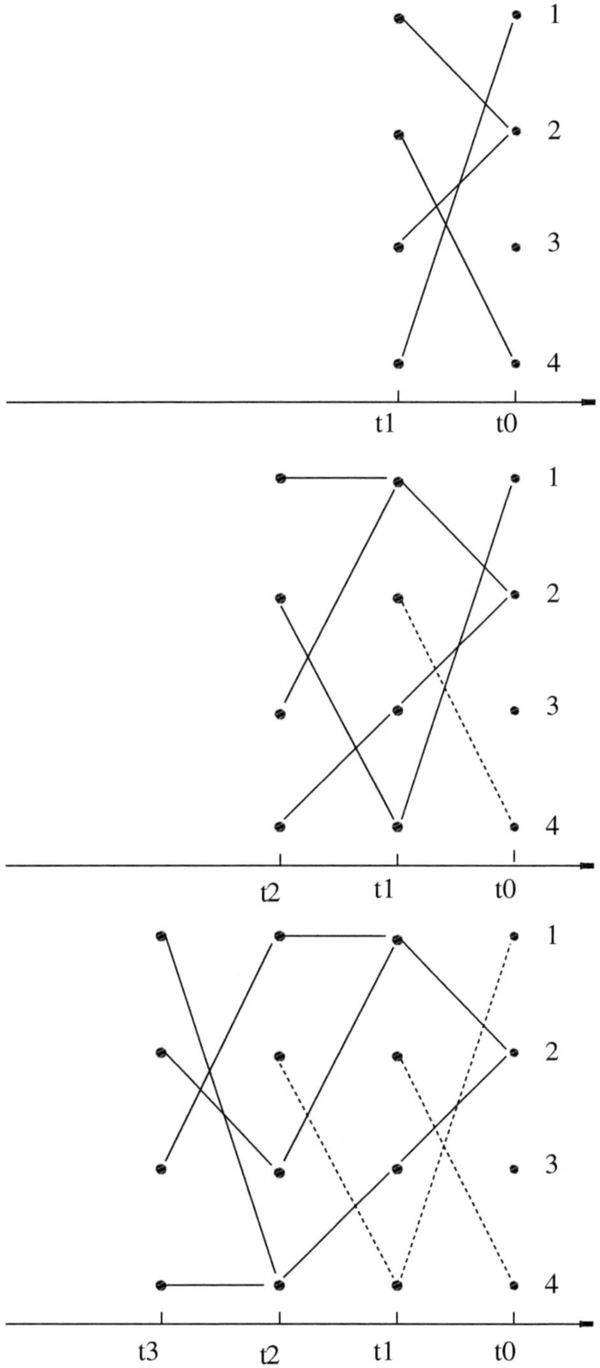

Fig. 4.4 Examples of trajectories of the Propp-Wilson algorithm in a 4-state space: the coupling time is $T_* = -3$, the coupling state is $i_* = 2$; the retained paths of the algorithm are shown as (——).

is the coupling state for the coupling time $T = -2$; or, this is still not the case and we iterate the algorithm between $t = -3$ and $t = -2$. In this way, we continue until we find the first coupling time. In the given example, the coupling time is $T = -3$ and the coupling state is $i_* = 2$, i.e., we had to go back 3 steps into the past in order that all paths issued from $((i,t), i \in \Omega)$ couple at time $t = 0$.

If, independent of the initial state at $T < 0$ the r-chains couple at i_* at time 0, T is called a coupling time from the past and i_* a coupling state. The result (4.14) shows that if $T' < T$, then T' is also a coupling time with the same coupling state. When

$$T_* = \sup\{-t < 0 : F^0_{-t}(\cdot) \text{ is constant}\}$$

is the first coupling time, we denote by $F^0_{-\infty} = F^0_{-T_*} = F^0_{-M}$ this shared coupling state. The exact simulation result is the following:

Proposition 4.3. *The coupling time T_* is almost surely finite. The coupling state $F^0_{-T_*}$ follows the distribution π.*

Proof: As the chain is ergodic, there exists some $k \geq 1$ such that for all i, j, $P^k(i,j) > 0$. Thus, F^t_{t-k} has some probability $\varepsilon > 0$ to be constant. As the variables F^0_{-k}, F^{-k}_{-2k}, F^{-2k}_{-3k}, ... are i.i.d. with probability $\varepsilon > 0$ of being constant, the Borel-Cantelli Lemma gives that at least one of these events occurs with probability 1: $P(T_* < \infty) = 1$.

As the sequence $\mathscr{S}_t = \{f_{-t}(i), i \in I\}$, $t = -1, -2, \dots$ is i.i.d., $F^{-1}_{-\infty}$ and $F^0_{-\infty}$ have the same distribution ν. Furthermore, we know that $F^{-1}_{-\infty}P = F^0_{-\infty}$. We have thus $\nu P = \nu$, that is, $\nu = \pi$ due to uniqueness of the invariant distribution. □

$\mathscr{S}_1 = \{f_{-1}(i), i \in \Omega\}$ is the starting point for the simulation. As the stopping rule of the procedure is implicitly given, there is no initiation bias in this kind of simulation.

4.6.2 Two improvements to the algorithm

To get the map F^0_t, $(-t) \times r$ operations have to be performed ($-t > 0$ steps back in time and at each instant, r simulations). This is unrealistic as r, the number of points in Ω is generally quite large. Not only that, a stack of length r is needed. Propp and Wilson proposed two improvements to get around these difficulties: the first is a simulator $\mathscr{S}_t = \{f_{-t}(i), i \in \Omega\}$ obtained from a unique uniform starting point U_{-t}. The second, useful when Ω is endowed with a partial ordering, is to construct a "monotone" algorithm for which it suffices to test the partial coupling $F^0_T(\underline{i}) = F^0_T(\overline{i})$ at two extremum states, as it turns out this effectively tests coupling for all states.

A unique starting point for defining $\mathscr{S}_{-t}$

Let $(U_{-t}, t \geq 1)$ be a uniform i.i.d. sequence of variables on $[0,1]$ and $\Phi : E \times [0,1] \to [0,1]$ a measurable function generating the transition P (cf. Ex. 4.2):

$$\forall i, j: P\{\Phi(i, U_{-1}) = j\} = P(i, j).$$

Thus, $f_{-t}(i) = \Phi(i, U_t)$ generates $P(i, \cdot)$. At each instant, only one simulation is necessary but all values $\Phi(i, U_t)$, $i \in \Omega$, must be calculated and compared. Having a *monotone chain* helps to overcome this problem.

Monotone Monte-Carlo algorithm

Suppose that Ω is a partially ordered set with respect to the relation $\prec$ that has a minimal element $\mathbf{0}$ and maximal element $\mathbf{1}$:

$$\forall x \in \Omega: \mathbf{0} \prec x \prec \mathbf{1}.$$

We say that algorithm $\mathscr{S}$ is *monotone* if the update rule preserves the partial order $\prec$:

$$\forall x, y \in \Omega \text{ s.t. } x \prec y, \forall u \in [0,1]: \Phi(x, u) \prec \Phi(y, u).$$

In this case, $F_{t_1}^{t_2}(x, u) = \Phi_{t_2-1}(\Phi_{t_2-2}(\ldots \Phi_{t_1}(x, u_{t_1}), \ldots, u_{t_2-2}), u_{t_2-1})$, where $u = (\ldots, u_{-1}, u_0)$ and the monotone property ensures that:

$$\forall x \prec y \text{ and } t_1 < t_2, F_{t_1}^{t_2}(x, u) \prec F_{t_1}^{t_2}(y, u).$$

In particular, if $u_{-T}, u_{-T+1}, \ldots, u_{-1}$ satisfy $F_{-T}^{0}(\mathbf{0}, u) = F_{-T}^{0}(\mathbf{1}, u)$, then $-T$ is a coupling time from the past: it suffices to follow the two trajectories starting from $\mathbf{0}$ and $\mathbf{1}$ in order to characterize the coupling time as all other trajectories $\{F_{-T}^{t}(x), -T \leq t \leq 0\}$ are found in between $\{F_{-T}^{t}(\mathbf{0}), -T \leq t \leq 0\}$ and $\{F_{-T}^{t}(\mathbf{1}), -T \leq t \leq 0\}$.

It is possible to perform the simulation algorithm in the following way: successively initialize the two chains at times $-k$, where $k = 1, 2, 2^2, 2^3, \ldots$, until the first value 2^k that satisfies $F_{-2^k}^{0}(\mathbf{0}, u) = F_{-2^k}^{0}(\mathbf{1}, u)$. Values of u_t are progressively stored in memory. The number of operations necessary is $2 \times (1+2+4+\ldots+2^k) = 2^{k+2}$. As $-T_* > 2^{k-1}$, at worst the number of operations is 4 times the optimal number of simulations $-2T_*$.

Example 4.4. Simulation of an attractive Ising model

Suppose $S = \{1, 2, \ldots, n\}$, $\Omega = \{-1, +1\}^S$ $(r = 2^n)$ is endowed with the ordering:

$$x \prec y \Leftrightarrow \{\forall i \in S, x_i \leq y_i\}.$$

There is a minimal state, $\mathbf{0} = (x_i = -1, i \in S)$ and a maximal state $\mathbf{1} = (x_i = +1, i \in S)$. We say the Ising distribution π is *attractive* if for all i,

$$\forall x \prec y \Rightarrow \pi_i(+1|x^i) \leq \pi_i(+1|y^i). \tag{4.15}$$

It is easy to see that if π is associated with the energy

$$U(x) = \sum_i \alpha_i x_i + \sum_{i<j} \beta_{i,j} x_i x_j,$$

then π is attractive if, for all i, j, $\beta_{i,j} \geq 0$. Attractive Ising models have monotone dynamics: choose some site i and note by $x \uparrow$ (resp. $x \downarrow$) the configuration $(+1, x^i)$ (resp. $(-1, x^i)$); (4.15) is equivalent to:

$$x \prec y \Rightarrow \frac{\pi(x \downarrow)}{\pi(x \downarrow) + \pi(x \uparrow)} \geq \frac{\pi(y \downarrow)}{\pi(y \downarrow) + \pi(y \uparrow)}.$$

The dynamics defined by:

$$f_t(x, u_t) = \begin{cases} f_t(x, u_t) = x \downarrow \text{ if } u_t < \frac{\pi(x\downarrow)}{\pi(x\downarrow) + \pi(x\uparrow)}, \\ f_t(x, u_t) = x \uparrow \text{ otherwise} \end{cases}$$

are monotone whenever π is attractive.

The Propp-Wilson algorithm takes advantage of the existence of a minimal element $\mathbf{0}$ and maximal element $\mathbf{1}$ under the order relation $\prec$ on Ω. This condition is not always satisfied: for example, if for some PP on S the inclusion relation over configurations has the empty configuration as minimal element, there is no maximal element. For such cases, (104; 130) generalize the CFTP algorithm by creating the "dominated CFTP" exact simulation algorithm for distributions defined on more general state spaces.

4.7 Simulating Gaussian random fields on $S \subseteq \mathbb{R}^d$

We are interested here in simulating centered *Gaussian random fields* Y on $S \subset \mathbb{R}^d$, where S is finite or continuous.

If $S = \{s_1, \ldots, s_m\} \subseteq \mathbb{R}^d$ is *finite*, $Y = (Y_{s_1}, Y_{s_2}, \ldots, Y_{s_m})$ is a Gaussian random vector; if its covariance $\Sigma = Cov(Y)$ is p.d., then there exists some lower diagonal matrix T of dimension $m \times m$ such that $\Sigma = T\,{}^tT$, i.e., the Cholesky decomposition of Σ. Thus, $Y = T\varepsilon \sim \mathcal{N}_m(0, \Sigma)$ if $\varepsilon \sim \mathcal{N}_m(0, I_m)$. The simulation method associated with this decomposition is the standard one, as long as we know how to calculate T, which is difficult if m is large; though, once T is given, it is simple to perform new simulations of Y using new sampled values of ε. As we saw before (cf. §4.3.2), an alternative method that is well adapted to Gaussian Markov random fields is to simulate using Gibbs sampling.

If S is a *continuous subset* $S \subseteq \mathbb{R}^d$, other methods useful in geostatistics enable us to simulate the random field. We now present some of these methods and invite the reader to consult Lantuéjoul's book (139) for a more comprehensive presentation of the subject.

4.7.1 Simulating stationary Gaussian random fields

Suppose that we know how to simulate over $S \subseteq \mathbb{R}^d$ a stationary random field X in L^2 that is not necessarily a Gaussian random field, though centered with variance 1 and with correlation function $\rho(\cdot)$. Let $\{X^{(i)}, i \in \mathbb{N}\}$ be an i.i.d. sequence of such random fields and

$$Y_s^{(m)} = \frac{1}{\sqrt{m}} \sum_{i=1}^{m} X_s^{(i)}.$$

Then, for large m, $Y^{(m)}$ gives approximately a stationary, centered Gaussian random field with correlation ρ. Now let us consider how to simulate the generating random field X. We note that the online help of the `RandomFields` package gives details on how to implement this.

The spectral method

The formula $C(h) = \int_{\mathbb{R}^d} e^{i\langle u,h\rangle} F(du)$ linking covariance and spectral measure shows that F is a probability if $Var(X_s) = 1$. If V, a random variable with distribution F and $U \sim \mathscr{U}(0,1)$ are independent, we consider the random field $X = (X_s)$ defined by:

$$X_s = \sqrt{2}\cos(\langle V,s\rangle + 2\pi U).$$

Since $\int_0^1 \cos(\langle V,s\rangle + 2\pi u)du = 0,\ E(X_s) = E(E(X_s|V)) = 0$ and

$$\begin{aligned} C(h) &= 2\int_{\mathbb{R}^d}\int_0^1 \cos(\langle v,s\rangle + 2\pi u)\cos(\langle v,(s+h)\rangle + 2\pi u)duF(dv) \\ &= \int_{\mathbb{R}^d} \cos(\langle v,h\rangle)F(dv). \end{aligned}$$

X is therefore a centered random field with covariance $C(\cdot)$. This leads us to the following algorithm:

1. Generate $u_1,\ldots,u_m \sim \mathscr{U}(0,1)$ and $v_1,\ldots,v_m \sim F$, all independently.
2. For $s \in S$, output values

$$Y_s^{(m)} = \sqrt{\frac{2}{m}} \sum_{i=1}^{m} \cos(\langle v_i,s\rangle + 2\pi u_i).$$

We choose m based on the quality of convergence of the 3rd and 4th order moments of $Y_s^{(m)}$ towards those of a Gaussian random variable. This simulation is feasible if we know how to get close to the spectral measure F. This is the case if F has bounded support or when F has a density that decreases rapidly to zero in the limit. If not, we can use the following method, known as the *turning bands* method.

The turning bands method

This method, suggested by Matheron (139) simulates an isotropic process on $\mathbb{R}^d$ starting from a stationary process on $\mathbb{R}^1$ (cf. §1.2.2). Let $\mathscr{S}_d = \{s \in \mathbb{R}^d : \|s\| = 1\}$ be the sphere with radius 1 in $\mathbb{R}^d$, Z a centered stationary process on $\mathbb{R}^1$ with covariance

C_Z and V a uniformly generated direction on $\mathscr{S}_d$. Set $Y_s = Z_{\langle s,V \rangle}$, $s \in S$. Y_s is centered as $E(Z_{\langle s,V \rangle}|V = v) = E(Z_{\langle s,v \rangle}) = 0$, with covariance

$$\begin{aligned} C_Y(h) &= E_V[E(Z_{\langle (s+h),V \rangle} Z_{\langle s,V \rangle}|V)] \\ &= E_V[C_Z(\langle h,V \rangle)] = \int_{\mathscr{S}_d} C_Z(\langle h,v \rangle)\tau(dv), \end{aligned}$$

where τ is the uniform distribution on $\mathscr{S}_d$. We thus consider the algorithm:

1. `Generate` m `directions` $v_1,\ldots,v_m \sim \mathscr{U}(\mathscr{S}_d)$.
2. `Generate` $z^{(1)},\ldots,z^{(m)}$, m `independent processes with correlations` $C_Z(\langle h,v_i \rangle)$, $i = 1,\ldots,m$.
3. `For` $s \in S$, `output values:` $Y_s^{(m)} = m^{-1/2}\sum_{i=1}^m z^{(i)}_{\langle s,v_i \rangle}$.

Denoting $C_Y(h) = C_1(\|h\|) = C(\langle h,v \rangle)$ with $C = C_Z$, we have that the relationship between C_1 and C for $d = 2$ and $d = 3$ is given by:

$$d = 2 : C(r) = \frac{1}{\pi}\int_0^{\pi} C_1(r\sin\theta)d\theta \quad \text{and} \quad C_1(r) = 1 + r\int_0^{\pi/2} \frac{dC}{dr}(r\sin\theta)d\theta; \tag{4.16}$$

$$d = 3\text{: } C(r) = \int_0^1 C_1(tr)dt \quad \text{and} \quad C_1(r) = \frac{d}{dr}(rC(r)). \tag{4.17}$$

For example, for the exponential covariance $C(h) = \sigma^2\exp\{-\|h\|/a\}$ on $\mathbb{R}^3$, $C_1(r) = \sigma^2(1 - r/a)\exp(-r/a), r \geq 0$.

In practice, C is given and we must start by calculating C_1 (cf. Chilès and Delfiner (43, p. 648) for relationships between C et C_1). A closer look at (4.16) and (4.17) explains why, when simulating random fields on $\mathbb{R}^2$ we would rather use the trace

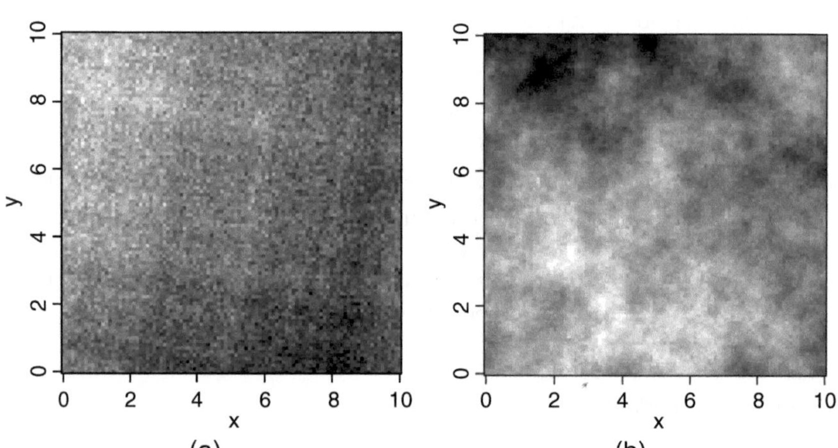

Fig. 4.5 Two simulated Gaussian random fields with exponential covariance obtained using m turning bands: (a) $m = 2$, (b) $m = 10$. The simulation is over the 100×100 grid.

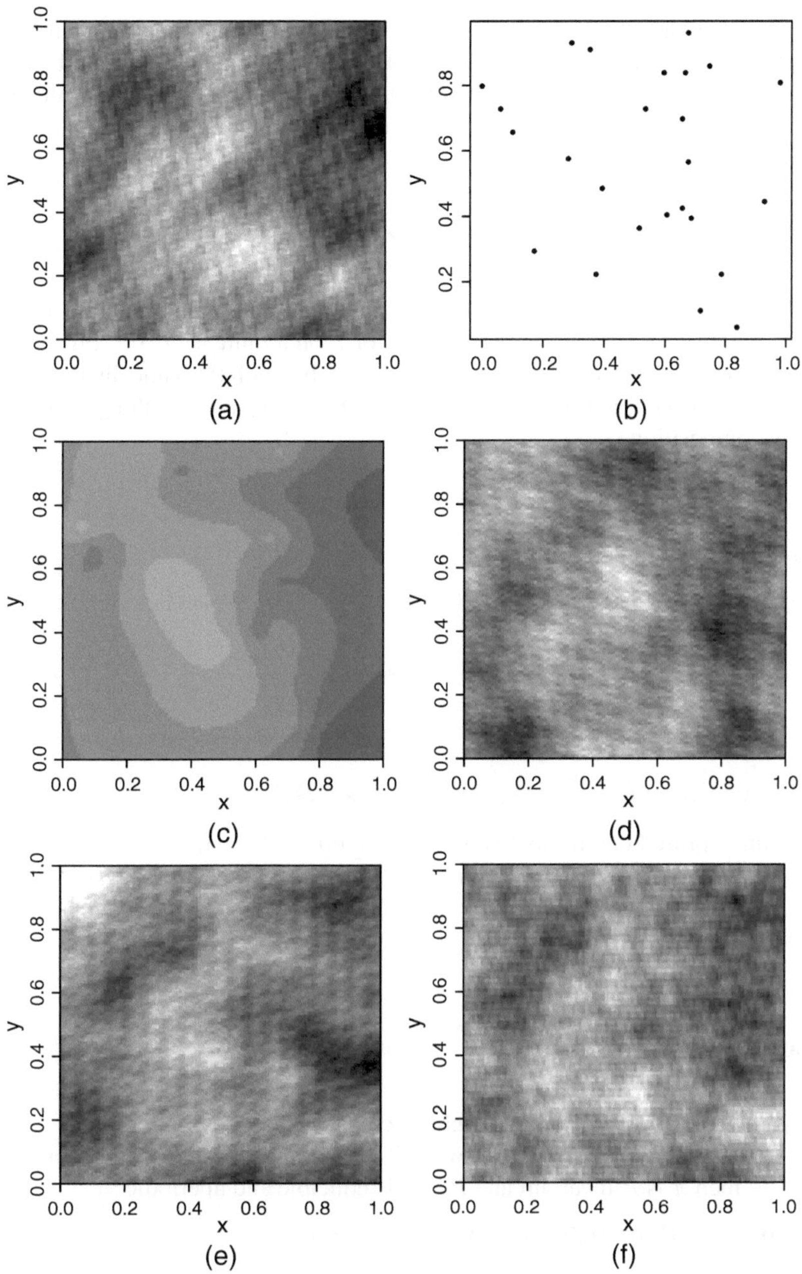

Fig. 4.6 (a) Simulation on a regular 100×100 grid over $[0,1]^2$ of a centered Gaussian process with covariance $C(h) = \exp(-\|h\|)$; (b) 25 sampled points; (c) kriging based on the 25 points of (b); (d) simulation conditional on these 25 values; (e) simulation conditional on 50 values; (f) simulation conditional on 100 values.

on $\mathbb{R}^2$ of a turning bands simulation on $\mathbb{R}^3$ than directly use turning bands on $\mathbb{R}^2$. For example, the simulations shown in Figure 4.5 are traces on $\mathbb{R}^2$ of turning bands simulations of a random field on $\mathbb{R}^3$ with exponential isotropic covariance with parameters $\sigma^2 = 4$ and $a = 10$.

4.7.2 Conditional Gaussian simulation

For Y a centered Gaussian random field on $\mathbb{R}^d$, we would like to generate Y on $S \cup C \subseteq \mathbb{R}^d$ conditional on some observation y_C of Y on a finite set C (cf. Fig. 4.6). To do this, we generate independently $X = \{X_s, s \in S\}$ but with the same distribution as Y, $\{\widehat{X}_s, s \in S\}$ the simple kriging of X at values $X_C = \{X_c, c \in C\}$ along with the following decompositions:

$$X_s = \widehat{X}_s + (X_s - \widehat{X}_s), \qquad Y_s = \widehat{Y}_s + (Y_s - \widehat{Y}_s), \qquad s \in S.$$

We propose the following algorithm:

1. `Given` Y_C, `calculate the simple kriging predictor` $\widehat{Y}_s$, $s \in S$.
2. `Generate` X `on` $S \cup C$ `with the same distribution as` Y, `independent of` Y.
3. `Given` X_C, `calculate the simple kriging predictor` $\widehat{X}_s$, $s \in S$.
4. `Output values` $\widetilde{Y}_s = \widehat{Y}_s + (X_s - \widehat{X}_s)$ `for` $s \in D$.

$\widetilde{Y}_s$ provides the required conditional simulation. In effect, $\widetilde{Y}_c = y_c + X_c - X_c = y_c$ if $s \in C$. Furthermore, $\{(X_s - \widehat{X}_s), s \in S\}$ and $\{(Y_s - \widehat{Y}_s), s \in S\}$ have the same distribution, that of the conditional residual of Y at s which is no other than the recentered distribution of Y conditional on y_C.

Exercises

4.1. Necessary conditions for Markov chain convergence.
Show that if a Markov chain with transition P is such that for any x and event A, $P^n(x,A) \to \pi$, then π is P-invariant and P is π-irreducible and aperiodic.

4.2. Simulating Markov chains on a discrete state space.

1. Suppose $P = (P_{ij})$ is a transition on $\Omega = \{1,2,\ldots\}$ and $\Phi : \Omega \times [0,1] \to \Omega$ is defined by $\Phi(i,u) = j$ if $u \in [\sum_{l=1}^{j-1} P_{i,l}, \sum_{l=1}^{j} P_{i,l}[$, $i,j \in \Omega$, with the convention $\sum_{l=1}^{0} P_{i,l} = 0$ and, if Ω is finite, the last half-open interval is right closed.
   Show that if (U_n) is an i.i.d. sequence of $\mathscr{U}(0,1)$ random variables, the sequence $\{X_0, X_{n+1} = \Phi(X_n, U_n), n \geq 1\}$ gives a simulation of a Markov chain with transition P.

2. Show that the chain on $\{1,2,3,4\}$ with transition $P(1,1) = P(1,2) = 1/2 = P(4,3) = P(4,4), P(2,1) = P(2,2) = P(2,3) = 1/3 = P(3,2) = P(3,3) = P(3,4)$ is irreducible and aperiodic. What is its invariant distribution π? Choose as initial value $X_0 = 1$ and simulate to obtain the chain. Verify that the empirical estimate $\widehat{\pi}_n$ (resp. $\widehat{P}^n_{i,j}$) of π (resp. of $P_{i,j}$) based on $X_1, X_2, \ldots, X_n$ converges to π (resp. $P_{i,j}$) when $n \longrightarrow \infty$.

4.3. Simulating hard-core models (cf. Example 4.1).
With the choice of the torus $S = \{1,2,\ldots,20\}^2$ and the 4-NN relation, simulate the uniform distribution over the hard-core configuration space. Calculate:

1. The mean number of occupied sites for a hard-core configuration.
2. The centered 90% confidence interval around this number.

4.4. Simulating bivariate Gibbs models.
Consider the 4-NN Gibbs process $Z_i = (X_i, Y_i) \in \{0,1\}^2$ on $S = \{1,2,\ldots,n\}^2$ with energy:

$$U(z) = \sum_{i\in S} \Phi_1(z_i) + \sum_{\langle i,j\rangle} \Phi_2(z_i, z_j),$$

where

$$\Phi_1(z_i) = \alpha x_i + \beta y_i + \gamma x_i y_i \text{ and } \Phi_2(z_i, z_j) = \delta x_i x_j + \eta y_i y_j.$$

1. Calculate the distributions $\pi_i(z_i|z^i)$, $\pi_i^1(x_i|x^i, y)$, $\pi_i^2(y_i|x, y^i)$.
2. Construct Gibbs sampling based on $\{\pi_i^1, \pi_i^2, i \in S\}$.

4.5. Matérn-I and Matérn-II models.

1. Let X be a homogeneous PPP(λ). A Matérn-I model (154; 48) can be obtained by deleting, in X, all pairs of points that are $\leq r$ apart:

$$X^I = \{s \in X : \forall s' \in X, s' \neq s, \|s - s'\| > r\}.$$

   Show that the intensity of the resulting process is $\lambda^* = \lambda \exp\{-\lambda\pi r^2\}$. Implement a simulation of this type of process on $[0,1]^2$ with $r = 0.02$ and $\lambda = 30$.
2. Suppose X is a homogeneous PPP(λ). To each point s of X associate an independent and continuous mark Y_s with density $h(\cdot)$. A Matérn-II model is obtained from X by deleting points $s \in X$ whenever there is some $s' \in X$ that is $\leq r$ away and if furthermore $Y_{s'} < Y_s$. Simulate such a process on $[0,1]^2$ with $r = 0.02$, $\lambda = 30$ and h the exponential distribution with mean 4.

4.6. Simulating bivariate Gaussian random fields.
Suppose we are interested in simulating a bivariate Gaussian distribution on S with energy:

$$U(x,y) = -\sum_{i\in S}(x_i^2 + y_i^2) - \beta \sum_{\langle i,j\rangle}(x_i y_j + x_j y_i), \qquad |\beta| < 1/2.$$

Calculate the conditional distributions $\pi_i(z_i|z^i)$, $\pi_i^1(x_i|x^i, y)$ and $\pi_i^2(y_i|x, y^i)$. Suggest two simulation methods.

4.7. Suppose Σ_n is the set of permutations of $\{1,2,\ldots,n\}$ and π the uniform distribution over Σ_n. Show that the chain with transition P on Σ_n that randomly permutes indices i and j is π-reversible and converges to π.

4.8. Simulating spatio-temporal dynamics.
Let $X=(X(t),t\geq 0)$ be a Markov chain on $\Omega=\{0,1\}^S$, $S=\{1,2,\ldots,n\}\subset\mathbb{Z}$, with transition from $x=x(t-1)$ to $y=x(t)$ of:

$$P(x,y)=Z(x)^{-1}\exp\{\sum_{i\in S}y_i[\alpha+\beta v_i(t)+\gamma w_i(t-1)]\},$$

with $v_i(t)=y_{i-1}+y_{i+1}$, $w_i(t-1)=x_{i-1}+x_i+x_{i+1}$ and $x_j=y_j=0$ if $j\notin S$.

1. Simulate these dynamics using Gibbs sampling.
2. With $\alpha=-\beta=2$ and $n=100$, study the evolution of stains $N_t=\{i\in S: X_i(t)=1\}$ as a function of γ.
3. Propose an analogous model on $S=\{1,2,\ldots,n\}^2\subset\mathbb{Z}^2$ and give a simulation of it.

4.9. Using spin-flips, simulate the 4-NN Ising model with parameters $\alpha=0$ and β on the torus $\{1,2,\ldots,64\}^2$, with the initial configuration having an equal number of $+1$ and -1 spins. Calculate the empirical correlation $\rho(\beta)$ at a distance 1 and construct the empirical curve $\beta\mapsto\rho(\beta)$. Answer the same questions using Gibbs sampling.

4.10. Simulating grayscale textures.
A Φ-model with grayscale textures $\{0,1,\ldots,G-1\}$ has energy:

$$U(x)=\theta\sum_{\langle i,j\rangle}\Phi_d(x_i-x_j),\qquad \Phi_d(u)=\frac{1}{1+(u/d)^2}.$$

For such models, the grayscale contrast increases with d and θ controls the spatial correlation. Simulate various textures using Gibbs sampling by varying the parameters θ and d as well as the number of levels of gray G.

4.11. Dynamics of site by site relaxation for simulation of π.
Consider the Ising model $\pi(x)=Z^{-1}\exp\{\beta\sum_{\langle i,j\rangle}x_ix_j\}$ on $\{-1,+1\}^S$, where $S=\{1,2,\ldots,n\}^2$ is associated with the 4-NN relation. We are interested in sequential dynamics where, at each step, the relaxation P_s occurs at *only one* site s, i.e., $P_s(x,y)=0$ if $x^s\neq y^s$. Furthermore, if $m=m(s)\in\{0,1,2,3,4\}$ counts the number of $+1$ neighbors of s, we impose that P_s depends only on m ($P_s(x,y)=P_s(x_s\to y_s|m)$) and is symmetric: ($P_s(-x_s\to -y_s|4-m)=P_s(x_s\to y_s|m)$).

1. Show that P_s depends on 5 parameters, $a_0=P_s(+1\to -1|m=2)$, $a_1=P_s(+1\to -1|m=3)$, $a_2=P_s(-1\to +1|m=3)$, $a_3=P_s(+1\to -1|m=4)$ and $a_4=P_s(-1\to +1|m=4)$.
2. Show that P_s is π-reversible if and only if $a_3=a_4\exp(-8\beta)$ and $a_1=a_2\exp(-4\beta)$. Thus, we note $a=(a_0,a_2,a_4)\in]0,1]^3$.

3. Give the values of a corresponding to the choices of Gibbs sampling and Metropolis dynamics.
4. Show that the dynamic associated with P_s for uniform s is ergodic.

4.12. Use Metropolis dynamics to simulate a Strauss process on $[0,1]^2$ with $n = 50$ points. Comment on the resulting configurations for $\gamma = 0.01$, 0.5, 1, 2, 10 and $r = 0.05$.

4.13. Reconstruction of a system of faultlines.
A faultline on $S = [0,1]^2$ is represented by a straight line d that cuts S. Each faultline d has an associated value $V_d : S \longrightarrow \{-1,+1\}$ that is constant on each half-space generated by d. Suppose that S is cut by n faultlines $R = \{d_1, d_2, \ldots, d_n\}$ (we know n but not R); these n faults generate a resulting value $V_R(z) = \sum_{i=1}^{n} V_{d_i}(z)$ for each $z \in S$. The following information is available to us: we have m wells located at m known locations in S, $\mathscr{X} = \{z_1, z_2, \ldots, z_m\}$ and we have observations $V_i = V_R(z_i)$ on $\mathscr{X}$. Our goal is to reconstruct R. We can consider this problem from the point of view of simulating n straight lines intersecting S under the following constraint (C):

$$C(R) = \sum_{i=1}^{m} (V_R(z_i) - V_i)^2 = 0.$$

Simulate R under the constraint (C) with the following prior distribution π: (i) the n straight lines are uniform i.i.d. and cut across S; (ii) independently, values associated with each are uniform i.i.d. on $\{-1,+1\}$.
Hints: (i) a straight line $d(x,y) = x\cos\theta + y\sin\theta - r = 0$ can be parametrized by (θ, r); characterize the subset $\Delta \subseteq [0, 2\pi[\times[0,+\infty[$ of straight lines that can cut S; a random straight line corresponds to the random choice of a point in Δ; (ii) the value associated with d can be characterized by $V_d(0) = \varepsilon_d \times d(0,0)$, where the ε_d are uniform i.i.d. on $\{-1,+1\}$ and independent of the randomly generated straight lines.

4.14. Suppose $S = \{s_1, s_2, \ldots\}$ is a discrete subset of $\mathbb{R}^2$. Using simple kriging, implement a recursive simulation algorithm on S for a centered Gaussian process with covariance $C(h) = e^{-\|h\|/4}$.

4.15. Simulating binary Markov textures.
Consider the 2-dimensional torus $S = \{1, 2, \ldots, n\}^2$ given the following neighbor relation: $(i,j) \sim (i',j')$ if either $i - i'$ is congruent to n and $j = j'$ or if $i = i'$ and $j - j'$ is congruent to n. Consider an Ising random field X on S with states $\{-1,+1\}$ for which translation-invariant potentials are associated with four families of pair potentials:

$$\Phi_{1,i,j}(x) = \beta_1 x_{i,j} x_{i+1,j}, \qquad \Phi_{2,i,j}(x) = \beta_2 x_{i,j} x_{i,j+1},$$

$$\Phi_{3,i,j}(x) = \gamma_1 x_{i,j} x_{i+1,j+1} \quad \text{and} \quad \Phi_{4,i,j}(x) = \gamma_2 x_{i,j} x_{i+1,j-1}.$$

X is an 8-NN Markov random field with parameter $\theta = (\beta_1, \beta_2, \gamma_1, \gamma_2)$.

1. Show that the marginal distribution at each site is uniform: $P(X_i = -1) = P(X_i = +1)$.

2. Describe an algorithm for simulating X using Gibbs sampling.

4.16. Confidence intervals for spatial correlations.
Consider the isotropic Ising model for the 2-dimensional torus $S = \{1,2,\dots,n\}^2$ with energy $U(x) = \beta \sum_{\langle s,t \rangle} x_s x_t$, where $\langle s,t \rangle$ represents the 4-NN relation (modulo n for boundary points).

1. Show that $E(X_i) = 0$ and $Var(X_i) = 1$.
2. Denote by $\rho(\beta) = E(X_{i,j}X_{i+1,j})$ the spatial correlation at distance 1. A natural estimator of this correlation is:

$$\widehat{\rho}(\beta) = \frac{1}{n^2} \sum_{(i,j)\in S} X_{i,j}X_{i+1,j},$$

   where $X_{n+1,j}$ is taken to be equivalent to $X_{1,j}$. Show that $\widehat{\rho}(\beta)$ is an unbiased estimator of $\rho(\beta)$.
3. As is the case for marginal distributions of Gibbs random fields, the mapping $\beta \mapsto \rho(\beta)$ is analytically unknown. One way to get near it is to generate N i.i.d. examples of X using an MCMC method, for example Gibbs sampling. Implement this procedure and find both the empirical distribution of $\widehat{\rho}(\beta)$ and the symmetric 95% bilateral confidence interval for $\rho(\beta)$. Implement the method for values from $\beta = 0$ to 1 in step sizes of 0.1.
4. Suppose now that we have independence $\beta = 0$. Show that $Var(\widehat{\rho}(0)) = n^{-2}$. Prove, noting $Z_{ij} = X_{i,j}X_{i+1,j}$, that Z_{ij} and $Z_{i'j'}$ are independent if $j \neq j'$ or if $|i-i'| > 1$. Deduce that it is reasonable to believe that, when we have independent $\{X_{ij}\}$, $\widehat{\rho}(\beta)$ is close to some variable $\mathcal{N}(0,n^{-2})$ for relatively large n. Test this using N samples of X.

4.17. Hierarchical modeling.
Suppose we have a random field $Y = \{Y_i,\ i \in S\}$ over a discrete subset S endowed with a symmetric neighbor graph $\mathcal{G}$ without loops. Consider the following hierarchical model:

1. $(Y_i|X_i)$, $i \in S$, conditionally independent Poisson random variables with mean $E(Y_i|X_i) = \exp(U_i + X_i)$.
2. $U_i \sim \mathcal{N}(0, \kappa_1)$ independent and $X = \{X_i, i \in S\}$ an intrinsic auto-Gaussian random field with conditional distributions (CAR model):

$$(X_i|X_{i^-}) \sim \mathcal{N}(|\partial i|^{-1} \sum_{j\in\partial i} X_j, (\kappa_2|\partial i|)^{-1}),$$

   where ∂i is the neighborhood of i.
3. $\kappa_i^{-1} \sim \Gamma(a_i, b_i)$ independent.

Suggest an MCMC algorithm to simulate the posterior distribution of X given $Y = y$.

Chapter 5
Statistics for spatial models

In this chapter we present the main statistical methods used to deal with the three types of data seen in earlier chapters. As well as general statistical methods that can be applied to various structures (maximum likelihood, minimum contrast, least squares, estimation of generalized linear models, the method of moments), we have specific techniques for each type of structure: variogram clouds in geostatistics, conditional pseudo-likelihood, Markov random field coding, nearest-neighbor distances, composite likelihood for PPs, etc. We will present each method in turn.

For further details and results, we suggest consulting the books cited in the text. We also recommend the online help for `R`, which can be freely downloaded from the website: `www.R-project.org` ((178) and cf. Appendix D). This well-documented help is frequently updated and contains useful references.

We also remark that when the model being examined is easily simulated, Monte Carlo techniques (testing, model validation) are useful in the absence of theoretical results.

When there are a large number of observations, we distinguish *two kinds of asymptotics* (Cressie, (48)). *Increasing domain asymptotics* are used when the number of observations increases with the size of the domain of observation. This approach is adopted when the observation sites (districts, measuring stations, parcels of agricultural land) are spatially distinct, such as in epidemiology, spatial geography, environmental modeling, ecology and agronomy. The other type, *infill asymptotics* are for when the number of observations increases inside a fixed and bounded domain S. This might be the case in mineral exploration or radiographic analysis (increasing image resolution). These two research areas remain relatively open as the "spatial" context here is more technical, probabilistic results are lacking (ergodicity, weak dependency, CLT) and/or it is difficult to verify hypotheses necessary for useful theorems. On the other hand, the extensive development of MCMC methods, valid in many cases, does not exactly encourage spending time on such difficult problems. Here we will only mention a few results related to increasing domain asymptotics and invite the reader to consult Stein's book (200) or (231) for more about infill asymptotics.

C. Gaetan, X. Guyon, *Spatial Statistics and Modeling*, Springer Series in Statistics,
DOI 10.1007/978-0-387-92257-7_5, © Springer Science+Business Media, LLC 2010

In spatial statistics, boundary effects are more important than in temporal statistics. For a time series ($d = 1$), the percentage of points on the boundary of the domain $S_N = \{1,2,\ldots,N\}$ is in the order of N^{-1}, which has no effect on the bias (the asymptotic distribution) of a renormalized classical estimator $\sqrt{N}(\widehat{\theta}_N - \theta)$. This is no longer true if $d \geq 2$: for example, for $d = 2$, the fraction of points on the boundary of the domain $S_N = \{1,2,\ldots,n\}^2$ with $N = n^2$ points is in the order of $1/\sqrt{N}$, leading to bias in the renormalized estimator. A nice solution, proposed by Tukey (216) consists of tapering the data on the boundary of the spatial domain (cf. §5.3.1), a further advantage is that this gives less importance to observations on the domain boundary, which are often not consistent with the postulated model. For spatial PPs, a correction for boundary effects was proposed by Ripley (184) (cf. §5.5.3).

We first present statistical methods for geostatistics, followed by those suited to second-order models or applicable to Markov random fields, and finish with a look at statistics for point processes. Four appendices containing additional information round off the chapter: Appendix A describes classical simulation methods; Appendix B provides details on ergodicity, the law of large numbers and the central limit theorem for spatial random fields; Appendix C develops the general methodology for minimum contrast estimation, as well as its asymptotic properties; technical proofs for several results in Chapter 5 are also collected here. Lastly, Appendix D presents useful software packages and gives examples of their use.

5.1 Estimation in geostatistics

5.1.1 Analyzing the variogram cloud

Let X be an intrinsic real-valued random field on $S \subset \mathbb{R}^d$ with constant mean $E(X_s) = \mu$ and variogram:

$$2\gamma(h) = E(X_{s+h} - X_s)^2 \text{ for all } s \in S.$$

We suppose that X is observed at n sites $\mathscr{O} = \{s_1,\ldots,s_n\}$ of S and we note $X(n) = {}^t(X_{s_1},\ldots,X_{s_n})$.

Suppose to begin with that γ is isotropic. The variogram cloud is the set of $n(n-1)/2$ points

$$\mathscr{N}_{\mathscr{O}} = \{(\|s_i - s_j\|, (X_{s_i} - X_{s_j})^2/2), i,j = 1,\ldots,n \text{ and } s_i \neq s_j\}$$

of the first quadrant of $\mathbb{R}^2$. As $(X_{s_i} - X_{s_j})^2/2$ is an unbiased estimator of $\gamma(s_i - s_j)$, this cloud is the correct representation of $h \mapsto 2\gamma(h)$. Note that pairs (s_i, s_j) of sites with large squared value $(X_{s_i} - X_{s_j})^2$ can turn up next to pairs the same distance apart but with small squared value (cf. Fig. 5.1-a). This may indicate local data outliers (175). Thus, in its initial form the cloud does not allow us to effectively analyze the variogram's characteristics such as its range or sill, nor does it let us look for the existence of a nugget effect. To correct for this, a smoothing of the cloud is superimposed onto the cloud itself; this smoothing is sometimes performed by moving averages, though more commonly by using a convolution kernel.

A priori, the variogram 2γ is not isotropic and it is prudent to consider several orientations of the vector h and evaluate the variogram cloud in each direction using the vectors $s_i - s_j$ "close" to that direction. In this way we can empirically detect possible anisotropy in the variogram. A common way to proceed is to use the 4 directions S, SE, E and NE with an angular tolerance of ± 22.5 degrees about each.

Example 5.1. Rainfall in the State of Parana (continued)

We return to Example 1.11 on rainfall $X = (X_s)$ in the State of Parana (Brazil). As the isotropic variogram cloud calculated on this raw data is very noisy, we perform smoothing using a Gaussian kernel with a smoothing parameter (here the standard deviation) of 100 (cf. Fig. 5.1-a). This smoothing, calculated in the 4 suggested directions, reveals obvious anisotropy (cf. Fig. 5.1-b). Note that this anisotropy could be due to a non-stationary mean: for example, if $X_{s,t} = \mu + as + \varepsilon_{s,t}$ where ε is intrinsic and isotropic, $2\gamma_X(s,0) = a^2 s^2 + 2\gamma_\varepsilon(s,0)$ and $2\gamma_X(0,t) = 2\gamma_\varepsilon(s,0)$; this shows that there can be confusion between first-order non-stationarity and non-isotropy.

As the variograms in the $0°$ and $45°$ directions appear to have quadratic shapes, it seems that an affine trend in these directions has been missed. As this non-stationarity seems to be confirmed by Figure 1.9, we propose an affine response surface, $E(X_s) = m(s) = \beta_0 + \beta_1 x + \beta_2 y$, $s = (x,y) \in \mathbb{R}^2$. However, the residuals calculated by OLS remain anisotropic (cf. Fig. 5.1-c). For a quadratic response surface $m(s) = \beta_0 + \beta_1 x + \beta_2 y + \beta_3 x^2 + \beta_4 xy + \beta_5 y^2$, the variogram cloud obtained suggests we should retain a model with stationary errors and isotropy close to white noise (cf. Fig. 5.1-d).

5.1.2 Empirically estimating the variogram

The natural empirical estimator of $2\gamma(h)$ is the moments estimator (Matheron (152)):

$$\widehat{\gamma}_n(h) = \frac{1}{\#N(h)} \sum_{(s_i,s_j)\in N(h)} (X_{s_i} - X_{s_j})^2, \qquad h \in \mathbb{R}^d. \tag{5.1}$$

In this formula, $N(h)$ is a lag class of pairs (s_i,s_j) at distance h ($h \in \mathbb{R}^d$) within a certain tolerance Δ. In the isotropic case, we take for example, with $r = \|h\| > 0$:

$$N(h) = \{(s_i,s_j) : r - \Delta \leq \left\|s_i - s_j\right\| \leq r + \Delta \; ; i,j = 1,\dots,n\}.$$

In practice, we estimate the variogram $2\gamma(\cdot)$ at a finite number k of lags:

$$\mathscr{H} = \{h_1, h_2, \dots, h_k\},$$

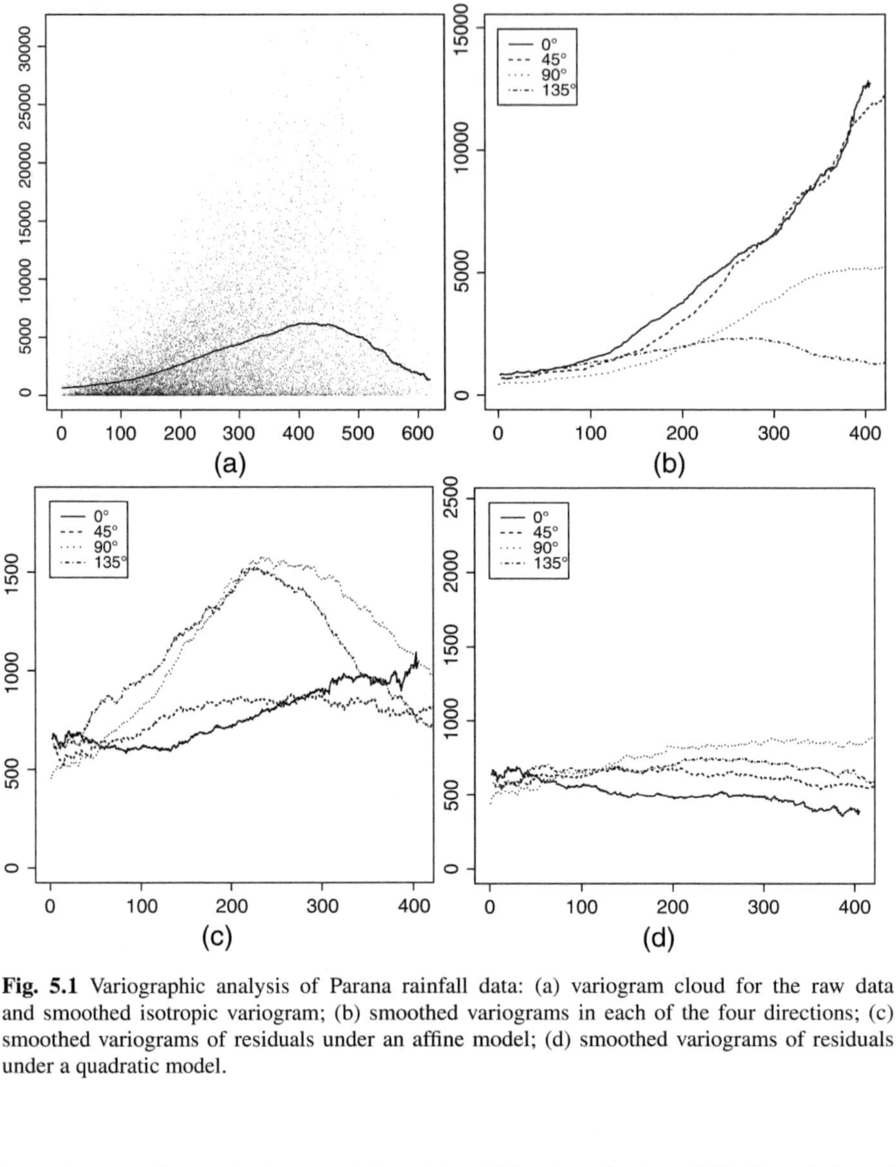

Fig. 5.1 Variographic analysis of Parana rainfall data: (a) variogram cloud for the raw data and smoothed isotropic variogram; (b) smoothed variograms in each of the four directions; (c) smoothed variograms of residuals under an affine model; (d) smoothed variograms of residuals under a quadratic model.

in such a way that each class contains at least 30 pairs of points (48). The choice of family of lags $\mathscr{H}$ must cover "adequately" the domain of $\gamma(\cdot)$ whilst ensuring each class has sufficient points (cf. Fig. 5.2-a). When estimating a parametric model (cf. §5.1.3), $\mathscr{H}$ must also allow us to identify the parameter. More generally, without supposing isotropy and for the vector $h = r(\cos\alpha, \sin\alpha)$ with direction α in $\mathbb{R}^2$, we take (cf. Fig. 5.2-b):

$$
\begin{aligned}
N(h) &= \{(s_i, s_j) : s_i - s_j \in \mathscr{V}_{\Delta,\delta}(h,\alpha)\ ;\ i,j = 1,\ldots,n\},\\
\mathscr{V}_{\Delta,\delta}(h) &= \{v = u(\cos\beta, \sin\beta) \in \mathbb{R}^2, \text{ where } |u - r| \leq \Delta \text{ and } |\beta - \alpha| \leq \delta\}.
\end{aligned}
$$

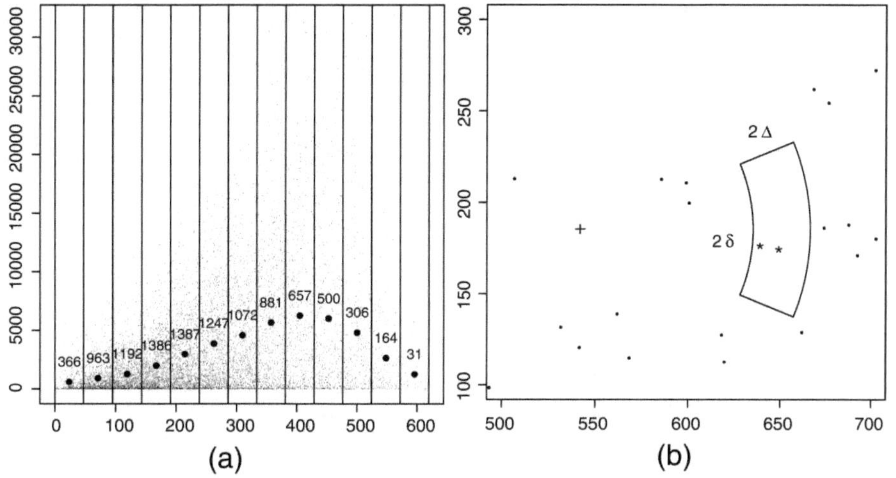

Fig. 5.2 (a) Variogram cloud and isotropic spacings for rainfall data from the State of Parana showing the number of pairs of points falling in each division; (•) indicates the empirical estimation of the variogram 2γ; (b) sites s_j $(*)$ such that $s_i - s_j \in \mathscr{V}_{\Delta,\delta}(h,\alpha)$ for a site s_i $(+)$, with $\|h\| = 109.15$, $\alpha = 0$, $\Delta = 15.59$ and $\delta = \pi/8$.

If X is second-order stationary, the covariance can be empirically estimated by

$$\widehat{C}_n(h) = \frac{1}{\#N(h)} \sum_{s_i,s_j \in N(h)} (X_{s_i} - \overline{X})(X_{s_j} - \overline{X}), \qquad h \in \mathbb{R}^d, \tag{5.2}$$

where $\widehat{\mu} = \overline{X} = n^{-1}\sum_{i=1}^{n} X_{s_i}$ is an unbiased estimator of the mean μ.

An advantage of $2\widehat{\gamma}_n(h)$ in comparison to $\widehat{C}_n(h)$ is that it does not require a prior estimate of the mean μ. Furthermore, under the intrinsic stationarity hypothesis, $2\widehat{\gamma}_n(h)$ is an unbiased estimator of $2\gamma(h)$, which is not the case for $\widehat{C}_n(h)$.

The following proposition gives the distribution of $2\widehat{\gamma}_n(h)$ when X is a Gaussian process (without the Gaussian hypothesis, see Prop. 5.3), which can be deduced from properties of quadratic forms of Gaussian vectors. In effect, $2\widehat{\gamma}(h)$ can be written ${}^tXA(h)X$ where $A(h)$ is an $n \times n$ symmetric p.s.d. matrix with coefficients $A_{s_i,s_j} = -1/\#N(h)$ if $s_i \neq s_j$ and $A_{s_i,s_i} = (\#N(h) - 1)/\#N(h)$ otherwise. As the matrix $A(h)$ is rank $\leq \sharp N(h)$, we note $\lambda_i(h)$ the $\sharp N(h)$ non-zero eigenvectors of $A(h)\Sigma$. Since $A(h)\mathbf{1} = 0$, we may suppose $\mu = 0$.

Proposition 5.1. *Distribution of empirical variograms for Gaussian processes*

If $X \sim \mathscr{N}_n(0,\Sigma)$, *then* $\widehat{\gamma}(h) \sim \sum_{i=1}^{\sharp N(h)} \lambda_i(h)\chi_{i1}^2$ *for some* χ^2 *i.i.d. with* 1*-df. In particular:*

$$E(\widehat{\gamma}(h)) = Trace(A(h)\Sigma) \ \ and \ \ Var(\widehat{\gamma}(h)) = 2 \times Trace\{A(h)\Sigma\}^2.$$

Proof. The proof is standard:

1. $Y = \Sigma^{-1/2}X$ follows a $\mathscr{N}_n(0,I_n)$ and $\widehat{\gamma}(h) = {}^tY\Gamma Y$ where $\Gamma = {}^t\Sigma^{1/2}A(h)\Sigma^{1/2}$.

2. If $\Gamma = {}^tPDP$ is the spectral decomposition of Γ (P orthogonal, D the matrix of eigenvalues (λ_i) of Γ), then ${}^tY\Gamma Y = \sum_{i=1}^n \lambda_i Z_i^2$ where the variables (Z_i) are i.i.d. standardized Gaussian random variables: in effect, ${}^tY\Gamma Y = {}^tZDZ$ if $Z = PY$ and $Z \sim \mathcal{N}_n(0, I_n)$.
3. The proposition follows from the fact that ${}^t\Sigma^{1/2}A(h)\Sigma^{1/2}$ and $A(h)\Sigma$ have the same eigenvalues and that $E(\chi^2_{i1}) = 1$ and $Var(\chi^2_{i1}) = 2$.

□

As the estimation $\widehat{\gamma}(h)$ is not very robust for large values of $X_{s_i} - X_{s_j}$, Cressie and Hawkins (49; 48, p. 74) propose the robustified estimator:

$$\overline{\gamma}_n(h) = \left\{0.457 + \frac{0.494}{\#N(h)}\right\}^{-1} \left\{\frac{1}{\#N(h)} \sum_{(s_i,s_j)\in N(h)} |X_{s_i} - X_{s_j}|^{1/2}\right\}^4 .$$

$|X_{s_i} - X_{s_j}|^{1/2}$, with expectation proportional to $\gamma(s_i - s_j)^{1/4}$, is in effect less sensitive to large values of $|X_{s_i} - X_{s_j}|$ than $(X_{s_i} - X_{s_j})^2$ and the denominator corrects asymptotically the bias for $2\overline{\gamma}_n$ when $\#N(h)$ is large. Moreover, Cressie and Hawkins show that the mean quadratic error of $\overline{\gamma}_n$ is less than that of $\widehat{\gamma}_n$.

5.1.3 Parametric estimation for variogram models

The variogram models $\gamma(\cdot;\theta)$ presented in Chapter 1 (cf. §1.3.3) depend on a parameter $\theta \in \mathbb{R}^p$ which is generally unknown. We now present two methods for estimating θ, least squares and maximum likelihood.

Least squares estimation

The estimation of θ by *ordinary least squares* (OLS) is a value

$$\widehat{\theta}_{OLS} = \underset{\alpha\in\Theta}{\text{argmin}} \sum_{i=1}^k (\widehat{\gamma}_n(h_i) - \gamma(h_i;\alpha))^2, \tag{5.3}$$

where k is the number of classes chosen for the empirical estimation $\widehat{\gamma}_n$ of γ at lags $\mathcal{H} = \{h_1, h_2, \dots, h_k\}$. In this expression, we have the option of replacing $\widehat{\gamma}_n(h_i)$ by the robustified estimator $\overline{\gamma}_n(h_i)$.

As in regression, the OLS method generally performs poorly as the $\widehat{\gamma}_n(h_i)$ are neither independent nor have the same variance. We might instead prefer to estimate θ by *generalized least squares* (GLS) :

$$\widehat{\theta}_{GLS} = \underset{\alpha\in\Theta}{\text{argmin}}\, {}^t(\widehat{\gamma}_n - \gamma(\alpha))\{Cov_\alpha(\widehat{\gamma}_n)\}^{-1}(\widehat{\gamma}_n - \gamma(\alpha)), \tag{5.4}$$

where $\widehat{\gamma}_n = {}^t(\widehat{\gamma}_n(h_1), \ldots, \widehat{\gamma}_n(h_k))$ and $\gamma(\alpha) = {}^t(\gamma(h_1, \alpha), \ldots, \gamma(h_k, \alpha))$. As calculating $Cov_\alpha(\widehat{\gamma}_n)$ is often difficult, the *weighted least squares* (WLS) method is a compromise between OLS and GLS where the squares are weighted by the variances of the $\widehat{\gamma}_n(h_i)$. For example, in the Gaussian case, since $Var_\alpha(\widehat{\gamma}_n(h_i)) \simeq 2\gamma^2(h_i, \alpha)/\#N(h_i)$, we obtain:

$$\widehat{\theta}_{WLS} = \underset{\alpha \in \Theta}{\operatorname{argmin}} \sum_{i=1}^{k} \frac{\#N(h_i)}{\gamma^2(h_i; \alpha)} (\widehat{\gamma}_n(h_i) - \gamma(h_i; \alpha))^2. \tag{5.5}$$

Simulation studies (232) show that the performance of the WLS estimator remains relatively satisfactory with respect to GLS.

These three methods operate under the same principle, *least squares estimation* (LSE): for $V_n(\alpha)$ a $k \times k$ p.d. symmetric matrix with known parametric form, we want to minimize the distance $U_n(\alpha)$ between $\gamma(\alpha)$ and $\widehat{\gamma}_n$:

$$\widehat{\theta}_{LSE} = \underset{\alpha \in \Theta}{\operatorname{argmin}}\, U_n(\alpha), \text{ where } U_n(\alpha) = {}^t(\widehat{\gamma}_n - \gamma(\alpha))V_n(\alpha)(\widehat{\gamma}_n - \gamma(\alpha)). \tag{5.6}$$

The LSE method is a special case of the minimum contrast estimator (cf. Appendix C). As the following proposition shows, the consistency (resp. asymptotic normality) of $\widehat{\theta}_n = \widehat{\theta}_{LSE}$ follows from the consistency (resp. asymptotic normality) of the empirical estimator $\widehat{\gamma}_n$. We note:

$$\Gamma(\alpha) = \frac{\partial}{\partial \alpha} \gamma(\alpha)$$

the $k \times p$ matrix of the p derivatives of the vector $\gamma(\alpha) \in \mathbb{R}^k$ and suppose:

(V-1) For all $\alpha_1 \neq \alpha_2$ in Θ, $\sum_{i=1}^k (2\gamma(h_i, \alpha_1) - 2\gamma(h_i, \alpha_2))^2 > 0$.
(V-2) θ is interior to Θ and $\alpha \mapsto \gamma(\alpha)$ is $\mathscr{C}^1$.
(V-3) $V_n(\alpha) \to V(\alpha)$ in P_α-probability where V is symmetric, p.d. and $\alpha \mapsto V(\alpha)$ is $\mathscr{C}^1$.

Proposition 5.2. *Convergence and asymptotic normality of the LSE (Lahiri, Lee and Cressie (138))*

Suppose conditions (V1-2-3) are satisfied and note θ the true unknown parameter value.

1. *If $\widehat{\gamma}_n \longrightarrow \gamma(\theta)$ P_θ-a.s., then $\widehat{\theta}_n \longrightarrow \theta$ a.s.*
2. *Suppose further that, for a sequence (a_n) tending to infinity,*

$$a_n(\widehat{\gamma}_n - \gamma(\theta)) \xrightarrow{d} \mathcal{N}_k(0, \Sigma(\theta)), \tag{5.7}$$

where $\Sigma(\theta)$ is p.d. and that the matrix $\Gamma(\theta)$ has full rank p. Then:

$$a_n(\widehat{\theta}_n - \theta) \xrightarrow{d} \mathcal{N}_p(0, \Delta(\theta)),$$

where

$$\Delta(\theta) = B(\theta)^t\Gamma(\theta)V(\theta)\Sigma(\theta)V(\theta)\Gamma(\theta)^t B(\theta),$$

with $B(\theta) = ({}^t\Gamma(\theta)V(\theta)\Gamma(\theta))^{-1}$.

Comments:

1. Convergence (5.7) is studied in the following section in the broader context of models with linear trends.
2. (V-1) is an identification condition for the parameters of the model: if $\alpha_1 \neq \alpha_2$, the k lags $\mathscr{H} = \{h_1, h_2, \ldots, h_k\}$ allow us to distinguish between two variograms $\gamma(\cdot,\alpha_1)$ and $\gamma(\cdot,\alpha_2)$. This condition requires that there are at least p identification vectors h_i: $k \geq p$.
3. The proof of the consistency of $\widehat{\theta}_n$ uses continuity of the functions $\alpha \mapsto \gamma(\alpha)$ and $\alpha \mapsto V(\alpha)$; differentiability is needed to establish the asymptotic normality of $\widehat{\theta}_n - \theta$.
4. For the OLS method ($V(\alpha) \equiv Id_k$), if we can choose $k = p$, then $\Delta(\theta) = \Gamma^{-1}(\theta)\Sigma(\theta)\,{}^t(\Gamma^{-1}(\theta))$.
5. The method is efficient if we take for $V_n(\alpha)$ the inverse of the variance matrix of $\widehat{\gamma}_n$. As this matrix is difficult to obtain, Lee and Lahiri (144) propose to estimate it by a subsampling procedure that remains asymptotically efficient.
6. This result allows us to construct a subhypothesis test on the parameter θ.

5.1.4 Estimating variograms when there is a trend

It remains to see under what conditions the empirical estimator $\widehat{\gamma}_n$ is asymptotically normal as supposed in (5.7). We present here a result of Lahiri, Lee and Cressie (138) which guarantees this property in the more general context of linear models:

$$X_s = {}^t z_s \delta + \varepsilon_s, \qquad \delta \in \mathbb{R}^p, \tag{5.8}$$

with a zero mean error intrinsic random field

$$E(\varepsilon_{s+h} - \varepsilon_s)^2 = 2\gamma(h,\theta), \qquad \theta \in \mathbb{R}^q.$$

This model, with parameter $(\delta,\theta) \in \mathbb{R}^{p+q}$, can be considered from two points of view. If δ is the parameter of interest as is the case in econometrics, we read (5.8) as a spatial regression. In this case, θ is an auxiliary parameter that must be pre-estimated (by $\widetilde{\theta}$) so that we can then efficiently estimate δ using GLS with the estimated variance $Var_{\widetilde{\theta}}(\varepsilon)$ (cf. §5.3.4).

If instead we are largely interested in the dependency parameter θ as is the case in spatial analysis, δ becomes the auxiliary parameter, estimated for example by OLS. The variogram is then estimated as follows:

1. Estimate δ by $\widehat{\delta}$ using a method that does not require knowledge of θ, for example OLS.
2. Calculate the residuals $\widehat{\varepsilon}_s = X_s - {}^t z_s \widehat{\delta}$.
3. Estimate the empirical variogram for $\widehat{\varepsilon}$ on $\mathscr{H}$ with (5.1).

We now fix the asymptotic framework to be used. Let $D_1 \subset \mathbb{R}^d$ be an open set with Lebesgue measure $d_1 > 0$ containing the origin and with regular boundary (∂D_1 a finite union of rectifiable surfaces with finite measure, for example $D_1 =]-1/2,1/2]^d$, or $D_1 = B(0,r)$, $r > 0$). We suppose that the domain of observation is:

$$D_n = (nD_1) \cap \mathbb{Z}^d.$$

In this case, $d_n = \sharp D_n \sim d_1 n^d$ and due to the geometry of D_1, $\sharp(\partial D_n) = o(\sharp D_n)$. Suppose now that the following conditions hold:

(VE-1) $\exists \eta > 0$ s.t. $E\,|\varepsilon_s|^{4+\eta} < \infty$; X is α-mixing (cf. B.2), satisfying:

$$\exists C < \infty \text{ and } \tau > \frac{(4+\eta)d}{\eta} \text{ such that, } \forall k,l,m : \alpha_{k,l}(m) \leq Cm^{-\tau}.$$

(VE-2) $\sup_{h \in \mathscr{H}} \sup\{\|z_{s+h} - z_s\| : s \in \mathbb{R}^d\} < \infty$.

(VE-3) $\|\widehat{\delta}_n - \delta\| = o_P(d_n^{-1/4})$, where $\widehat{\delta}_n$ is an estimator of δ.

Proposition 5.3. *Asymptotic normality of empirical variograms (138)*
Under conditions (VE-1-2-3),

$$d_n^{-1/2}(\widehat{\gamma}_n - \gamma(\theta)) \xrightarrow{d} \mathscr{N}_k(0, \Sigma(\theta)),$$

where

$$\Sigma_{l,r}(\theta) = \sum_{i \in \mathbb{Z}^d} cov_\theta([\varepsilon_{h_l} - \varepsilon_0]^2, [\varepsilon_{i+h_r} - \varepsilon_i]^2), \qquad l,r = 1,\ldots,k.$$

Comments:

1. (VE-1) ensures normality if the mean of X is constant. This is a consequence of the CLT for the random field of squares $\{[\varepsilon_{i+h_r} - \varepsilon_i]^2\}$ that is mixing (cf. §B.3 for this result and for certain other mixing random fields). If X is a stationary Gaussian random field with an exponentially decreasing covariance, (VE-1) is satisfied.
2. (VE-2) and (VE-3) ensure that if $\widehat{\delta}_n \to \delta$ at a sufficient rate, the previous result remains valid when working with residuals. A parallel can be drawn between this condition and a result of Dzhaparidze ((75); cf. §5.3.4) showing the efficiency of the estimator $\widehat{\psi}_1$ in the first step of the Newton-Raphson algorithm for resolving a system of equations $F(\widehat{\psi}) = 0$ under the condition that the initial estimator $\widehat{\psi}_0$ is consistent at a sufficient rate.
3. Some standard conditions on the sequence of regressors and the error model allow us to control the variance of the OLS estimator for β (cf. Prop. 5.7) and thus to verify condition (VE-3).
4. If we add conditions (V) to (VE) we obtain asymptotic normality for the LSE $\widehat{\theta}$ (5.8).
5. Lahiri, Lee and Cressie (138) examine a mixed asymptotics framework where, in conjunction with increasing domain asymptotics, the set of points in D_n also

becomes denser. For example, for successively finer resolutions with stepsize $h_n = m_n^{-1}$ in each direction, $m_n \in \mathbb{N}^*$ and $m_n \longrightarrow \infty$, the domain of observation becomes:

$$D_n^* = nD_1 \cap \{\mathbb{Z}/m_n\}^d.$$

Though we multiply the number of observation points by a factor of $(m_n)^d$, the rate of convergence remains the same as for increasing domain asymptotics, that is, at the rate $\sqrt{d_n} \sim \sqrt{\lambda(nD_1)}$, the asymptotic variance being:

$$\Sigma_{l,r}^*(\theta) = \int_{\mathbb{R}^d} cov_\theta([\varepsilon_{h_l} - \varepsilon_0]^2, [\varepsilon_{s+h_r} - \varepsilon_s]^2)ds, \qquad l,r = 1,\ldots,k.$$

This echoes the classical setting in diffusion statistics ($d = 1$), whether we measure diffusion at a discrete set of points ($D_n = \{1,2,\ldots,n\}$) at greater and greater definition with stepsize m_n^{-1} ($D_n^* = \{k/m_n, k = 1,\ldots,nm_n\}$), or continuously on $D_n = [1,n]$, $n \to \infty$; in both cases, we have convergence at the rate $\sqrt{n}$ and only the factor modulating this rate varies from one case to the next.

Maximum likelihood

If X is a Gaussian vector with mean μ and covariance $Var(X) = \Sigma(\theta)$, the log-likelihood is:

$$l_n(\theta) = -\frac{1}{2}\left\{\log|\Sigma(\theta)| + {}^t(X-\mu)\Sigma^{-1}(\theta)(X-\mu)\right\}. \qquad (5.9)$$

To maximize (5.9), we iterate the calculation of the determinant $|\Sigma(\theta)|$ and the inverse of the covariance $\Sigma^{-1}(\theta)$, which requires $O(n^3)$ operations (149). If $\widehat{\theta}$ is the ML estimation of θ, that of the variogram is $2\gamma(h,\widehat{\theta})$.

The asymptotic properties of the maximum likelihood estimator in the Gaussian case are presented in §5.3.4, which deals with the estimation of Gaussian spatial regressions.

5.1.5 Validating variogram models

Choosing a variogram model from within a family of parametric models can be done using the LS Akaike criterion (4; 114) (cf. §C.3). Once a model $\mathscr{M}$ has been chosen, it must be validated. One way to do so is to immerse $\mathscr{M}$ in a dominating model $\mathscr{M}^{\max}$ and test $\mathscr{M} \subset \mathscr{M}^{\max}$. Two "nonparametric" alternatives are as follows.

Cross-validation

The underlying idea in cross-validation is to put aside each observation in turn and use kriging (cf. §1.9) to predict its value using the other observations without again estimating the variogram model. For each site we then have an observed value X_{s_i} and predicted value $\widehat{X}_{s_i}$. Validation is via the mean square normalized error criteria:

$$MSNE = \frac{1}{n}\sum_{i=1}^{n}\frac{(X_{s_i}-\widehat{X}_{s_i})^2}{\tilde{\sigma}_{s_i}^2},$$

where $\tilde{\sigma}_{s_i}^2$ is the kriging variance. If the variogram model is correctly identified and well-estimated, then the MSNE should be close to 1. If as a first approximation we suppose the normalized residuals $(X_{s_i}-\tilde{X}_{s_i})/\tilde{\sigma}_{s_i}$ are independent Gaussian variables with variance 1, the model can be validated if, noting $q(n,\beta)$ the β-quantile of a χ_n^2,

$$q(n,\alpha/2) \le nMSNE \le q(n,1-\alpha/2).$$

Looking at a graph of the renormalized residuals can be useful to validate the model. A homogeneous spatial distribution of the positive and negative residuals indicates an adequate model (cf. Fig. 5.5-a). In contrast, Fig. 5.5-b indicates residual heteroscedasticity.

Validation via Monte Carlo methods: parametric bootstrap

Let $\widehat{\theta}$ be an estimator of θ for the variogram $\gamma(\cdot,\theta)$ being validated. Generate m i.i.d. samples $(X^{(j)},\ j=1,\dots,m)$ of an intrinsic random field on $S=\{s_1,s_2,\dots,s_n\}$ with variogram $\gamma(\cdot;\widehat{\theta})$. For each sample j, calculate the empirical estimates $\{\widehat{\gamma}_n^{(j)}(h_i), i=1,\dots,k\}$ on $\mathscr{H}$ as well as their empirical lower bounds $\widehat{\gamma}_{inf}(h_i)=\min_{j=1,\dots,m}\widehat{\gamma}_n^{(j)}(h_i)$ and upper bounds $\widehat{\gamma}_{sup}(h_i)=\max_{j=1,\dots,m}\widehat{\gamma}_n^{(j)}(h_i)$, $i=1,\dots,k$. Allotting the same weight to each sample, these two bounds define an approximate empirical confidence interval of level $1-2/(m+1)$:

$$P(\widehat{\gamma}_n^{(j)}(h_i)<\widehat{\gamma}_{inf}(h_i))=P(\widehat{\gamma}_n^{(j)}(h_i)>\widehat{\gamma}_{sup}(h_i))\le\frac{1}{m+1}.$$

Next we represent graphically the functions $h_i \to \widehat{\gamma}_{inf}(h_i)$ and $h_i \to \widehat{\gamma}_{sup}(h_i)$, $i=1,\dots,k$. If the initial empirical estimates $\widehat{\gamma}_n(h_i)$ are contained within the confidence envelope $\{[\widehat{\gamma}_{inf}(h_i),\widehat{\gamma}_{sup}(h_i)], i=1,\dots,k\}$, we can reasonably conclude that X is an intrinsic process with variogram $\gamma(\cdot,\theta)$ (cf. Fig. 5.4-b). Otherwise, we reject the model $\gamma(\cdot,\theta)$. By examining the reasons for doing so, we may find a path to an alternative model. Nevertheless note that this procedure is relatively conservative and tends not to reject the hypothesis we seek to validate.

Example 5.2. The radioactive cloud from Chernobyl and daily rainfall in Switzerland

We consider daily cumulative rainfall data from the Swiss meteorological service measured on May 8, 1986, the day Chernobyl's radioactive cloud traveled across Europe. As daily rainfall is a good indicator of the effect of radioactive fallout, these

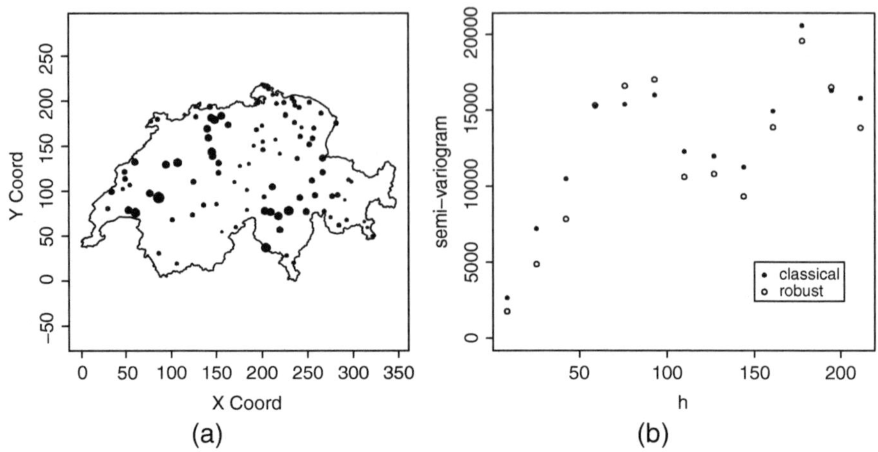

Fig. 5.3 (a) Rainfall data for 100 Swiss weather stations: size of symbols is proportional to rainfall intensity; (b) 2 nonparametric estimates of the semivariogram for 13 classes of distance.

data allowed contamination risk to be evaluated after the Chernobyl disaster. These data were used in a competition where researchers were invited to propose models and prediction methods. 100 data points from the Swiss network (cf. Fig. 5.3) were made available (dataset `sic.100` in the `geoR` package) out of a total of 467, the challenge was to best predict the remaining 367 values (known but hidden) with respect to the square error criteria (70).

While estimators (5.1) and (5.2) do not guarantee the c.n.d. condition that variograms must satisfy (cf. Prop. 1.2), they nevertheless allow us to estimate a parametric model (§1.3.3): Table 5.1 gives estimations obtained using `geoR` with a Matérn variogram with three parameters (a, σ^2, ν) (cf. §1.3.3).

Fig. 5.4-a gives a comparison of semivariograms obtained using OLS, WLS and ML under a Gaussian hypothesis on the distribution. The MSNE (cf. Table 5.1) obtained using cross-validation suggests validity of the Matérn model and WLS estimation. Confirmation is given in Fig. 5.4-b, showing that the 13 empirical estimations of γ are all contained within the empirical 95% confidence interval obtained from 40 simulations of the estimated Matérn model. Figures 5.5-a and 5.5-b, showing the distribution of residuals, also support this model.

Table 5.1 Parametric OLS, WLS and ML estimates of the Matérn semivariogram for the Swiss rainfall data and the MSNE cross-validation error.

	$\widehat{a}$	$\widehat{\sigma}^2$	$\widehat{\nu}$	MSNE
OLS	17.20	15135.53	1.21	1.37
WLS	18.19	15000.57	1.00	1.01
ML	13.40	13664.45	1.31	1.09

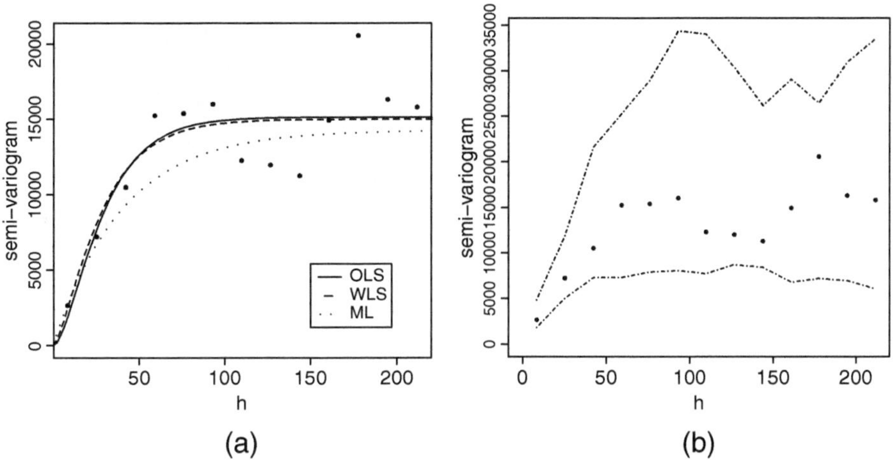

Fig. 5.4 (a) Three parametric estimates of the semivariogram; (b) empirical estimates of γ compared with the upper and lower confidence levels (dotted lines) obtained by generating 40 samples from the estimated Matérn model (parameters $\widehat{a} = 18.19$, $\widehat{\sigma}^2 = 15000.57$ and $\widehat{\nu} = 1.0$).

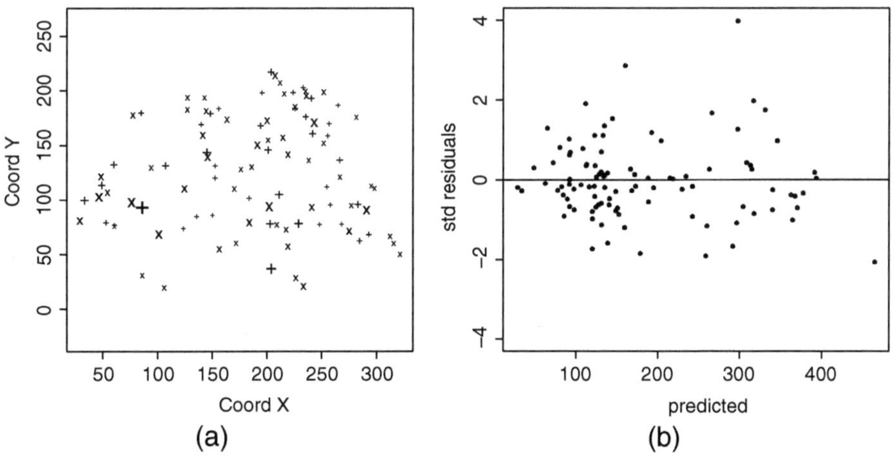

Fig. 5.5 Swiss rainfall data: (a) spatial distribution of positive (+) and negative (x) normalized residuals; (b) graph of normalized residuals as a function of predicted values.

Lastly, Fig. 5.6 gives the map of predictions (a) by kriging for rainfall over the whole landmass and (b) its standard deviation, both calculated using the estimated model.

Example 5.3. Rainfall in the State of Parana (cont.)

We now continue our modeling of rainfall in the State of Parana. The variographic analysis carried out in Example 5.1 led to the supposition of either an affine

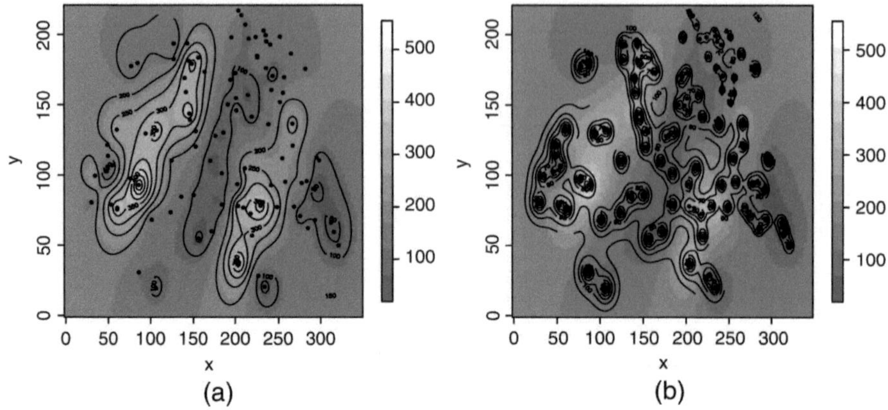

Fig. 5.6 Kriging of Swiss rainfall data: (a) predictions $\widehat{X}_s$ and (b) standard deviation of prediction error.

Table 5.2 Results of Gaussian maximum likelihood estimation for the rainfall data from the State of Parana. l is the value of the log of the Gaussian likelihood and m the number of parameters.

Model	l	m	*AIC*	*MSNE*
A	-663.9	6	1340	0.98
B	-660.0	8	1336	1.14
C	-660.2	9	1338	0.99

response surface with anisotropic variogram or a quadratic response surface and a white noise variogram. We now propose to choose among three models:

1. Model A: $m(s) = \beta_0 + \beta_1 x + \beta_2 y$ and exponential isotropic covariance with nugget effect:
$$C(h) = \sigma^2 \exp(-\|h\|/\phi) + \tau^2 \mathbf{1}_0(h).$$
2. Model B: $m(s) = \beta_0 + \beta_1 x + \beta_2 y$ and exponential covariance with nugget effect and geometric anisotropy:
$$C(h) = \sigma^2 \exp(-\|A(\psi,\lambda)h\|/\phi) + \tau^2 \mathbf{1}_0(h),$$
where $A(\psi,\lambda)$ is the rotation around the origin in $\mathbb{R}^2$ of angle ψ followed by a dilation of $0 \le 1/\lambda \le 1$ along the new y axis.
3. Model C: $m(s) = \beta_0 + \beta_1 x + \beta_2 y + \beta_3 x^2 + \beta_4 xy + \beta_5 y^2$ and
$$C(h) = \sigma^2 \exp(-\|h\|/\phi) + \tau^2 \mathbf{1}_0(h).$$

We remark that A is a submodel of B and C. If we let l represent the Gaussian log-likelihood of the observations and m the number of model parameters, we get the results in Table 5.2. Formally speaking (i.e., by applying the likelihood ratio test), model B is better than A whereas A and C are not significantly different. Using the AIC criteria leads to the choice of the anisotropic model B, whereas the MSNE criteria (measuring the mean quadratic error (by kriging) between the predicted values and real values) suggests keeping model C. Models B and C are thus both potential choices.

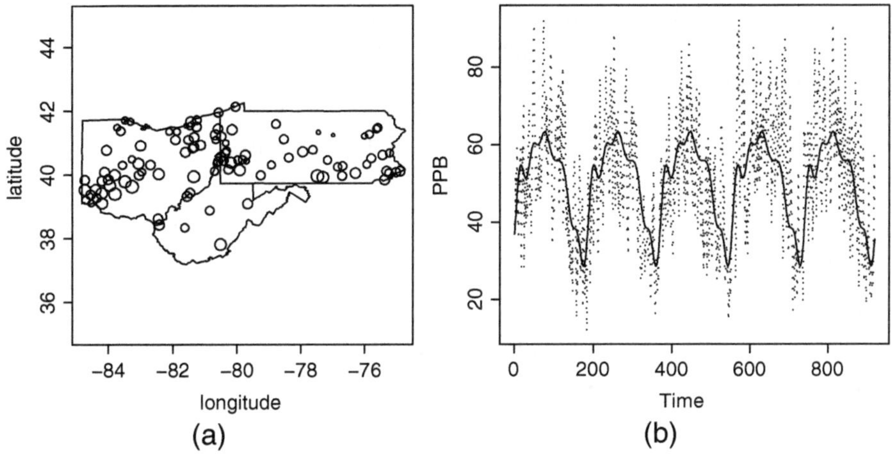

Fig. 5.7 Ozone level data at 106 stations in 3 U.S. states: (a) spatial distribution of stations: the size of symbols is proportional to the mean annual level in parts per billion (PPB); (b) temporal evolution of the mean over the 106 stations and estimated periodic components.

Example 5.4. Analyzing spatio-temporal data for ozone levels

The data we refer to here are maxima of hourly means over eight consecutive hours of ozone levels measured at 106 stations in the U.S. states of Ohio, Pennsylvania and Virginia (cf. Fig. 5.7-a) from May 1 to October 31, 1995–1999, giving 184 observations per site per year. The original data, available at `http://www.image.ucar.edu/GSP/Data/O3.shtml`, were centered by removing a seasonal component from the data at each station s (cf. Fig. 5.7-b),

$$\mu_{s,t} = \alpha_s + \sum_{j=1}^{10} [\beta_j \cos(2\pi jt/184) + \gamma_j \sin(2\pi jt/184)].$$

We limit ourselves here to an analysis of May–September data of the final year, putting aside the October data to use for testing and validating the model. We are left with 15,133 observations after excluding missing data.

Selection of a spatio-temporal model begins by a visual inspection of the empirical estimation of the spatio-temporal variogram $2\gamma(h,u) = Var(X_{s+h,t+u} - X_{s,t})$, $(h,u) \in \mathbb{R}^2 \times \mathbb{R}$. In the case of a second-order stationary process, $\gamma(h,u) = (C(0,0) - C(h,u))$. The moment estimator for stationary and isotropic processes is:

$$2\widehat{\gamma}_n(h,u) = \frac{1}{\#N(h,u)} \sum_{(s_i,s_j;t_i,t_j)\in N(h,u)} (X_{s_i,t_i} - X_{s_j,t_j})^2, \qquad h \in \mathbb{R}^d,$$

with

$$N(h,u) = \{(s_i,s_j;t_i,t_j) : \ \|h\| - \Delta_S \leq \|s_i - s_j\| \leq \|h\| + \Delta_S,$$
$$|u| - \Delta_T \leq |t_i - t_j| \leq |u| + \Delta_T \ ; i,j = 1,\ldots,n\}.$$

We plot (cf. Fig. 5.8) the empirical estimates of γ for various choices of spacing h_k, u_l. Fig. 5.8-a shows that the instantaneous spatial correlation decreases with distance, as does the temporal correlation at individual sites. We therefore consider the following four models of spatio-temporal covariance $C(h,u)$ that are both flexible and coherent with respect to the plotted empirical variograms:

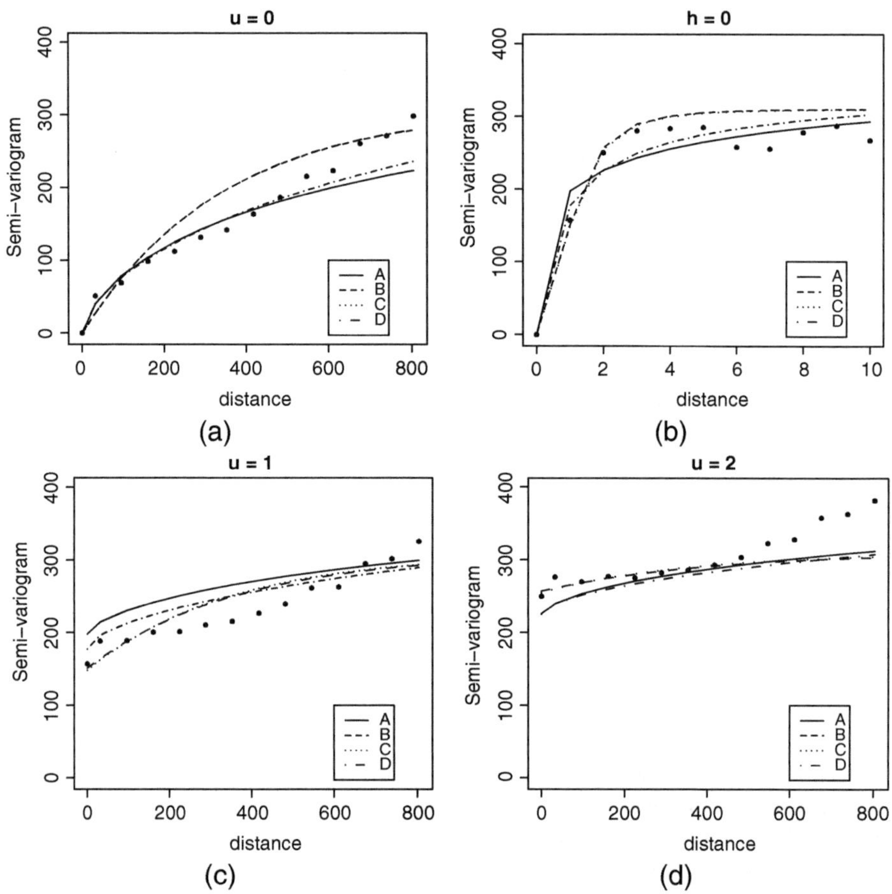

Fig. 5.8 Empirical (•) and parametric estimates of the spatio-temporal semivariogram of the ozone level data for models A, B, C and D and distance classes: (a) $\gamma_n(h_k,0)$; (b) $\gamma_n(0,u_l)$; (c) $\gamma_n(h_k,1)$; (d) $\gamma_n(h_k,2)$.

-Model A:

$$C(h,u) = \sigma^2(1 + |u/\psi|^a + \|h/\phi\|^b)^{-3/2},$$

with $0 < a, b \leq 2$, $\phi, \psi > 0$ (a special case of model (1.18)).
-Model B:

$$C(h,u) = \sigma^2(1+|u/\psi|^a)^{-3/2}\exp\left\{-\frac{\|h/\phi\|}{(1+|u/\psi|^a)^{b/2}}\right\},$$

with $0 < a \leq 2$, $0 \leq b \leq 1$ (this is model (1.17) from Ex. 1.7).
-Model C: model B with $b = 0$ (C is thus a separable model).
-Model D:

$$C(h,u) = \sigma^2(1+|u/\psi|^a)^{-b}M_c\left(\frac{\|h/\phi\|}{(1+|u/\psi|^a)^{b/2}}\right),$$

with $0 < a \leq 2$, $0 \leq b \leq 1$, where $M_c(v) = (2^{b-1}\Gamma(c))^{-1}v^c\mathcal{K}_c(v)$, $c > 0$ is the Matérn model. Model D comes from the class (1.16).

We estimate parameters using WLS, minimizing the criteria:

$$W(\theta) = \sum_{k=1}^{m_t}\sum_{l=1}^{m_s}\frac{|N(h_k,u_l)|}{\gamma(h_k,u_l;\theta)^2}\left(\widehat{\gamma}(h_k,u_l) - \gamma(h_k,u_l;\theta)\right)^2, \tag{5.10}$$

where m_s (resp. m_t) is the number of spatial (resp. temporal) divisions used for a given fit. This is the most commonly used method as calculating Gaussian likelihoods (5.9) is costly (requiring $O(N^3)$ operations for N observations; note however that one way to get around this is to approximate the likelihood (126; 201; 79)).

We consider two approaches to choosing the model. The first simply uses $W(\widehat{\theta})$, the WLS criteria (5.10). The second is based on the quality of predictions of the model: using simple kriging for October 1999, we calculate the prediction $\widehat{x}_{s,t}$ for one day for each station using data from the three previous days. Quality of prediction on the 3075 available observations is evaluated using the mean square error $MSE = \sum_{s,t}(x_{s,t} - \widehat{x}_{s,t})^2/3075$. Table 5.3 shows the results: the non-separable models are better, in particular D. Under the prediction criteria, C is the best model.

Table 5.3 Parametric estimates of the semivariogram using the WLS method on the ozone level data.

Model	$\widehat{a}$	$\widehat{b}$	$\widehat{c}$	$\widehat{\sigma}^2$	$\widehat{\phi}$	$\widehat{\psi}$	$W(\widehat{\theta})$	MSE
A	0.67	0.32	–	461.87	1920.20	11.58	134379.3	161.46
B	2.00	0.21	–	310.15	347.13	1.34	200141.4	151.84
C	2.00	–	–	309.70	347.30	1.36	200303.0	152.03
D	1.69	0.23	0.29	392.69	1507.78	0.77	110462.1	155.61

5.2 Autocorrelation on spatial networks

In this section we suppose that X is a real-valued random field defined over a discrete network $S \subset \mathbb{R}^d$ that is not necessarily regular, endowed with an *influence graph* $\mathscr{R}$ (*a priori* directed), $(i,j) \in \mathscr{R}$ signifying that j influences i, $j \neq i$.

Furthermore, suppose we have a positive and bounded *weights matrix* $W = \{w_{ij}, (i,j) \in \mathscr{R}\}$ that quantifies the amount of influence j has on i, such that for all i, $w_{ii} = 0$ and $w_{ij} = 0$ if $(i,j) \notin \mathscr{R}$. The choice of W, left to an expert, is an important step in modeling $\mathscr{R}$ and depends on the problem under consideration. When i represents a geographical region (district, county, state, etc.), Cliff and Ord (45) propose to let w_{ij} depend on quantities such as the Euclidean distance $d(i,j) = \|i-j\|$, the percentage $f_{i(j)}$ of the border of i shared with j, the strength $c_{i,j}$ of communication networks passing between i and j when we have urban areas, etc. For example,

$$w_{ij} = f_{i(j)}^{b} d(i,j)^{-a}$$

with $a, b > 0$ gives large weights w_{ij} for neighboring states, even more so if i shares a large part of its border with j. W is the contiguity matrix of the graph $\mathscr{R}$ if $w_{ij} = 1$ for $(i,j) \in \mathscr{R}$ and 0 otherwise. A normalized contiguity matrix is one associated with the normalized weights $w_{ij}^{*} = w_{ij} / \sum_{k \in \partial i} w_{ik}$.

Classically we distinguish between two indices for measuring global spatial dependency on a network $(S, \mathscr{R})$: Moran's index calculates a spatial correlation and Geary's index a spatial variogram.

5.2.1 Moran's index

Let X be a second-order real-valued random field observed on a subset $D_n \subset S$ of cardinality n. Suppose to begin with that X is centered and note $\sigma_i^2 = Var(X_i)$. Let W be an $n \times n$ matrix with known weights. A W-measure of global spatial autocovariance is defined by:

$$C_n = \sum_{i,j \in D_n} w_{i,j} X_i X_j.$$

This autocovariance has to be normalized to provide an autocorrelation.

Under the hypothesis (H_0) of *spatial independence* of the X_i, it is easy to see that:

$$Var(C_n) = \sum_{i,j \in D_n} (w_{ij}^2 + w_{ij} w_{ji}) \sigma_i^2 \sigma_j^2.$$

Thus, under (H_0), $I_n = \{Var(C_n)\}^{-1/2} C_n$ is a centered variable with variance 1. If the σ_i^2 are known, we can get an asymptotic test of (H_0) if we have a central limit theorem for C_n. If the variances are known up to a multiplicative constant, $\sigma_i^2 = a_i \sigma^2$, $a_i > 0$ known, we can estimate σ^2 by $\widehat{\sigma}^2 = \frac{1}{n} \sum_{i \in D_n} X_i^2 / a_i$ and then work with the index C_n renormalized by its estimated standard deviation.

Moran's index is the generalization of I_n to the case where X has constant but unknown mean and variance, $E(X_i) = \mu$ and $Var(X_i) = \sigma^2$. Estimating these parameters respectively by $\overline{X} = n^{-1} \sum_{i \in D_n} X_i$ and $\widehat{\sigma}^2 = n^{-1} \sum_{i \in D_n} (X_i - \overline{X})^2$, *Moran's index* is given by:

$$I_n^M = \frac{n \sum_{i,j \in D_n} w_{ij}(X_i - \overline{X})(X_j - \overline{X})}{s_{0,n} \sum_{i \in D_n} (X_i - \overline{X})^2}, \tag{5.11}$$

where $s_{0n} = \sum_{i,j \in D_n} w_{ij}$. Setting $s_{1n} = \sum_{i,j \in D_n} (w_{ij}^2 + w_{ij} w_{ji})$, we have

$$I_n^M = \frac{s_{1n}^{1/2}}{s_{0n}} \{\widehat{Var}(C_n)\}^{-1/2} C_n = \frac{s_{1n}^{1/2}}{s_{0n}} \widehat{I}_n.$$

As s_{0n} and s_{1n} are of order n and under (H_0), $E(C_n) = 0$ and $\widehat{\sigma}^2 \xrightarrow{Pr} \sigma^2$, we have under (H_0) :

$$E(I_n^M) = o(1) \quad \text{and} \quad Var(I_n^M) = \frac{s_{1n}}{s_{0n}^2}(1 + o(1)).$$

Though it looks like a correlation coefficient, I_n^M can be outside the interval $[-1, +1]$ (cf. Ex. 5.2). The more similar the values are at neighboring sites, the larger I_n^M is, and vice versa. In the first case we talk of aggregation or spatial cooperation and the second case of repulsion or spatial competition. If we are dealing with independent variables, I_n^M will be close to 0.

We remark that Moran's index can also be calculated if data are binary or ordinal (45).

Moran's index at distance d

Suppose that W is the spatial contiguity matrix of graph $\mathscr{R}$,

$$w_{ij} = 1 \qquad \text{if } (i,j) \in \mathscr{R}, \qquad w_{ij} = 0 \quad \text{otherwise.}$$

For integer $d \geq 1$, we say that j is a d-neighbor of i if a sequence $i_1 = i$, i_2, ..., $i_d = j$ exists such that $w_{i_l, i_{l+1}} \neq 0$ for $l = 1, \dots, d-1$ and if this path has minimal length. Thus, i_2 is a neighbor of i, i_3 of i_2, ... and so on up to j being a neighbor of i_{d-1}. To this relation we can associate the neighbor graph $\mathscr{R}^{(d)}$ and its contiguity matrix $W^{(d)}$ for the distance d. Relative to $W^{(d)}$, we can define Moran indices $I^M(d)$ at distance $d \geq 1$ in the same way as in (5.11). Note that $W^{(d)} \equiv \{W^d\}^*$, where $M_{ij}^* = 0$ (resp. 1) if $M_{ij} = 0$ (resp. $M_{ij} \neq 0$).

In order to test the spatial independence hypothesis (H_0), we can either use an asymptotic test or permutation test.

5.2.2 Asymptotic test of spatial independence

Suppose that S is an infinite set of sites no closer to each other than some minimum distance > 0. Let (D_n) be a strictly increasing sequence of finite subsets of $S \subset \mathbb{R}^d$ and W a known bounded weights matrix on the graph $\mathscr{R} \subset S^2$ having range R to neighborhoods of uniformly bounded size,

$$W = (w_{ij}, i,j \in S) \quad \text{with} \quad w_{ii} = 0 \quad \text{and} \quad w_{ij} = 0 \text{ if } \|i - j\| > R,$$
$$\exists M < \infty \text{ such that } \forall i,j : |w_{ij}| \leq M \quad \text{and} \quad \sharp\partial i \leq M.$$

The following asymptotic normality result does not require X to be a Gaussian random field.

Proposition 5.4. *Asymptotic distribution of Moran's index under* (H_0)
Suppose that:

$$\exists \delta > 0 \text{ such that } \sup_{i \in S} E(|X_i|^{4+2\delta}) < \infty \text{ and } \liminf_n \frac{s_{1n}}{n} > 0.$$

Then, under the spatial independence hypothesis (H_0)*, Moran's index is asymptotically Gaussian:*

$$\frac{s_{0n}}{\sqrt{s_{1n}}} I_n^M \xrightarrow{d} \mathcal{N}(0,1).$$

Proof: We prove the result in the simplified context of centered variables X_i. It suffices to show asymptotic normality of C_n. Indeed:

$$C_n = \sum_{\widetilde{S}_n} Z_i, \text{ where } Z_i = X_i V_i, \ V_i = \sum_j w_{ij} X_j$$

and $\widetilde{S}_n = \{i \in D_n \text{ s.t. } \partial i \subseteq D_n\}$. It is true that under (H_0), variables $Z_i = X_i V_i$ are $2R$-dependent, i.e., independent as soon as $\|i-j\| > 2R$. Furthermore, their moments of order $2+\delta$ are uniformly bounded. Thus, Z satisfies the conditions of the CLT for mixing random fields (cf. §B.3). Under (H_0), I_n is therefore asymptotically a Gaussian random variable. □

Exact calculation of expectation and variance of I_n^M *for Gaussian random fields*

If the X_i are i.i.d. $\mathcal{N}(\mu, \sigma^2)$, then the previous result can be stated more precisely as we are able to analytically calculate the moments of I_n^M. Under (H_0) and Gaussian hypothesis (45), we have:

$$E(I_n^M) = -\frac{1}{n-1} \quad \text{and} \quad Var(I_n^M) = \frac{n^2 s_{1n} - n s_{2n} + 3 s_{0n}^2}{(n^2-1) s_{0n}^2} - \frac{1}{(n-1)^2},$$

where $s_{2n} = \sum_{i \in D_n} (w_{i\cdot} + w_{\cdot i})^2$, $w_{i\cdot} = \sum_{i \in D_n} w_{ij}$ and $w_{\cdot j} = \sum_{j \in D_n} w_{ij}$. Calculation of these two moments relies on the following result of Pitman:

Proposition 5.5. *Let* $X = (X_1, X_2, \ldots, X_n)$ *be sampled from* $\mathcal{N}(0,1)$*,* $h(X)$ *a homogeneous real-valued function of* X *with degree* 0 *and* $Q(X) = \sum_{i=1}^n X_i^2$*. Then the variables* $h(X)$ *and* Q *are independent.*

Proof. If $v < 1/2$, we have

$$\begin{aligned} M(u,v) &= \mathbb{E}(\exp\{iuh(X) + vQ(X)\}) \\ &= (2\pi)^{-n/2} \int_{\mathbb{R}^n} \exp\{iuh(x) - \frac{1}{2}(1-2v)Q(x)\}\, dx_1 dx_2 \ldots dx_n. \end{aligned}$$

Make the following change of variable: for $j = 1,\dots,n$, $x_j = y_j\sqrt{1-2v}$. As h is homogeneous and has degree 0, $h(x) = h(y)$ and

$$\begin{aligned} M(u,v) &= (2\pi)^{-n/2} \times (1-2v)^{-n/2} \int_{\mathbb{R}^n} \exp\{iuh(y) - (1/2)Q(y)\} dy_1 dy_2 \dots dy_n \\ &= (1-2v)^{-n/2} \mathbb{E}(\exp\{iuh(X)\}) \end{aligned}$$

$h(X)$ and $Q(X)$ are therefore independent. □

This result allows us to calculate moments for Moran's index $I = P/Q$, a ratio of quadratic forms: I, homogeneous with degree 0 is independent of $Q(X)$ and thus, for all $p \geq 1$,

$$E(P^p) = E(I^p Q^p) = E(I^p)E(Q^p), \text{ i.e., } E(I^p) = \frac{E(P^p)}{E(Q^p)}.$$

It only remains to calculate the moments of the quadratic forms in the numerator and denominator. We remark that this result remains true if $(X_1, X_2, \dots, X_n)$ is a sample from $\mathcal{N}(\mu, \sigma^2)$. It suffices to replace $\sum_{i=1}^n X_i^2$ by $Q(X) = \sum_{i=1}^n (X_i - \overline{X})^2$. To calculate the moments, we note that in the present case $Q(X)$ is distributed according to the random variable $\sigma^2 \chi^2_{n-1}$.

5.2.3 Geary's index

Geary's index measures spatial dependency in the same way as variograms:

$$I_n^G = \frac{(n-1)\sum_{i,j \in D_n} w_{ij}(X_i - X_j)^2}{2 s_{0n} \sum_{i \in D_n} (X_i - \overline{X})^2}.$$

The more similar values are at neighboring points, the smaller I^G, and vice versa. I^G is sensitive to large differences between neighboring points in the same way that I^M is sensitive to extreme values of X.

Under the hypotheses of Proposition 5.4 and under (H_0), I_n^G is asymptotically a Gaussian random variable:

$$\sqrt{\frac{s_{0n}}{2s_{1n} + s_{2n}}}(I_n^G - 1) \sim \mathcal{N}(0,1).$$

When X is a Gaussian random field, we have:

$$\mathbb{E}(I_n^G) = 1, \qquad Var(I_n^G) = \frac{(2s_{1n} + s_{2n})(n-1) - 4s_{2n}}{2(n+1)s_{0n}}.$$

5.2.4 Permutation test for spatial independence

Generally speaking, the permutational distribution of a real-valued statistic $I(X)$ of $X = (X_i, i = 1, \ldots, n)$ conditional on n observed values $(x_i, i = 1, \ldots n)$ is uniform over the set of values $I_\sigma = I(x_\sigma)$ of I for the $n!$ permutations σ of $\{1, 2, \ldots, n\}$. The associated bilateral (resp. unilateral) significance level is:

$$p_a = \frac{1}{n!} \sum_\sigma \mathbf{1}\{|I_\sigma| > a\} \qquad \text{(resp. } p_a^* = \frac{1}{n!} \sum_\sigma \mathbf{1}\{I_\sigma > a\}\text{)}.$$

If enumerating all possible permutations is impossible, we use instead Monte Carlo methods and choose at random, for m relatively large, m permutations $\{\sigma_1, \sigma_2, \ldots, \sigma_m\}$ for which we calculate the values I_σ and the associated Monte Carlo significance levels p_a^{MC} and p_a^{MC*}.

To test the independence hypothesis (H_0): the X_i are i.i.d. without knowing the common distribution of the X_i, we could use the permutational distribution of Moran's index and/or Geary's index. This test can be justified by the fact that under (H_0) permuting the $\{X_i\}$ does not change the global distribution of X,

$$(X_i, i = 1, \ldots, n) \sim (X_{\sigma(i)}, i = 1, \ldots, n).$$

The advantage of using a permutation method is that it provides a non-asymptotic test without hypotheses on the distribution of X.

If the n observed values $(x_i, i = 1, \ldots, n)$ are all different, the expectation $\mathbb{E}_P$ of Moran's index I^M under the permutational distribution is, under (H_0) (cf. Ex. 5.3):

$$\mathbb{E}_P(I_n^M) = -\frac{1}{n-1}.$$

Example 5.5. Percentage of people with blood group A in 26 Irish counties

Fig. 5.9 shows the percentage of people with blood group A in Ireland's 26 counties (cf. (45) and the `eire` dataset in the `spdep` package).

Fig. 5.9-a clearly shows similarities between neighbors, that is, counties sharing a common border (cf. Fig. 5.9-b). We calculate (cf. Table 5.4) Moran's index and Geary's index $t^a = (I^a - E(I^a))/\sqrt{Var(I^a)}$, $a = M, G$ (cf. `spdep`) for this neighbor relation and $w_{ij} = 1/|\partial i|$ for $(i, j) \in \mathscr{R}$, 0 otherwise, as well as significance levels p^a (for the Gaussian distribution) and p_{MC}^a (for the permutational distribution with $m = 1000$). The results confirm this spatial dependency.

Other permutation test statistics for spatial independence

Statistics other than Moran's index and Geary's index can also be used to test spatial independence. In the spirit of Peyrard et al. (173), consider the random field $X = \{X_{i,j}, (i, j) \in S\}$ observed over the regular grid $S = \{(i, j) : i = 1, \ldots, I; j = 1, \ldots, J\}$ with X measuring the severity of the illness of a plant species in S. Even though

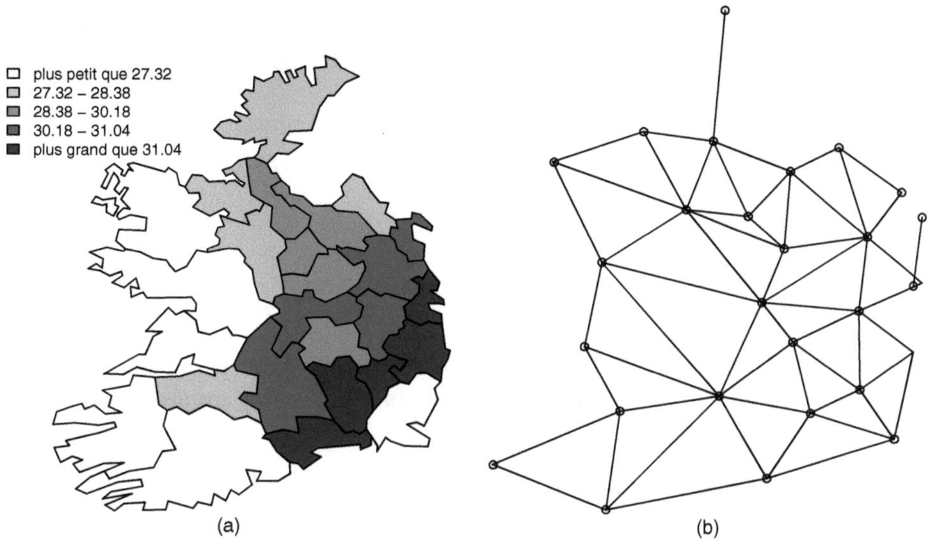

Fig. 5.9 (a) Percentage of the population with blood group A in each of the 26 counties of Ireland; (b) (symmetric) influence graph $\mathscr{R}$ over the 26 counties.

Table 5.4 Percentage with blood group A: Moran and Geary indices and their associated statistics.

	Index	t^a	p^a	p^a_{MC}
Moran	0.554	4.663	0	0.001
Geary	0.380	-4.547	0	0.001

S is regular, this does not necessarily mean that the grid is equally spaced. If the distance between rows i is larger than between columns j we might consider as our statistic for evaluating spatial dependency the variogram for distances d along rows, estimated by:

$$\Gamma(d) = \widehat{\gamma}(0,d) = \frac{1}{I(J-1)} \sum_{i=1}^{I} \sum_{j=1}^{J-d} (X_{i,j} - X_{i,j+d})^2, \qquad d \geq 1.$$

To test the *overall independence* (H_0^G) among all sites, we can take into account permutations σ on the set S of all sites and the associated values

$$\Gamma_\sigma(d) = \frac{1}{I(J-1)} \sum_{i=1}^{I} \sum_{j=1}^{J-d} (X_{\sigma(i,j)} - X_{\sigma(i,j+d)})^2, \qquad d \geq 1.$$

A Monte Carlo test of (H_0^G) can be performed for any distance d by constructing a confidence interval for $\gamma(0,d)$ using predetermined quantiles obtained using m global permutations σ chosen at random (cf. Fig. 5.11-a). If $\widehat{\gamma}(0,d)$ is outside this interval, we reject (H_0^G).

To test *independence between rows i*, (H_0^L), we calculate variograms for distances d along columns and compare them to those calculated using only permutations σ of row indices i:

$$\Delta(d) = \widehat{\gamma}(d,0) = \frac{1}{J(I-1)} \sum_{i=1}^{I-d} \sum_{j=1}^{J} (X_{i+d,j} - X_{i,j})^2, \qquad d \geq 1,$$

$$\widehat{\gamma}_\sigma(d,0) = \frac{1}{J(I-1)} \sum_{i=1}^{I-d} \sum_{j=1}^{J} (X_{\sigma(i)+d,j} - X_{\sigma(i),j})^2, \qquad d \geq 1.$$

The Monte Carlo test of (H_0^L) is thus constructed using m row permutations σ chosen at random (cf. Fig. 5.11-b). Other types of permutation are also possible for testing this hypothesis.

Example 5.6. Decaying lavender

Various permutation methods are applied to tests (spatial homogeneity, independence of pairs of random fields X and Y, conditional independence, etc.) in (173) for exploring data on regular grids. We now take a look at one of their studies.

A field has $I = 11$ rows and $J = 80$ columns of lavender plants. Consecutive plants in the same row are 20cm apart whereas rows themselves are spaced 1m apart. The field has a certain quantity of decaying plants (caused by mycoplasma), we give a score from 0 (healthy, black pixel) to 5 (extremely sick, white pixel) for the state of health of each plant (cf. Fig. 5.10).

Test results are given for $m = 200$ permutations and the confidence levels 2.5% and 97.5%. For the test of global independence (H_0^G) (cf. Fig. 5.11-a), $\Gamma(d)$ almost

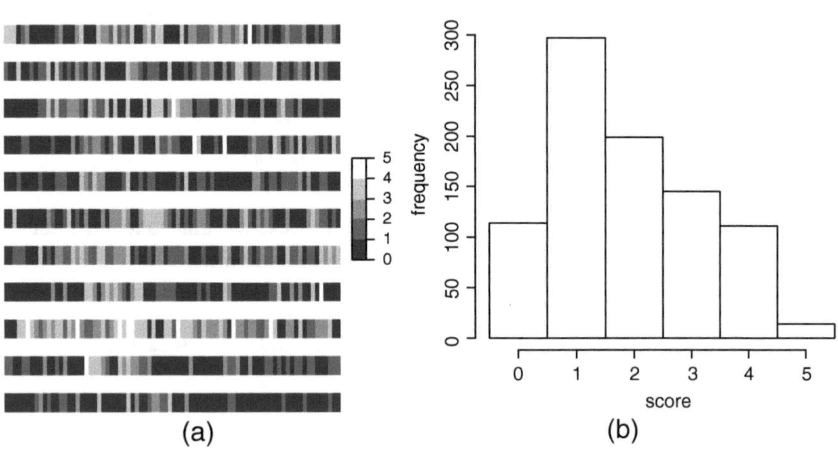

Fig. 5.10 Spatial distribution of scores (1–5) denoting the amount of decay of lavender plants grown on the Sault Plateau; (b) histogram of decay scores (Peyrard et al. (173)).

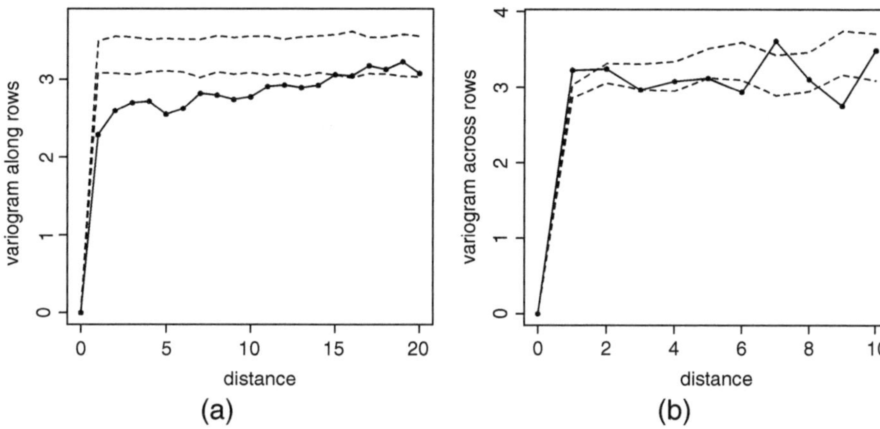

Fig. 5.11 (a) Permutation test of total independence; (b) test of independence of rows. Dashed lines correspond to 2.5% and 97.5% confidence levels.

always lies outside of the constructed confidence region. Thus, there is not global independence. For the second test (H_0^L) of independence between rows (cf. Fig. 5.11-b), $\Delta(d)$ is out of the confidence region for $d = 1, 6, 7$ and 9 and is found on its extremes for $d = 2, 3$ and 5. Here too we reject the hypothesis of independence between rows.

5.3 Statistics for second-order random fields

Suppose that X is a real-valued second-order random field defined on a discrete network S. First we examine the case of stationary X on $S = \mathbb{Z}^d$, followed by X as an AR field on a not necessarily regular network and last, the spatial regression case $X = Z\delta + \varepsilon$, where ε is a second-order centered spatial error random field. We present asymptotic results for when X is a Gaussian random field.

5.3.1 Estimating stationary models on $\mathbb{Z}^d$

Suppose that X is a centered and stationary second-order real-valued random field over $\mathbb{Z}^d$. To make things simpler, suppose that observations $X(n)$ of X are made on the cube $D_n = \{1, 2, \ldots, n\}^d$, with the number of observations being $N = n^d = \sharp(D_n)$. Except for having to deal with boundary effects, the results we present here generalize naturally those obtained for stationary time series ($d = 1$).

Empirical covariance and tapered data

The empirical covariance at distance $k \in \mathbb{Z}^d$ is:

$$\widehat{C}_n(k) = \frac{1}{N} \sum_{i,i+k \in D_n} X_i X_{i+k}, \qquad k \in \mathbb{Z}^d. \tag{5.12}$$

Normalizing by the same N^{-1} for all k means that $\widehat{C}_n(\cdot)$ becomes a p.s.d. function with support $\Delta(n) = \{i - j : i, j \in D_n\}$.

In spatial statistics, *boundary effects* on D_n increase as the dimension d of the network increases: in effect, for a fixed number of observations $N = n^d$, the percentage of points of D_n found on the boundary scales with $dn^{-1} = dN^{-1/d}$, thus increasing as d does. Hence, for empirical covariances, a simple calculation shows that:

$$\lim_n \sqrt{N} E(\widehat{C}_n(k) - C(k)) = \begin{cases} 0 & \text{if } d = 1 \\ -\{|k_1| + |k_2|\} & \text{if } d = 2 \\ +\infty & \text{if } d > 2 \text{ and } k \neq 0 \end{cases}.$$

In $\mathbb{Z}$, the boundary effects of order $n^{-1} = N^{-1}$ have no consequence on asymptotic bias. In $\mathbb{Z}^2$ they are of the order of $N^{-1/2}$ and begin to have a significant effect. For $d \geq 3$ they have a dominant effect.

To remove this bias, an initial solution is to replace the normalization term N^{-1} in the expression for the empirical covariance by $N(k)^{-1}$, where $N(k)$ is the number of pairs (X_i, X_{i+k}) found in (5.12). This said, there is no longer any guarantee that the estimated covariance is p.s.d.

To deal with such difficulties, Tukey (216) defined data $X^w(n)$ *tapered* on the boundary D_n by some tapering function w. For $d = 1,2,3$ and a suitable tapering function, such estimators no longer have the previously described problems, and manage to retain their efficiency. Tapering can also improve statistical analyses as it decreases weights of boundary points which are often not representative of the model being studied.

We now give a definition of tapering. Let $w : [0,1] \longrightarrow [0,1]$, $w(0) = 0$, $w(1) = 1$ be an increasing *tapering profile* of class $\mathscr{C}^2$. The w-tapering function that tapers $100(1-\rho)\%$ of boundary points is defined for $0 \leq \rho \leq 1$ by:

$$h(u) = \begin{cases} w(2u/\rho) & \text{if } 0 \leq u \leq \rho/2 \\ 1 & \text{if } \rho/2 \leq u \leq 1/2 \end{cases} \quad \text{and}$$

$$h(u) = h(1-u) \qquad \text{if} \quad 1/2 \leq u \leq 1.$$

For example, the Tukey-Hanning tapering function is defined as $w(u) = 2^{-1}(1 - \cos \pi u)$. The tapered data $X^w(n)$ are given by:

$$X_i^w = a_n(i) X_i, \text{ where } a_n(i) = \prod_{k=1}^{d} h\left(\frac{i_k - 0.5}{n}\right), \qquad i \in D_n.$$

The tapered empirical covariance $\widehat{C_n^w}$ can be found using the tapered data $X^w(n)$. For $d = 1,2,3$ and the choice $\rho_n = o(n^{-1/4})$, the bias of a tapered estimator is negligible (Dalhaus and Künsch (57); (96, Ch. 4)).

Whittle's Gaussian pseudo-likelihood

Suppose that X is parametrized by its spectral density f_θ, where θ is in the interior of a compact Θ in $\mathbb{R}^p$. The tapered spectrogram

$$I_n^w(\lambda) = \frac{1}{(2\pi)^d} \sum_k \widehat{C}_n^w(k) e^{i\lambda k}$$

is none other than the estimation of $f_\theta(\lambda)$ associated with the tapered empirical covariances. As with time series, $I_n^w(\lambda)$ is a poor estimator of $f_\theta(\lambda)$ for a fixed frequency. On the other hand, in integral form, the (tapered) spectrogram leads to good estimations. In particular, for a Gaussian random field X, Whittle (222) showed that a good approximation of the log-likelihood of $X(n)$ is, up to an additive constant, equal to $-2U_n(\theta)$, where

$$U_n(\alpha) = \frac{1}{(2\pi)^d} \int_{T^d} \left\{ \log f_\alpha(\lambda) + \frac{I_n^w(\lambda)}{f_\alpha(\lambda)} \right\} d\lambda, \qquad T = [0, 2\pi[. \tag{5.13}$$

Furthermore, whether or not X is a Gaussian random field, minimizing U_n leads to a good estimation of θ under reasonable hypotheses. $-2U_n(\theta)$ is known as *Whittle's pseudo log-likelihood* or the *Gaussian pseudo log-likelihood* of X. When the $(c_k(\theta), k \in \mathbb{Z}^d)$ are Fourier coefficients of f_θ^{-1}, another way to write U_n can be obtained from the relationship:

$$(2\pi)^{-d} \int_{T^d} I_n^w(\lambda) f_\alpha^{-1}(\lambda) d\lambda \ = \ \sum_k c_k(\theta) \widehat{C}_n^w(k).$$

Suppose that $\widehat{\theta}_n = \underset{\alpha \in \Theta}{\operatorname{argmin}}\, U_n(\alpha)$ is a minimum contrast estimator (cf. Appendix C). We now give asymptotic properties of $\widehat{\theta}_n$ when X is a Gaussian random field. For this, denote:

$$\Gamma(\theta) = (2\pi)^{-d} \int_{T^d} \{(\log f)_\theta^{(1)\,t} (\log f)_\theta^{(1)}\}(\lambda) d\lambda,$$

$$e(h) = \left[\int_0^1 h^4(u) du\right] \left[\int_0^1 h^2(u) du\right]^{-2}$$

and suppose that

(W1) There exists $0 < m \leq M < \infty$ such that $m \leq f_\theta \leq M$.
(W2) θ is identifiable, i.e., $\theta \mapsto f_\theta(.)$ is one-to-one.
(W3) $f_\theta(\lambda)$ is infinitely differentiable at λ and that $f_\theta^{(2)}$ exists and is continuous at (θ, λ).

Theorem 5.1. *(57; 96) If X is a stationary Gaussian random field satisfying (W), then $\widehat{\theta}_n \xrightarrow{Pr} \theta$. Furthermore, if $\Gamma^{-1}(\theta)$ exists, we have for dimensions $d \leq 3$,*

$$n^{d/2}(\widehat{\theta}_n - \theta) \xrightarrow{d} \mathcal{N}_p(0, e^d(h)\Gamma^{-1}(\theta)).$$

Comments:

1. Hypotheses (W) are satisfied if X is an identifiable ARMA, SAR or CAR model.
2. Hypothesis (W3) implies that X is exponentially α-mixing: in effect, if f is analytic, the covariances of X decrease exponentially and as X is a Gaussian random field, this implies that X is exponentially mixing (cf. Appendix B, §B.2). This weak dependency combined with the CLT for α-mixing random fields (§B.3) allows us to obtain asymptotic normality of the gradient $U_n^{(1)}(\theta)$ of the contrast (condition (N2) in §C.2.2). The hypothesis $f \in \mathscr{C}^\infty$ at λ can also be weakened.
3. This result also holds without tapering ($e(h) = 1$) when $d = 1$. In dimension $d = 2$ and 3, it is possible to choose a h_n-tapering such that $e(h_n) \downarrow 1$. In these cases, $\widehat{\theta}_n$ is asymptotically efficient.
4. A test of a subhypothesis defined with constraint $(H_0) : C(\theta) = 0$, where $C : \mathbb{R}^k \longrightarrow \mathbb{R}^l$ is regular and of class $\mathscr{C}^2$ can be performed upon noticing that $C(\widehat{\theta}_n)$ is asymptotically a centered Gaussian random variable under (H_0).
5. If X is not a Gaussian random field, we still have asymptotic normality of $\widehat{\theta}_n$ if X is 4th order stationary such that X_0 has a moment of order $4+\delta$ for some $\delta > 0$ and if the α-mixing coefficient of X decreases fast enough (57; 96). If so, the asymptotic variance of $\widehat{\theta}_n$ is

$$Var(n^{d/2}\widehat{\theta}_n) \sim e^d(h)\Gamma^{-1}(\theta)[\Gamma(\theta) + B(\theta)]\Gamma^{-1}(\theta),$$

with

$$B(\theta) = \frac{(2\pi)^d}{4} \int_{T^{2d}} \frac{f_{4,\theta}(\lambda, -\lambda, \mu)}{f_\theta(\lambda) f_\theta(\mu)} (\log f_\theta)^{(1)}(\lambda)\, {}^t(\log f_\theta)^{(1)}(\mu) d\lambda d\mu,$$

where $f_{4,\theta}$ is the spectral density of the 4th order cumulants (cf. (96)), zero if X is a Gaussian random field.

Identifying Gaussian $CAR(M)$ models with penalized Whittle contrast

Suppose $M \subset (\mathbb{Z}^d)^+$ is the positive half-plane of $\mathbb{Z}^d$ with respect to the lexicographic order and $m = \sharp M$. If X is a Gaussian $CAR(P_0)$ model where $P_0 \subseteq M$, the goal is to identify P_0 in the CAR family with supports $P \subseteq M$, $\sharp P = p$. Identification by penalized contrast (cf. Appendix C, §C.3) consists of estimating P_0 by:

$$\widehat{P}_n = \underset{P \subseteq M}{\text{argmin}} \left\{ U_n(\widehat{\theta}_P) + \frac{c_n}{n} p \right\},$$

where $\widehat{\theta}_P = \underset{\theta \in \Theta_P}{\text{argmin}}\, U_n(\theta)$. Θ_P is the set of $\theta \in \mathbb{R}^p$ such that the spectral density f_θ of the associated $CAR(P)$ model is everywhere strictly positive. Here, U_n is Whittle's contrast and c_n the rate of penalization brought to act on the model dimension. From

(102, Prop. 8) we have the following identification result: if

$$2C_0 \log\log n \le c_n \le c_0 n$$

for some small enough $c_0 > 0$ and large enough $C_0 < \infty$, then

$$\widehat{P}_n \longrightarrow P_0 \text{ in } P_{\theta_0}\text{-probability.}$$

5.3.2 Estimating autoregressive models

A centered real-valued second-order random field observed on $S = \{1, 2, \ldots, n\}$ is a centered random vector $X = X(n) = {}^t(X_1, X_2, \ldots, X_n) \in \mathbb{R}^n$ characterized by its covariance matrix $\Sigma = Cov(X)$. Let us take a look at two spatial autoregressive cases (cf. §1.7.4):

1. *Simultaneous autoregressive* (SAR) *model*:

$$X_i = \sum_{j \in S: j \neq i} a_{i,j} X_j + \varepsilon_i,$$

   where ε is a WN with variance σ^2. If A is the matrix $A_{i,i} = 0$ and $A_{i,j} = a_{i,j}$ if $i \neq j$, $i, j \in S$, then $X = AX + \varepsilon$ is well-defined as long as $(I - A)$ is invertible, with covariance:

$$\Sigma = \sigma^2 \{ {}^t(I-A)(I-A) \}^{-1}. \tag{5.14}$$

   For estimation, it is necessary to be *sure of the identifiability* of parameters A of a SAR.
2. *Conditional autoregressive* (CAR) *model*:

$$X_i = \sum_{j: j \neq i} c_{i,j} X_j + e_i, \quad \text{where } Var(e_i) = \sigma_i^2 > 0 \text{ and } Cov(X_i, e_j) = 0 \text{ if } i \neq j.$$

   If C is the matrix $C_{i,i} = 0$ and $C_{i,j} = c_{i,j}$ if $i \neq j$, $i, j \in S$ and if D is the diagonal matrix with coefficients $D_{i,i} = \sigma_i^2$, the model is given by $(I - C)X = e$ and

$$\Sigma = D(I - C)^{-1}. \tag{5.15}$$

   In the estimation procedure, we must not forget to impose the constraints: for all $i \neq j : c_{i,j}\sigma_j^2 = c_{j,i}\sigma_i^2$. If $\sigma_i^2 = \sigma^2$ for all i (resp. if X is stationary), these constraints become $c_{i,j} = c_{j,i}$ (resp. $c_{i-j} = c_{j-i}$).

SAR models (resp. CAR models) are *linear models* in A (resp. C). Furthermore, SAR models are CAR models, with $C = {}^tA + A + {}^tAA$, but here we lose linearity in A. For such models, two estimation methods can be envisaged:

1. Maximum likelihood if X is a Gaussian random field, with covariance of either (5.14) or (5.15) depending on the situation. If the model is not Gaussian, this

"Gaussian" likelihood remains a good estimating function. This method is further examined in Section 5.3.4 in the context of spatial regression estimation.

2. Ordinary Least Squares estimation (OLS), following naturally from the fact that AR models are linear models in $\theta = A$ for SAR models and in $\theta = C$ for CAR models. OLS minimizes $\|\varepsilon(\theta)\|^2$ for SAR models and $\|e(\theta)\|^2$ for CAR models. One advantage of OLS is that it gives an explicit estimator that is easy to calculate.

Nevertheless, in OLS estimation an important difference exists between SAR and CAR modeling:

Proposition 5.6. *OLS estimation of SAR models is not in general consistent. CAR model OLS estimation is consistent.*

Proof. Lack of convergence of OLS for SARs is a classical result for linear models $Y = Z\theta + \varepsilon$ when errors ε are correlated with regressors Z (simultaneous equations in econometrics). To prove this point, consider the bilateral SAR model on $\mathbb{Z}^1$ given by

$$X_t = a(X_{t-1} + X_{t+1}) + \varepsilon_t, \qquad |a| < 1/2.$$

If X is observed over $\{1, 2, \ldots, n\}$, the OLS estimation of a is

$$\widehat{a}_n = \frac{\sum_{t=2}^{n-1} X_t(X_{t-1} + X_{t+1})}{\sum_{t=2}^{n-1} (X_{t-1} + X_{t+1})^2}.$$

Denote $r(\cdot)$ the covariance of X. As X is ergodic, it is easy to show that, if $a \neq 0$,

$$\widehat{a}_n \longrightarrow \frac{r_1 - a(r_0 + r_2)}{r_0 + r_2} \neq a.$$

Consistency of OLS for CAR models is a consequence of standard properties of linear models: here, conditional errors e_i are not correlated with the variable $Z(i) = \{X_j,\ j \neq i\}$ characterizing X_i, $i = 1, \ldots, n$. □

Asymptotic properties of OLS for Gaussian CAR models with error variances all equal to σ^2 will be given in §5.4.2. In such cases, OLS coincides with the maximum of the conditional pseudo-likelihood as X is also a Markov Gaussian random field (cf. §5.4.2).

5.3.3 Maximum likelihood estimation

When X is a centered Gaussian vector in $\mathbb{R}^n$ with covariance $\Sigma(\theta)$, the maximum likelihood estimator $\widehat{\theta}_n$ of θ maximizes

$$l_n(\theta) = -\frac{1}{2}\left\{\log|\Sigma(\theta)| + {}^t X \Sigma^{-1}(\theta) X\right\}.$$

Calculating $\widehat{\theta}_n$ requires an iterative optimization technique. The numerical difficulty lies in calculating the determinant $|\Sigma(\theta)|$ and the quadratic form ${}^tX\Sigma^{-1}(\theta)X$. When using AR modeling, we know the parametric form of the inverse covariance Σ^{-1} and thus the form of ${}^tX\Sigma^{-1}(\theta)X$:

1. For SARs: ${}^tX\Sigma^{-1}(\theta)X=\sigma^{-2}\|(I-A)X\|^2$; the ML estimation is:

$$\widehat{\sigma}^2=\sigma^2(\widehat{\theta}_n)=n^{-1}\left\|(I-A(\widehat{\theta}_n))X\right\|^2, \text{ where}$$

$$\widehat{\theta}_n=\underset{\theta}{\operatorname{argmin}}\{-2n^{-1}\log|I-A(\theta)|+\log(\sigma^2(\theta))\}.$$

   An iterative calculation of $|\Sigma(\theta)|$ is quite straightforward for some models. For example, in the case of SARs with a single parameter ρ: $(I-\rho W)X=\varepsilon$, as W is symmetric and $I-\rho W$ invertible, $|\Sigma(\theta)|=\sigma^{2n}|I-\rho W|^{-2}$ and $|I-\rho W|=\prod_{i=1}^n(1-\rho w_i)$. Eigenvalues $w_1\le w_2\le\ldots\le w_n$ do not have to be recalculated during the series of iterations and $\Sigma(\widehat{\rho})$ is p.d. if: (a) $\widehat{\rho}<w_n^{-1}$ when $0\le w_1$, (b) $\widehat{\rho}>w_1^{-1}$ when $w_n\le 0$ and (c) $w_1^{-1}<\widehat{\rho}<w_n^{-1}$ when $w_1<0<w_n$.
2. For CAR models, ${}^tX\Sigma^{-1}(\theta)X=\sigma_e^{-2}\,{}^tX(I-C)X$ if $\sigma_i^2=\sigma_e^2$ for $i=1,\ldots,n$:

$$\widehat{\sigma}_e^2=\sigma_e^2(\widehat{\theta}_n)=n^{-1}{}^tX(I-A(\widehat{\theta}_n))X, \text{ where}$$

$$\widehat{\theta}_n=\underset{\theta}{\operatorname{argmin}}\{-2n^{-1}\log|I-A(\theta)|+\log(\sigma_e^2(\theta))\}.$$

   For CAR models with one parameter ρ: $(I-\rho W)X=e$, $|\Sigma(\theta)|=\sigma_e^{2n}|I-\rho W|^{-1}$ and $|I-\rho W|=\prod_{i=1}^n(1-\rho w_i)$.

5.3.4 Spatial regression estimation

Spatial regressions (cf. §1.8) are written, for a centered second-order random field $\varepsilon=(\varepsilon_s)$,

$$X=Z\delta+\varepsilon, \tag{5.16}$$

where the $n\times q$ matrix Z is the design matrix and $\delta\in\mathbb{R}^q$.

Linear models of the trend $E(X)$ are provided by analysis of variance models (quantitative exogenous Z with one or more factors), regression models (quantitative Z in $\mathbb{R}^q$), response surfaces $E(X_s)=f(s)$ and analysis of covariance models (Z with a mixed form). For example:

1. Additive models with two factors $I\times J$ $(q=I+J-1)$:

$$E(X_{i,j})=\mu+\alpha_i+\beta_j,\qquad i=1,\ldots,I \text{ and } j=1,\ldots,J,$$

$$\alpha_.=0,\qquad \beta_.=0.$$

2. Quadratic response surfaces in $(x,y)\in\mathbb{R}^2$ $(q=6)$:

$$E(X_{x,y}) = \mu + ax + by + cx^2 + dy^2 + exy.$$

3. For exogenous z_i,
$$E(X_i) = {}^t g(z_i)\delta,$$
where $g : E \longrightarrow \mathbb{R}^q$ is known.
4. The model combining a response surface, a regression and an analysis of variance:
$$E(X_s \mid i, z) = \mu_i + ax + by + {}^t g(z)\delta.$$

Regression estimation using ordinary least squares

The OLS estimate of δ is:

$$\widetilde{\delta} = ({}^tZZ)^{-1}\, {}^tZX.$$

This estimator is unbiased with variance, if $\Sigma = Var(\varepsilon)$, of

$$Var(\widetilde{\delta}) = \Delta = ({}^tZZ)^{-1}\, {}^tZ\Sigma Z\, ({}^tZZ)^{-1}.$$

When X is a Gaussian random field, $\widetilde{\delta} \sim \mathcal{N}_q(\delta, \Delta)$. We note that:

1. $\widetilde{\delta}$ is not efficient (the GLS estimator is the one that is, cf. (5.18) and §5.3.4).
2. If for all i, $Var(\varepsilon_i) = \sigma^2$, the standard estimator of σ^2 based on the residuals $\widetilde{\varepsilon} = X - Z\widetilde{\delta}$ is in general biased.
3. The usual statistic for tests F of subhypotheses dealing with δ deduced from this estimation does not generally follow a Fisher distribution.
4. The interest in OLS is that it is a good estimator that does not require knowledge of the spatial structure Σ. It can serve as an initial estimator in an iterative estimation procedure (cf. §5.3.4).

If $\widetilde{\delta}$ is consistent, OLS allows us to estimate Σ and thus makes a GLS procedure "feasible." We now give conditions ensuring consistency and give the rate of convergence. For A a p.s.d. matrix, note $\lambda_M(A)$ (resp. $\lambda_m(A)$) the largest (resp. smallest) eigenvalue of A. By letting our representation of matrices Z as in (5.16) and $\Sigma = Cov(\varepsilon)$ depend on n, we have the following property:

Proposition 5.7. *The OLS estimator $\widetilde{\delta}_n$ is consistent if the following two conditions hold:*

(i) *$\lambda_M(\Sigma_n)$ is uniformly bounded in n.*
(ii) *$\lambda_m({}^tZ_nZ_n) \rightarrow \infty$ or $({}^tZ_nZ_n)^{-1} \rightarrow 0$, if $n \rightarrow \infty$.*

Proof: As $\widetilde{\delta}$ is unbiased, it suffices to show that the trace (noted tr) of $Var(\widetilde{\delta})$ tends to 0. Using the identity $tr(AB) = tr(BA)$, we have for $A = ({}^tZZ)^{-1}\, {}^tZ$ and $B = \Sigma Z\, ({}^tZZ)^{-1}$:

$$tr(Var(\widetilde{\delta})) = tr(\Sigma Z({}^tZZ)^{-2}\,{}^tZ) \le \lambda_M(\Sigma) tr(Z({}^tZZ)^{-2}\,{}^tZ)$$
$$= \lambda_M(\Sigma) tr(({}^tZZ)^{-1}) \le q\frac{\lambda_M(\Sigma)}{\lambda_m({}^tZZ)} \underset{n\to\infty}{\longrightarrow} 0.$$

The first inequality is a consequence of, for two symmetric $n \times n$ p.s.d. matrices Σ and V, the upper bound:

$$tr(\Sigma V) \le \lambda_M(\Sigma) tr(V).$$

□

Comments:

1. Condition (ii) relies on the covariates (Z_n). If for example $({}^tZ_nZ_n)/n \longrightarrow Q$, where Q is p.d., then (ii) is satisfied and the variance goes to zero as n^{-1}. This condition is satisfied if, for $n = m \times r$, Z_n is the m-fold repetition of an experimental design R_0 of dimension $r \times q$ and full rank q ($Q = ({}^tR_0R_0)/m$).
2. (i) is related to the *spatial model* Σ and is satisfied if the covariance γ of X is such that:

$$\sup_{i\in S} \sum_{j\in S} |\gamma(i,j)| < \infty.$$

This is true if X is an ARMA, SAR or stationary CAR model.

Proof: First, write $\Sigma_n \equiv \Sigma = (\gamma(i,j))$. Let $v = (v_1, v_2, \dots, v_n)$ be a non-zero eigenvector associated with an eigenvalue $\lambda \ge 0$ of Σ, with i_0 satisfying $\left|v_{i_0}\right| = \max_i |v_i|$. $\left|v_{i_0}\right|$ is > 0 and the result $\Sigma v = \lambda v$ means that for coordinate i_0,

$$\lambda v_{i_0} = \sum_{j\in S} \gamma(i_0, j) v_j.$$

By dividing both sides by v_{i_0}, the triangle inequality can be applied to the previous formula, giving:

$$\sup\{\lambda\colon \text{eigenvalue of } \Sigma\} \le \sup_{i\in S} \sum_{j\in S} |\gamma(i,j)| < \infty.$$

□

Regression estimation by quasi-generalized least squares

We now consider model (5.16) and suppose initially that $\Sigma = Cov(X) = \sigma^2 R$ is known and invertible. By premultiplying by $R^{-1/2}$: $X^* = R^{-1/2}X$ and $Z^* = R^{-1/2}Z$, model (5.16) becomes a regression with white noise errors ε^*:

$$X^* = Z^*\delta + \varepsilon^*, \tag{5.17}$$

with $E(\varepsilon^*) = 0$ and $Var(\varepsilon^*) = \sigma^2 I$.

Model (5.17) is thus a standard linear model and we know (Gauss-Markov theorem) that the best linear estimator, unbiased and with minimal variance for δ is none other than the Generalized Least Squares (GLS) estimator:

$$\widehat{\delta}_{MCG} = ({}^tZ^*Z^*)^{-1}\,{}^tZ^*X^* = ({}^tZ\Sigma^{-1}Z)^{-1}\,{}^tZ\Sigma^{-1}X, \tag{5.18}$$

with $Var(\widehat{\delta}_{MCG}) = ({}^tZ\Sigma^{-1}Z)^{-1}$. If X is a Gaussian random field, the GLS estimate coincides with the maximum likelihood one.

This said, in general Σ is not known. If $\Sigma = \Sigma(\theta)$ has a known parametric form, then as $\theta \in \mathbb{R}^p$ is unknown, estimating $\eta = (\delta, \theta) \in \mathbb{R}^{q+p}$ using *quasi-generalized least squares* (QGLS) uses the following algorithm:

1. Estimate δ by OLS: $\tilde{\delta} = ({}^tZZ)^{-1}\,{}^tZX$.
2. Calculate the OLS residuals: $\tilde{\varepsilon} = X - Z\tilde{\delta}$.
3. Based on $\tilde{\varepsilon}$, estimate $\tilde{\theta}$ using a previously developed method (for example (5.3) and (5.4) for variograms).
4. Estimate Σ by $\tilde{\Sigma} = \Sigma(\tilde{\theta})$, then δ by $GLS(\tilde{\Sigma})$.

Steps 2–4 can be iterated until convergence of estimations, giving $\widehat{\delta}_{QGLS}$. Under certain conditions (75; 6), if the OLS estimation $\tilde{\delta}$ of δ converges sufficiently quickly (for example, $n^{1/4}$, cf. Prop. 5.7 and (75)), one iteration suffices as the GLS and QGLS are asymptotically equivalent:

$$\lim_n \sqrt{n}(\widehat{\delta}_{QGLS} - \widehat{\delta}_{GLS}) = 0 \text{ in probability}$$

and

$$\sqrt{n}(\widehat{\delta}_{QGLS} - \delta) \sim \mathcal{N}_p(0, \lim_n n\,{}^t(Z_n\Sigma_n^{-1}Z_n)^{-1}).$$

QGLS are therefore, for large n, good estimators. Let us now take a look at the Gaussian regression case.

ML for Gaussian spatial regression

Suppose that $X \sim \mathcal{N}_n(Z\delta, \Sigma(\theta))$. If $\Sigma(\theta) = \sigma^2 Q(\tau)$, $\theta = (\sigma^2, \tau)$, the negative of the log-likelihood of $X(n)$ can be written, up to an additive constant,

$$\begin{aligned} 2l(\delta,\theta) &= \log|\Sigma(\theta)| + {}^t(X - Z\delta)\Sigma^{-1}(\theta)(X - Z\delta) \\ &= n\log\sigma^2 + \log|Q(\tau)| + {}^t(X - Z\delta)Q^{-1}(\tau)(X - Z\delta)/\sigma^2. \end{aligned}$$

The ML estimation with known τ is thus:

$$\widehat{\delta} = ({}^tZQ^{-1}(\tau)Z)^{-1}\,{}^tZQ^{-1}(\tau)X, \tag{5.19}$$

$$\widehat{\sigma}^2 = (X - Z\widehat{\delta})Q^{-1}(\tau)(X - Z\widehat{\delta})/n. \tag{5.20}$$

The profile $l^*(\tau)$ of the log-likelihood at τ can be obtained by replacing (δ, σ^2) by their estimates in $l(\delta, (\sigma^2, \tau))$:

$$l^*(\tau) = -\frac{n}{2}\{\log[{}^t X Q^{-1}(\tau)(I - P(\tau))X] + \log|Q(\tau)| + n(1 - \log n)\}, \quad (5.21)$$

$$P(\tau) = Z({}^t Z Q^{-1}(\tau) Z)^{-1 t} Z Q^{-1}(\tau), \quad (5.22)$$

where P is the orthogonal projection with respect to the norm $\|\cdot\|_{Q^{-1}}$ onto the space generated by the columns of Z. We therefore estimate τ with $\widehat{\tau}$ by maximizing (5.21), then δ and σ^2 using (5.19) and (5.20) with $\widehat{\tau}$.

With the help of a result of Sweeting (210) on the asymptotic normality of ML, Mardia and Marshall (149) and Cressie (48, p. 484–485) show that under certain conditions, the ML estimator of the regression $X \sim \mathcal{N}_n(Z\delta, \Sigma(\theta))$, $\widehat{\eta} = (\widehat{\delta}, \widehat{\theta})$ of η is consistent and asymptotically Gaussian, with asymptotic variance given by the inverse of the Fisher information of the model $J(\delta,\theta) = E_{\delta,\theta}\{l^{(2)}(\delta,\theta)\}$.

Define the following matrices:

$$\Sigma_i = \partial\Sigma/\partial\theta_i,\ \Sigma^j = \partial\Sigma^{-1}/\partial\theta_j,\ \Sigma_{ij} = \partial^2\Sigma/\partial\theta_i\partial\theta_j,\ \Sigma^{ij} = \partial^2\Sigma^{-1}/\partial\theta_i\partial\theta_j$$

and $t_{ij} = tr(\Sigma^{-1}\Sigma_i\Sigma^{-1}\Sigma_j)$ for $i,j = 1,\dots,p$. We also denote by $(\lambda_l, l = 1,\dots,n)$ (resp. $(|\lambda_{i;l}|; i = 1,\dots,p$ and $l = 1,\dots,n)$, $(|\lambda_{i,j;l}|; i,j = 1,\dots,p$ and $l = 1,\dots,n))$ the eigenvalues of Σ (resp. Σ_i, Σ_{ij}) put into increasing order and $\|G\| = tr(G\,{}^tG)$ the Euclidean norm of matrix G.

Theorem 5.2. *(Mardia-Marshall (149))*

Suppose that the spatial regression $X \sim \mathcal{N}_n(Z\delta, \Sigma(\theta))$, $\delta \in \mathbb{R}^q$ and $\theta \in \mathbb{R}^p$ satisfies the following conditions:

(MM-1) $\lambda_n \to e < \infty$, $|\lambda_{i;n}| \to e_i < \infty$ *and* $|\lambda_{i,j;n}| \to e_{ij} < \infty$ *for* $i,j = 1,\dots,p$.
(MM-2) $\|\Sigma\|^{-2} = O(n^{-1/2-\kappa})$ *for some* $\kappa > 0$.
(MM-3) $t_{ij}/\sqrt{t_i t_j} \longrightarrow a_{ij}$ *for* $i,j = 1,\dots,p$ *and* $A = (a_{ij})$ *is regular.*
(MM-4) $({}^tZZ)^{-1} \longrightarrow 0$.

Then:

$$J^{1/2}(\delta,\theta)\{(\widehat{\delta},\widehat{\theta}) - (\delta,\theta)\} \xrightarrow{d} \mathcal{N}_{q+p}(0, I_{q+p}),$$

where $J(\delta,\theta) = \begin{pmatrix} J_\delta & 0 \\ 0 & J_\theta \end{pmatrix}$ for $J_\delta = {}^tZ\Sigma^{-1}(\theta)Z$ and $(J_\theta)_{ij} = (2^{-1}\mathrm{tr}(-\Sigma^j\Sigma_i))$.

Comments:

1. Calculation of the analytic form of $J(\delta,\theta)$ is given in §C.4.1.
2. $J(\delta,\theta)$ depends only on θ; in practice, we must estimate $J(\delta,\theta)$ either by replacing θ with $\widehat{\theta}$ or by using the observed information matrix $l^{(2)}\,(\widehat{\delta},\widehat{\theta})$.
3. $\widehat{\delta}$ and $\widehat{\theta}$ are asymptotically independent.
4. As $\Sigma^j = -\Sigma^{-1}\Sigma_j\Sigma^{-1}$ and $\Sigma_i = -\Sigma\Sigma^i\Sigma$,

$$(J_\theta)_{ij} = 2^{-1}\mathrm{tr}(\Sigma^{-1}\Sigma_j\Sigma^{-1}\Sigma_i) = 2^{-1}\mathrm{tr}(\Sigma^j\Sigma\Sigma^i\Sigma) = -2^{-1}\mathrm{tr}(\Sigma^i\Sigma_j).$$

We use the first expression for J_θ if $\Sigma(\theta)$ has a known parametric form (covariance or variogram model), the second if $\Sigma^{-1}(\theta)$ has a known parametric form (SAR and CAR models).

5. This asymptotic normality result allows us to construct tests and both joint and individual confidence regions for (δ,θ). For example, the confidence region with approximate level α for δ is, with $q(p;\alpha)$ denoting the α-quantile of a χ^2_p,

$$\{\delta : (\widehat{\delta}-\delta)J_{\widehat{\delta}}(\widehat{\delta}-\delta) \leq q(p;\alpha)\}.$$

We now give an example of identifying $J(\delta,\theta)$: Ord (167) considers the spatial regression with SAR errors:

$$X = Z\delta + \varepsilon, \qquad \varepsilon = (I-\theta W)e, \qquad e \sim \mathcal{N}_n(0,\sigma^2 I),$$

where $\theta \in \mathbb{R}$ and W is a known weights matrix. The model parameters are $(\delta,(\sigma^2,\theta)) \in \mathbb{R}^{q+2}$. Denoting $F = I-\theta W$, $G = WF^{-1}$ and

$$\nu = -\sum_1^n \frac{w_i^2}{1-\theta w_i^2},$$

where (w_i) are eigenvalues of W, the information matrix $J(\delta,\theta)$ is block diagonal with coefficients:

$$J(\delta) = \frac{1}{\sigma^2}\,{}^t(FZ)FZ, \qquad J(\sigma^2) = \frac{n}{2\sigma^4},$$

$$J(\theta) = tr({}^tGG-\nu), \qquad J(\sigma^2,\theta) = \frac{tr(G)}{\sigma^2}.$$

Mardia and Marshall (149) also deal with the case of stationary errors on $\mathbb{Z}^d$ with covariance:

$$\gamma(i,j;\theta) = \sigma^2\rho(i-j;\psi).$$

Denote $\rho_i = \partial\rho/\partial\theta_i$ and $\rho_{ij} = \partial^2\rho/\partial\theta_i\partial\theta_j$ for $i,j=1,\ldots,p$. If ρ, the ρ_i and the ρ_{ij} are summable over $\mathbb{Z}^d$ and if X is observed on $\{1,2,\ldots,n\}^d$, then under conditions (MM-3-4), the result of Theorem 5.2 is also true.

Corollary 5.1. *(149) Suppose that X, observed on $\{1,\ldots,n\}^d \subset \mathbb{Z}^d$ has stationary Gaussian errors and that correlations ρ and their derivatives ρ_i and ρ_{ij}, $i,j=1,\ldots,p$ are summable on $\mathbb{Z}^d$. Then, under conditions (MM-3) and (MM-4), the result of the previous theorem remains true.*

Example 5.7. Percentage of people with blood group A in the 26 counties of Ireland (cont.)

For the data displayed in Fig. 5.9, Moran's test indicated spatial dependency for the "percentage" variable. This dependency could perhaps be a product of other explicative variables that might account for these percentages, such as for example the percentage being higher in regions colonized by anglo-saxons. To examine this, we consider two explicative variables, one quantitative, `towns`, the number of towns per unit area, the other qualitative, `pale`, indicating whether or not the county was under anglo-saxon control.

Table 5.5 Results of ML estimations of various spatial regressions for the `eire` data: (a) two covariate model, `towns` and `pale` with i.i.d. Gaussian errors; (b) model with only the variable `pale` and SAR errors (standard errors shown in parentheses.

	Models	
Coefficient	(a)	(b)
`Intercept`	27.573 (0.545)	28.232 (1.066)
`towns`	−0.360 (2.967)	–
`pale`	4.342 (1.085)	2.434 (0.764)
ρ	–	0.684 (0.148)

Regressing the percentage onto these two variables with i.i.d. errors (model (a)) shows that only `pale` is significant (cf. Table 5.5, model (a)).

The spatial correlation between residuals remains: for Moran's index of the residuals $I^R = {}^t rWr/{}^t rr$, we have (7, pg. 102) that if there is spatial independence, $T = (I_R - E(I_R))/\sqrt{Var(I_R)} \sim \mathcal{N}(0,1)$ with $E(I_R) = tr(MW)/(n-p)$ and

$$Var(I_R) = \frac{tr(MWM({}^tW + W)) + \{tr(MW)\}^2}{(n-p)(n-p+2)} - \{E(I_R)\}^2,$$

where $M = I - P$, with P being the orthogonal projection onto the space generated by covariates Z. As the value of T was calculated as $t = 1.748$, the correlation between residuals remains significant at the 10% level. We therefore prefer the regression model (b) with only the covariate `pale` and the SAR error model:

$$r = \rho W r + \varepsilon,$$

where W is the spatial contiguity matrix of the neighbor graph. ML results (cf. Table 5.5, model (b)) show that the dependency parameter ρ is significantly different to zero.

Example 5.8. Influence of industrial locations and tobacco use on lung cancer

This example considers a spatial epidemiology study by Richardson et al. (182) (cf. `cancer-poumon` on the website) looking at lung cancer mortality and its links with metallurgy (*metal*), engineering (*meca*) and textile (*textile*) industries on the one hand and cigarette use (*tabac*) on the other.

Whenever epidemiological data shows spatial autocorrelation, the interpretation is not the same when considering infectious diseases (local spatial diffusion processes) as opposed to chronic illnesses, the case here. In the present situation, correlation may originate in covariates that themselves exhibit spatial dependency. If some of these variables are observable, we can introduce them into the regression while simultaneously trying to propose a parametrically "parsimonious" model. If the residuals in this regression are still correlated, one reason may be that these residuals "include" other hidden spatial risk factors (geographic variation, environmental influences, etc).

The explicative variable Y (cf. Fig. 5.12) is the standardized mortality rate for lung cancer (number of deaths per cancer/number of inhabitants) for men aged between 35 and 74 in the two years 1968–1969, measured in 82 French regions (data from INSERM-CépicDc IFR69).

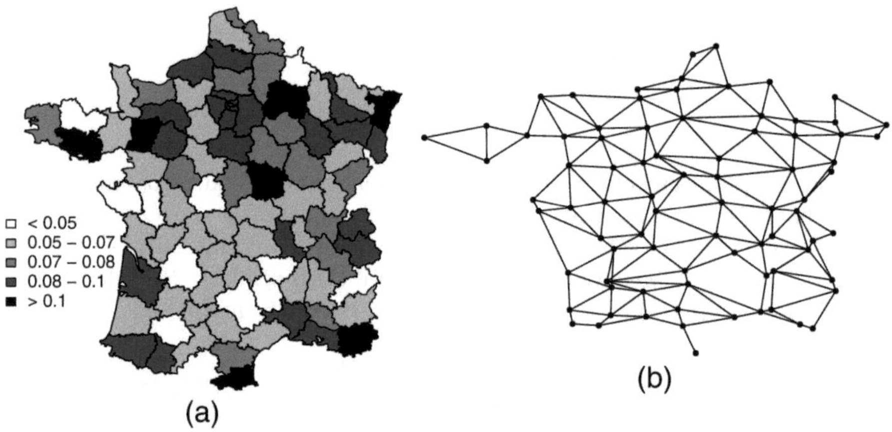

Fig. 5.12 (a) Rate (in percent) of standardized mortality due to lung cancer in 82 French regions; (b) (symmetric) influence graph $\mathscr{R}$ over the 82 regions.

We take data for cigarette sales (SEITA) from 1953, with the 15 year lag allowing us to take into account the influence of tobacco use on lung cancer. The regression we consider is

$$Y = X\beta + u, \qquad Y \text{ and } u \in \mathbb{R}^{82},$$

denoted (A) if we only take into account the 3 industrial variables and (B) if we also include the variable *tabac*. We propose five spatial models (with parameter θ) for errors: the first two involve ARs on networks and the three others have spatial covariance $\Sigma = Cov(u)$ with isotropic parametric structures:

1. Gaussian CAR model $u_i = c\sum_{j\in\partial i} w_{ij}x_j + e_i$, $Var(e_i) = \sigma_e^2$, where W is the spatial contiguity matrix of the 82 regions.
2. Gaussian SAR model $u_i = b\,\sum_{j\in\partial i} w_{ij}^* x_j + \varepsilon_i$, $Var(\varepsilon_i) = \sigma_\varepsilon^2$, where W^* is the matrix W but with each row normalized to 1.
3. Disk model: $\Sigma_{i,j} = \sigma^2 f_a(\|i-j\|_2)$.
4. Matérn model with parameters (a, σ^2, ν) (cf. §1.3.3).
5. Exponential nugget effect model: if $i \neq j$, $\Sigma_{i,j} = \sigma^2\gamma\exp\{-\lambda\,\|i-j\|_2\}$, $\gamma \leq 1$ and $= \sigma^2$ otherwise.

The disk model is the 2-dimensional equivalent of the spherical model (cf. §1.3.3) with the sphere of radius a replaced by a disk of the same radius.

Table 5.6 Improvement in fit by using ML (χ^2) and prediction (PRESS) for 5 error models and regressions (A) and (B) compared with OLS.

Regression	A	A	B	B
Model	χ^2	PRESS	χ^2	PRESS
OLS	–	1.90×10^{-6}	–	0.96×10^{-6}
CAR	$\chi_1^2=25.6$	1.44×10^{-6}	$\chi_1^2=2.1$	0.94×10^{-6}
SAR	$\chi_1^2=32.0$	1.37×10^{-6}	$\chi_1^2=8.2$	0.87×10^{-6}
Disk	$\chi_1^2=9.7$	1.40×10^{-6}	$\chi_1^2=9.4$	0.88×10^{-6}
Expo+nugget	$\chi_2^2=30.4$	1.34×10^{-6}	$\chi_2^2=8.4$	0.86×10^{-6}
Matérn	$\chi_2^2=25.5$	1.41×10^{-6}	$\chi_2^2=8.6$	0.86×10^{-6}

Table 5.7 ML estimation of the parameters of the 5 spatial models under (A) and (B) (s.e. for standard error).

Regression	A	A	B	B
Model	Parameters	s.e.	Parameters	s.e.
CAR	$\widehat{c}=0.175$	0.005	$\widehat{c}=0.077$	0.057
SAR	$\widehat{b}=0.613$	0.103	$\widehat{b}=0.353$	0.138
disk	$\widehat{a}=94.31$	5.66	$\widehat{a}=35.73$	4.92
expo+nugget	$\widehat{\gamma}=0.745$	0.091	$\widehat{\gamma}=0.554$	0.265
	$\widehat{\lambda}=0.0035$	0.0018	$\widehat{\lambda}=0.012$	0.008
Matérn	$\widehat{\nu}=0.305$	–	$\widehat{\nu}=0.228$	–
	$\widehat{a}=112.36$	16.41	$\widehat{a}=75.19$	23.74

Note that the 5 models are not nested and the first three are 2-dimensional, the last two 3-dimensional.

Moran's index calculated for the normalized weights matrix W^* indicates that all variables are spatially correlated (0.53 for Y, 0.58 for *tabac*, 0.41 for *metal* and 0.26 for *textile*) except *meca* (0.12). Also, the OLS regression residuals (A) and (B) are significantly correlated.

Table 5.6 shows the gain in log-likelihood of each of the residual models for (A) and (B) compared with OLS (i.i.d. errors u_i). This gain is also evaluated using the prediction sum of squares (PRESS) criterion, the sum of squares of conditional errors (ε_i) of the model, estimated by $\hat{\varepsilon}_i = u_i - \sum_{j:j\neq i}\frac{\hat{\sigma}^{ij}}{\hat{\sigma}^{ii}}u_j$ where $(\hat{\sigma}^{ij})^{-1} = \widehat{\Sigma}(\widehat{\beta},\widehat{\theta})$.

As the quantile of a χ_1^2 at 95% is 3.841 (resp. 5.911 for a χ_2^2), all the spatial models offer a significant improvement over the i.i.d. error model except for the CAR-(B) model. The improvement in prediction is in the order of 30% for (A) and 10% for (B), with the *tabac* variable as expected decreasing the value of the χ^2 statistic and the PRESS criteria.

Table 5.7 gives ML estimates of the spatial parameters as well as standard errors. All parameters are significant for (A) but show weaker dependency for (B). For the CAR-(B) model, c is no longer significant. For the SAR model associated with W^*, b can be interpreted as the weight of influence of the neighborhood. Parameter $2\widehat{a} \simeq 188$ km of the disk model represents the distance above which correlations disappear. Parameter $\widehat{\nu} = 0.305$ of the estimated Matérn model suggests a linear decrease at the origin.

Table 5.8 Estimates (and standard errors) of regression coefficients (A) and (B) for three error models: (i) independence, (ii) SAR and (iii) Matérn.

Regression and model u	Metallurgy	Engineering	Textile
(A) and *OLS*	2.46 (0.39)	1.88 (0.63)	1.12 (0.53)
(A) and *SAR*	1.39 (0.42)	2.11 (0.55)	0.38 (0.49)
(A) and Matérn	1.35 (0.43)	1.90 (0.57)	0.38 (0.50)
(B) and *OLS*	1.50 (0.27)	1.29 (0.46)	0.84 (0.38)
(B) and *SAR*	1.11 (0.32)	1.37 (0.45)	0.61 (0.39)
(B) and Matérn	1.05 (0.32)	1.24 (0.44)	0.62 (0.38)

Table 5.8 gives estimations of regressions (A) and (B) for three error models, (i) independent (OLS), (ii) SAR and (iii) Matérn covariance. We see that the type of spatial model heavily influences parameter estimates but not their precision: for example, for (A) and *metal*, spatial modeling reduces by half the slope ($\widehat{\beta}_{MCO}/\widehat{\beta}_{SAR} = 1.77$) and the associated statistic t ($t_{\beta_{MCO}}/t_{\beta_{SAR}} = 1.95$). Also, spatial modeling renders the variable *textile* non-significant. As we might have expected, taking into account the variable *tabac* significantly decreases the values of the estimates as well as their standard errors.

In conclusion, taking into account the spatial structure of errors significantly influences our estimations of regressions (A) and (B) and the variables *metal* and *meca* are significant, though not *textile*.

5.4 Markov random field estimation

Let X be a Markov random field on the discrete set of sites $S \subset \mathbb{R}^d$ taking values in $\Omega = E^S$, where S is endowed with a symmetric neighbor graph $\mathscr{G}$. Suppose that the distribution P_θ of X is known via its conditional specifications π_θ (cf. §2.2.1), these in turn associated with a parametric potential $\Phi_\theta = \{\Phi_{A,\theta}, A \in \mathscr{S}\}$, where θ is an interior point of a compact $\Theta \subset \mathbb{R}^p$. Denote by $\mathscr{G}(\pi_\theta)$ the set of distributions on Ω having specification π_θ and $\mathscr{G}_s(\pi_\theta)$ those which are stationary when $S = \mathbb{Z}^d$. Suppose that the graph $\mathscr{G}$ does not depend on θ and that X is observed on $D_n \cup \partial D_n \subset S$, where ∂D_n is the neighborhood boundary of D_n.

We now present three procedures for estimating both θ and certain asymptotic properties: maximum likelihood (ML), maximum conditional pseudo-likelihood (MCPL) and C-coding estimation. We also give results on the identification of the support of the neighborhood of Markov random fields.

5.4.1 Maximum likelihood

If X has distribution $P_\theta \in \mathcal{G}(\pi_\theta)$, the distribution of X on D_n conditional on $x(\partial D_n)$ has energy density H_n (cf. §2.2.1):

$$\pi_n(x;\theta) = Z_n^{-1}(x_{\partial D_n};\theta)\exp\{H_n(x;\theta)\},$$

where $H_n(x;\theta) = \sum_{A:A\cap D_n \neq \emptyset} \Phi_A(x;\theta)$.

Consistency of ML: $S = \mathbb{Z}^d$ with invariant π_θ

Suppose that $S = \mathbb{Z}^d$ and that π_θ is translation-invariant and belongs to the exponential family $H_n(x;\alpha) = {}^t\alpha\, h_n(x)$,

$$h_{k,n}(x) = \sum_{i\in D_n:(i+A_k)\cap D_n \neq \emptyset} \Phi_{A_k}(\tau_i(x)), \qquad \text{for } k = 1,\dots,p, \tag{5.23}$$

where τ_i is the translation on $\mathbb{Z}^d$ defined by: $\forall j$, $(\tau_i(x))_j = x_{i+j}$. With this notation, the $\Phi = \{\Phi_{A_k}, k = 1,\dots,p\}$ are p measurable and *bounded* generating potentials and $\alpha = {}^t(\alpha_1,\alpha_2,\dots,\alpha_p) \in \mathbb{R}^p$ is the model parameter. Let $\widehat{\theta}_n = Arg\max_{\alpha\in\Theta} \pi_n(x;\alpha)$ be the ML estimator on Θ. To simplify things, we suppose that $D_n = [-n,n]^d$. We then have the following result:

Theorem 5.3. *Suppose that X has translation-invariant specification (5.23) π_θ and is stationary. Then, if the generating potentials are measurable and bounded and if the parametrization in α of $\pi_{0,\alpha}$, the conditional distribution at 0, is well-defined, the ML estimation $\widehat{\theta}_n$ is consistent.*

A proof of this result is given in Appendix C, §C.4.2.

Comments

1. One of the difficulties in using ML is having to calculate the normalization constant $Z_n(x_{\partial D_n};\alpha)$. In §5.5.6 we give an MCMC method allowing an approximate calculation of $Z_n(x_{\partial D_n};\theta)$ for Gibbs processes.
2. Consistency still holds if conditioning (here, on $x_{\partial D_n}$) occurs with respect to an arbitrary exterior condition $y_{\partial D_n}$ or even a sequence of external conditions (ML *with fixed boundary conditions*). It is similarly the case for ML *with free boundary conditions* which make the contribution of potentials disappear as soon as their support spreads out of D_n (in this case, the energy is $H_n(x;\theta) = \sum_{A\subset D_n} \Phi_A(x;\theta)$).
3. Let us give a sufficient condition ensuring the parametrization $\alpha \mapsto \pi_{0,\alpha}$ is well-defined, where:

$$\pi_{0,\alpha}(x|v) = Z^{-1}(v;\alpha)\exp({}^t\alpha\, h(x,v)), \text{ with } h_k(x,v) = \Phi_{A_k}(x,v).$$

Without putting restrictions on the generality of the model, suppose that the identifiability constraint for potentials Φ_{A_k} is satisfied (cf. (2.3)). Then, the representation in terms of α of $\pi_{0,\alpha}$ is well-defined if:

$$\exists w_i = (x_i, v_i),\, i = 1, \ldots, K \ \text{ such that } \ H = (h(w_1), \ldots, h(w_K)) \text{ is of rank } p. \tag{5.24}$$

In effect, if $\alpha \neq \theta$, $\exists (x,v)$ such that ${}^t(\alpha - \theta)h(x,v) \neq 0$ and we can deduce that

$$\frac{\pi_{0,\alpha}(x|v)}{\pi_{0,\alpha}(a|v)} = \exp({}^t\alpha\, h(x,v)) \neq \exp({}^t\theta\, h(x,v)) = \frac{\pi_{0,\theta}(x|v)}{\pi_{0,\theta}(a|v)}.$$

Consider for example the 8-NN binary model ($E = \{0,1\}$) on $\mathbb{Z}^2$ with conditional energy

$$H_0(x_0, v) = x_0\{\alpha + \beta_1(x_1 + x_5) + \beta_{,2}(x_3 + x_7) + \gamma_1(x_2 + x_6) + \gamma(x_4 + x_8)\},$$

where $1, 2, \ldots, 8$ is the enumeration in the trigonometric sense of the 8-NN of 0. Noting $x = (x_0, x_1, \ldots, x_8)$, it is easy to see that the following 5 configurations satisfy (5.24): $(1,\mathbf{0})$, $(1,1,\mathbf{0})$, $(1,0,0,1,\mathbf{0})$, $(1,0,1,\mathbf{0})$ and $(1,0,0,0,1,\mathbf{0})$ (here, $\mathbf{0}$ is the vector of 0s defined so that each configuration belongs to $\mathbb{R}^9$).

4. ML consistency is retained if π_θ is translation-invariant, without requiring hypotheses of belonging to an exponential family as long as the potentials $\phi_{A,\alpha}$, $0 \in A$ are uniformly continuous in α, this uniformly in x (96).
5. When the state space E of X is compact, Comets (46) showed consistency of ML *without stationarity hypotheses* on X. The proof uses a large deviations inequality for Gibbs random fields (46; 85; 96) as well as compactness of $\mathcal{G}_s(\pi_\theta)$. The interest in this result comes from the fact that it makes no hypotheses on the distribution of X except requiring translation-invariance of its conditional specification. Furthermore, (46) showed that convergence occurs at an exponential rate, i.e.,

$$\forall \delta > 0, \exists \delta_1 > 0 \text{ s.t. for large } n,\ P_\theta(\left\|\widehat{\theta}_n - \theta\right\| \geq \delta) \leq C \exp -\delta_1(\sharp D_n).$$

6. Obtaining general results on the asymptotic distribution of the ML estimator for Gibbs random fields is difficult. One of the reasons is that as the distribution $P_\theta \in \mathcal{G}(\pi_\theta)$ of X is only known via its conditional specifications, we do not known if there is a phase transition or not if the distribution P_θ of X is weakly dependent, for example if P_θ is α-mixing (cf. §B.2). If P_θ is characterized by its specifications and is α-mixing, then $\widehat{\theta}_n$ is asymptotically Gaussian if the mixing is fast enough (96).

Asymptotic normality of ML and subhypothesis tests when X is weakly dependent

Let $d_n = \sharp D_n$ and suppose that $\sharp \partial D_n = o(d_n)$. Asymptotic normality of the ML estimator $\widehat{\theta}_n$ can be shown under conditions where X is weakly dependent and the potentials $\Phi_A(\alpha), \alpha \in \Theta$ are bounded $\mathscr{C}^2$ functions (96, Th. 5.3.2). The asymptotic variance of $\widehat{\theta}_n$ can be characterized in the following way. Define the mean energy at site i by:

$$\overline{\Phi}_i(x;\theta) = \sum_{A:i\in A} \frac{\Phi_A(x;\theta)}{\sharp A}$$

and for μ_θ, the distribution of X under θ, Fisher's information matrix:

$$I_n(\theta) = \sum_{i,j\in D_n} Cov_{\mu_\theta}(\overline{\Phi}_i^{(1)}(x;\theta), \overline{\Phi}_j^{(1)}(x;\theta)).$$

Then, if there exists a non-random positive-definite symmetric matrix I_0 such that $\liminf_n d_n^{-1} I_n(\theta) \geq I_0$, then for large n,

$$(\widehat{\theta}_n - \theta) \sim \mathscr{N}_p(0, I_n(\theta)^{-1}).$$

A value for $I_n(\theta)$ can be found using Monte Carlo methods under $\widehat{\theta}_n$.

Denote by (H_p) the hypothesis: $\theta \in \Theta \subset \mathbb{R}^p$ and let (H_q) be a subhypothesis of class $\mathscr{C}^2$ and dimension q parametrized by $\varphi \in \Lambda \subset \mathbb{R}^q$. Letting $l_n = \log \pi_n(x)$ represent the log-likelihood conditional on $x(\partial D_n)$, we have the following result for the log-likelihood ratio statistic: under (H_q),

$$2\{l_n(\widehat{\theta}_n) - l_n(\widehat{\varphi}_n)\} \xrightarrow{d} \chi^2_{p-q}.$$

5.4.2 Besag's conditional pseudo-likelihood

The conditional pseudo-likelihood (CPL) of a Markov random field X is the product over sites $i \in D_n$ of the conditional densities at i. The log-pseudo-likelihood and estimation of the CPL maximum are respectively:

$$l_n^{CPL}(\theta) = \sum_{i\in D_n} \log \pi_i(x_i \mid x_{\partial i}, \theta) \;\text{ and }\; \widehat{\theta}_n^{CPL} = \operatorname*{argmax}_{\alpha\in\Theta} l_n^{CPL}(\alpha).$$

This estimation method, proposed by Besag in 1974 (25) is easier to put into practice than ML as it avoids having to calculate the normalizing constant of a Markov random field. Also, when it can be calculated, efficiency of CPL compared with ML is quite good if the spatial correlation is relatively weak (cf. §5.4.4). As an illustration, let us give details of this method in two particular contexts.

Estimating grayscale textures

Let X be a Markov texture on $\mathbb{Z}^2$, with states $E = \{0,1\}$, potentials $\Phi_{\{i\}}(x) = \beta_0 x_i$ and $\Phi^k_{\{i,j\}}(x) = \beta_k x_i x_j$ if $i - j = \pm k$, $i,j \in \mathbb{Z}^2$, where $k \in L \subset \mathbb{Z}^2$ for finite and symmetric L, $0 \notin L$. If X is observed over the square $D_n \cup \partial D_n$, the CPL is given by:

$$L_n^{PL} = \prod_{i \in D_n} \frac{\exp x_i(\beta_0 + \sum_{k \neq 0} \beta_k x_{i+k})}{1 + \exp(\beta_0 + \sum_{k \neq 0} \beta_k x_{i+k})}.$$

Fig. 5.13 gives three examples of binary real textures studied by Cross and Jain (51). These textures were modeled using Markov random fields and then estimated using

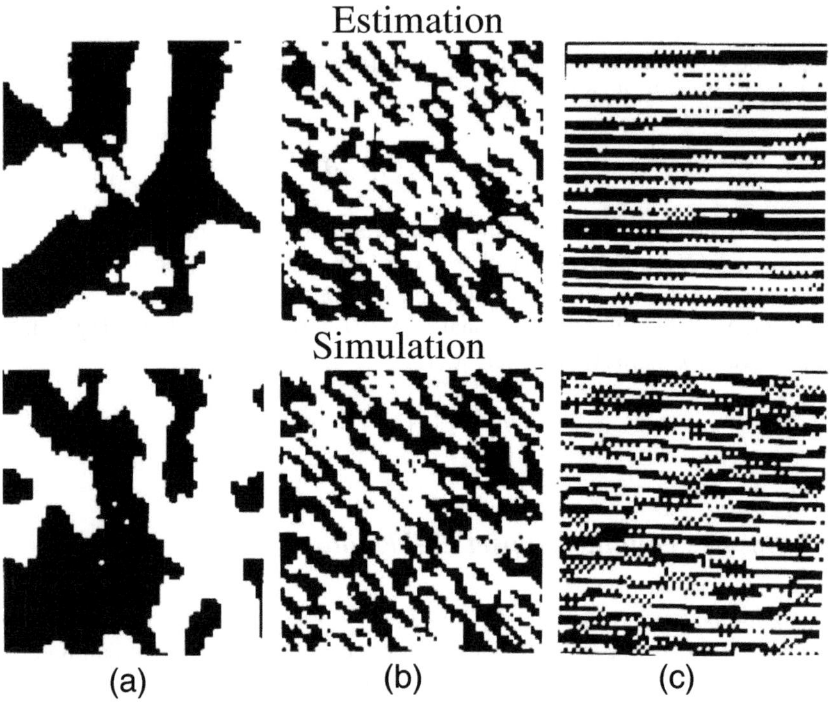

Fig. 5.13 64×64 grid of binary real textures: estimated and simulated. (a) pebbles, (b) cork, (c) curtain (Cross and Jain (51), ©1983 IEEE).

CPL. Then, the estimated models were simulated using Gibbs sampling. The visual similarity between the original textures and those simulated shows both the quality of modeling and also that of estimation by CPL.

Gaussian CAR model with one parameter

Consider the following CAR model with one real-valued parameter β:

$$X_i = \beta \sum_{j: j \neq i} w_{ij} X_j + e_i, \; Var(X_i \mid \cdot) = \kappa, \qquad i = 1, \ldots, n,$$

where W is a symmetric matrix with zeros on the diagonal. An associated model exists if $C = I - \beta W$ is positive definite, i.e., if $\beta\,|\lambda| < 1$ for all eigenvalues λ of W. In the Gaussian case, CPL estimation of β is identical to OLS on the conditional errors e: in effect, $X_i|x_{\partial i} \sim \mathcal{N}(\beta v_i, \kappa)$, where $v_i = \sum_{j:\langle i,j\rangle} w_{ij} x_j$ and the negative of the conditional log-likelihood is:

$$l_n^{CPL}(\beta) = constant + \frac{\sharp D_n}{2}\log\kappa + \frac{RSS_n(\beta)}{2\kappa},$$

where

$$RSS(\beta) = \sum_{i \in D_n} (X_i - \beta v_i)^2.$$

As this model is linear in β, we obtain:

$$\widehat{\beta} = \frac{{}^t XWX}{{}^t XW^2X}, \qquad \widehat{\kappa} = \frac{{}^t XX - ({}^t XWX)^2/{}^t XWX}{n}.$$

As for large n, residuals e_i are not correlated with v_i for $i = 1, \ldots, n$, we have (26): $E(\widehat{\beta}_n) \sim \beta$, $E(\widehat{\kappa}) \sim \kappa$, $Var(\widehat{\beta}_n) = 2tr(W^2)\{tr(C^{-1}W^2)\}^{-2}$ and $Var(\widehat{\kappa}) = 2\kappa^2 tr(C^2)/n^2$.

We are now going to present three asymptotic results dealing with maximum CPL estimation: the first is a consistency result for random fields X defined on not necessarily regular networks S; the next two are asymptotic normality results for when the specification on $\mathbb{Z}^d$ is translation-invariant, first in the bounded potentials case and then for Gaussian potentials.

Consistency of CPL estimation

Suppose that S is a *not necessarily regular* countable discrete network, $\mathcal{G}$ a symmetric graph on S and X a Markov random field on $(S, \mathcal{G})$ taking values in $\Omega = \prod_{i \in S} E_i$ (different state spaces for different sites are allowed). X, with distribution P_θ and specification π_θ (here, θ is in the interior of a compact $\Theta \subset \mathbb{R}^p$) is observed over a strictly increasing sequence $(D_n \cup \partial D_n)$, $d_n = \sharp\{D_n\}$. We now give several definitions and conditions.

Coding subset

- C is a *coding subset* of S if, for all $i \neq j$ in C, $j \notin \partial i$.
- A is a *strong coding subset* if, for all $i \neq j$ in A, we have $\partial i \cap \partial j = \emptyset$.

Let π_i and $\pi_{\partial i}$ represent the specifications at i and ∂i,

$$m_i(\theta, \alpha; x_{\partial i}) = -E_\theta^{x_{\partial i}} \left\{ \frac{\log(\pi_i(X_i|x_{\partial i}, \alpha)}{\pi_i(X_i|x_{\partial i}, \theta)} \right\} \; (\geq 0)$$

and suppose the following conditions (C):

(C1) *On the graph* $(S,\mathscr{G})$: there exists a coding subset C of S which is a disjoint union of K strong coding subsets $\{C_k\}$ such that, defining $A_n = A \cap D_n$, $c_n = \sharp(C_n)$ and $c_{n,1} = \sharp(C_{1,n})$:

(i) $\liminf_n c_{1,n}/c_n > 0$ and $\liminf_n c_n/d_n > 0$.

(ii) $\chi = \prod_{i\in\partial i} E_i$ is a fixed space of neighborhood configurations when $i \in C_1$.

(C2) *Lower bounds for* π_α: $\exists c > 0$ such that $\forall i \in C_1$, $\forall x_i, x_{\partial i}, x_{\partial\partial i}$ and $\alpha \in \Theta$ we have $\pi_i(x_i|x_{\partial i},\alpha)$ and $\pi_{\partial i}(x_{\partial i}|x_{\partial\partial i},\alpha) \geq c$, where π_i is uniformly (in i, x_i and $x_{\partial i}$) continuous at α.

(C3) *Identifiability at* θ *of* π_θ: there exists $m(\theta,\alpha;z) \geq 0$, $(\alpha,z) \in \Theta \times \chi$, λ-integrable for all α such that:

(i) $m_i(\theta,\alpha;z) \geq m(\theta,\alpha;z)$ if $i \in C_1$.

(ii) $\alpha \mapsto K(\theta,\alpha) = \int_\chi m(\theta,\alpha;z)\lambda(dz)$ is continuous and has a unique minimum at $\alpha = \theta$.

Theorem 5.4. *Consistency of CPL estimation (96)*

Let X be a Markov random field on S satisfying conditions (C1–3). Then the maximum CPL estimator on D_n is consistent.

Proof. We start by showing consistency of the C_1-coding estimator (cf. (5.25)). The property providing the key to the proof is the *independence of* $\{X_i, i \in C_1\}$ *conditional on* x^{C_1}, the random field outside C_1. First, we prove the following *subergodicity* result:

Lemma 5.1. *(83; 121; 96) Let A be a measurable subset of χ and the empirical frequency of A on C_1 defined by*

$$F_n(A;C_1) = \frac{1}{c_{1,n}} \sum_{i\in C_{1,n}} \mathbf{1}(x_{\partial i} \in A).$$

Then:

$$\liminf_n F_n(A;C_1) \geq \frac{c}{2}\lambda(A) \ P_\theta\text{-a.s.}$$

Proof of the Lemma: From (C2), variables $\mathbf{1}(X_{\partial i} \in A)$ have expectation $\geq c\lambda(A)$, variance ≤ 1 and, conditional on $x^{\partial C_1}$, are independent. The strong law of large numbers for independent variables in L^2, conditional on $x^{\partial C_1}$, gives (Breiman (32, Th. 3.27)):

$$\liminf_n F_n(A;C_1) \geq \frac{c}{2}\lambda(A) \quad P_\theta^{x^{\partial C_1}}\text{-a.s.}$$

As the upper bound is independent of $x^{\partial C_1}$, it is true P_θ-a.s. □

Proof of Theorem (cont.): Consider now the coding C_1-contrast:

$$U_n^{C_1}(\alpha) = -\frac{1}{c_{1,n}} \sum_{i\in C_{1,n}} \log \pi_i(x_i \mid x_{\partial i};\alpha) \qquad (5.25)$$

and write $U_n^{C_1}(\alpha) - U_n^{C_1}(\theta) = A_n + B_n$, where:

$$A_n = -\frac{1}{c_{1,n}} \sum_{i \in C_{1,n}} \left\{ \log \frac{\pi_i(x_i \mid x_{\partial i};\alpha)}{\pi_i(x_i \mid x_{\partial i};\theta)} + m_i(\theta,\alpha;x_{\partial i}) \right\}.$$

A_n is the sum of centered variables with bounded variance (condition (C2)) and independently conditional on x^{C_1}, $\lim_n A_n = 0$ P_θ-a.s. We deduce P_θ-a.s. the following sequence of inequalities:

$$\begin{aligned}
\liminf_n (U_n^{C_1}(\alpha) - U_n^{C_1}(\theta)) &= \liminf_n B_n \qquad (5.26)\\
&\geq \liminf_n \int_\chi m(\theta,\alpha;z) F_n(C_1,dz) \quad \text{(C3)}\\
&\geq \int_\chi m(\theta,\alpha;z) \liminf_n F_n(C_1,dz) \quad (\text{as } m \geq 0)\\
&\geq \frac{c}{2} \int_\chi m(\theta,\alpha;z)\lambda(dz) \mathrel{\hat=} K(\theta,\alpha) \quad \text{(Lemma (5.1))}.
\end{aligned}$$

As the coding estimator is associated with $U_n^C(\alpha) = \sum_{k=1}^K c_{k,n} c_n^{-1} U_n^{C_k}(\alpha)$, its consistency follows from (5.26), conditions (C) and the corollary of the general consistency properties of minimum contrast methods (cf. Appendix C, §C.2). Consistency of the CPL estimator is a consequence of the same corollary.

□

Conditions (C1–3) have to be verified case by case. Let us take a look at two examples, one associated with irregular graphs $(S,\mathscr{G})$, the other with translation-invariant random fields on $S = \mathbb{Z}^d$.

Example 5.9. Ising models on irregular graphs

Consider an Ising model with states $\{-1,+1\}$ on a graph $(S,\mathscr{G})$ such that (C1) holds, with the specification for $i \in C$ expressed as:

$$\pi_i(x_i \mid x_{\partial i};\alpha) = \frac{\exp x_i v_i(x_{\partial i};\alpha)}{2\, ch\{v_i(x_{\partial i};\alpha)\}},$$

where $v_i(x_{\partial i};\alpha) = \beta u_i + \gamma \sum_{j \in \partial i} w_{ij} x_j$ with known weights (u_i), (w_{ij}) and symmetric w satisfying for all $i : \sum_{\partial i} w_{ij} = 1$. The model parameter is $\alpha = {}^t(\beta,\gamma)$. Suppose that weights u and w are bounded. As the potentials are bounded, it is easy to see that (C2) is satisfied. Furthermore, we see that $m_i(\theta,\alpha;x_{\partial i}) = m(v_i(\theta), v_i(\alpha))$, where

$$m(a,b) = (a-b)th(a) - \log \frac{ch(a)}{ch(b)} \geq 0,$$

with $m(a,b)$ equal to 0 if and only if $a = b$. If $\theta = (\beta_0, \gamma_0)$,

$$v_i(x_{\partial i};\theta) - v_i(x_{\partial i};\alpha) = (\beta_0 - \beta)u_i + (\gamma_0 - \gamma) \sum_{j \in \partial i} w_{ij} x_j.$$

If there exists $\delta > 0$ such that $\inf_{i \in C} u_i \geq \delta$, (C3) is satisfied: in effect, for the constant configuration $x^*_{\partial i}$ on ∂i (-1 if $(\beta_0 - \beta)(\gamma_0 - \gamma) < 0$ and $+1$ otherwise),

$$\left| v_i(x^*_{\partial i};\theta) - v_i(x^*_{\partial i};\alpha) \right| \geq \delta \left\| \theta - \alpha \right\|_1 .$$

Example 5.10. Markov random fields on $\mathbb{Z}^d$ with invariant specification

Let X be a Markov chain on $\mathbb{Z}^d$ with invariant specification that belongs to the exponential family (5.23), where the potentials $\{\Phi_{A_k}, k = 1, \dots, K\}$ are measurable and uniformly bounded. If the parametrization of $\alpha \mapsto \pi_{\{0\},\alpha}$ is identifiable, then conditions (C1–3) are satisfied and the CPL method is consistent. In effect, if $R = \sup\{\|i\|_1, i \in \partial 0\}$, $C = \{a\mathbb{Z}\}^d$ is a coding set if a is an integer $\geq 2R$. Also, the choice of b-translations of $C_1 = \{2a\mathbb{Z}\}^d$ for the 2^d vectors $b = (b_1, b_2, \dots, b_d)$, where $b_i = 0$ or a, $i = 1, \dots, d$, defines a partition of C into 2^d strong coding subsets for which (C1) is satisfied. As for (C2), the result comes directly from the uniform upper bound in (x, α) of the energy: $|H_\Lambda(x,\alpha)| \leq c(\Lambda)$.

Asymptotic normality of CPL

Markov random fields with translation-invariant specification: Suppose that X is a Markov random field on $S = \mathbb{Z}^d$ with distribution $P_\theta \in \mathcal{G}(\pi_\theta)$ that has a translation-invariant specification and belongs to the exponential family (5.23). Write:

$$J_n(\theta) = \sum_{i \in D_n} \sum_{j \in \partial i \cap D_n} Y_i(\theta)\, {}^tY_j(\theta), \qquad I_n(\theta) = \sum_{i \in D_n} Z_i(\theta), \tag{5.27}$$

with $Y_i(\theta) = \{\log \pi_i(x_i/x_{\partial i};\theta)\}^{(1)}_\theta$ and $Z_i(\theta) = -\{\log \pi(x_i/x_{\partial i};\theta)\}^{(2)}_{\theta^2}$.

Theorem 5.5. *Asymptotic normality of the CPL estimator (Comets and Janzura (47))*

Suppose that X is a Markov random field on $S = \mathbb{Z}^d$ with translation-invariant specification that belongs to the exponential family (5.23), with measurable and bounded generating potentials $\{\Phi_{A_k}\}$. Then, if $\pi_{i,\theta}$ is identifiable at θ,

$$\{J_n(\widehat{\theta}_n)\}^{-1/2} I_n(\widehat{\theta}_n)\{\widehat{\theta}_n - \theta\} \xrightarrow{d} \mathcal{N}_p(0, I_p).$$

The interest in this result is that it needs no hypotheses on the global distribution of X, nor any of the following: uniqueness of P_θ in $\mathcal{G}(\pi_\theta)$, stationarity and/or ergodicity or weak dependency. The random normalization $\{I_n(\widehat{\theta}_n)\}^{-1/2} J_n(\widehat{\theta}_n)$ does not necessarily become stable as n increases but gives asymptotic normality to $\{\widehat{\theta}_n - \theta\}$. This itself results from a CLT dealing with conditionally centered functionals of Markov random fields (cf. Appendix B, §B.4). The functional used here is the gradient $Y_i(\theta)$ of the conditional log-density (cf. (47) for the general case and (99) for ergodic random fields).

Stationary Gaussian CAR model on $\mathbb{Z}^d$

Suppose we have the Gaussian CAR model: $X_i = \sum_{j \in L^+} c_j (X_{i-j} + X_{i+j}) + e_i$, $i \in D_n$, $Var(e_i) = \sigma^2$. This can also be written in matrix form as

$$X(n) = \mathscr{X}_n c + e(n),$$

where $\mathscr{X}_n$ is an $n \times p$ matrix with i^{th} row $(\mathscr{X}(i) = (X_{i-j} + X_{i+j}), j \in L^+)$, $i \in D_n$ and $c = (c_j, j \in L^+) \doteq \theta \in \mathbb{R}^p$. The maximum CPL estimate $\widehat{c}_n$, here equal to the OLS estimate can be explicitly derived because the model is linear:

$$\widehat{c}_n = (\mathscr{X}_n \mathscr{X}_n)^{-1} \mathscr{X}_n X(n) \quad \text{and} \quad \widehat{\sigma}_n^2 = \frac{1}{n-p} \sum_{i \in D_n} e_i^2(\widehat{c}_n).$$

Setting $V_j = Cov(\mathscr{X}(0), \mathscr{X}(j))$, denote:

$$G = \sigma^2 \{V_0 + \sigma^2 I - 2 \sum_{L^+} c_j V_j\} \quad \text{and} \quad \Delta = V_0^{-1} G V_0^{-1}$$

and suppose that $\sum_{L^+} \left| c_j \right| < 1/2$. Under this hypothesis, the distribution of X is thus completely characterized and is that of an ergodic Gaussian field with exponentially decreasing covariance. X is therefore exponentially α-mixing (cf. Appendix B, §B.2; (96)) and so we can deduce:

$$\sqrt{n}(\widehat{c}_n - c) \xrightarrow{d} \mathscr{N}_p(0, \Delta).$$

This result can be extended to other models where $c = (c_s, s \in L^+)$ is constant on subsets of L^+, such as submodels exhibiting isotropy.

We note that asymptotic normality of the CPL estimator can be obtained without invariance conditions on the specification of X as long as X is α-mixing with the required mixing rate (96, §5.3.2).

Characterizing the asymptotic variance of CPL estimations

The CPL estimation $\widehat{\theta}_n^{CPL}$ can be obtained using software for generalized linear model (GLM) estimation as long as the conditional specifications $\{\pi_i(\cdot \mid x_{\partial i}, \theta), i \in S\}$ follow log-linear models. In effect, in these cases the neighbor covariances $x_{\partial i}$ are added to standard covariates of GLMs and, when optimizing the estimation functional, there is no difference between CPL and GLM likelihoods when variables are independent.

However, *the standard errors given by a "GLM" routine are incorrect.* In effect, unlike the standard GLM case, the variables $(X_i \mid x_{\partial i})$ are not independent and arguments invoked to find the variance of the ML estimator for GLMs are not longer valid. Let us now show how to find the asymptotic variance of $\widehat{\theta}_n^{CPL}$.

As already stated in Comets and Janzura's result, this asymptotic variance is expressed using two pseudo-information matrices I_n and J_n (cf. (5.27) and Appendix C, Hypotheses (C.2.2)).

Consider now the case of a weakly dependent Markov random field X (96, Th. 5.3.3). For the conditional energies:

$$h_i(x_i \mid x_{\partial i}, \alpha) = \sum_{A: i \in A} \Phi_A(x; \alpha), \qquad i \in S,$$

define the two pseudo-information matrices $I_n(\theta)$ and $J_n(\theta)$:

$$J_n(\theta) = \sum_{i,j \in D_n} Cov_{\mu_\theta}(h_i^{(1)}(\theta), h_j^{(1)}(\theta)),$$

$$I_n(\theta) = \sum_{i \in D_n} E_{\mu_\theta}\{Var_\theta^{X_{\partial i}} h_i(X_i \mid X_{\partial i}, \theta)\}. \tag{5.28}$$

If there exists a non-random symmetric p.d. matrix K_0 such that $\liminf_n d_n^{-1} J_n(\theta)$ and $\liminf_n d_n^{-1} I_n(\theta) \geq K_0$, then, for large n:

$$Var(\widehat{\theta}_n^{CPL}) \simeq I_n^{-1}(\theta) J_n(\theta) I_n^{-1}(\theta).$$

A value for the variance of $\widehat{\theta}_n^{CPL}$ can be obtained in three ways:

1. Using result (47) after calculating the two pseudo-information matrices (5.27).
2. Estimating the expectations and variances in (5.28) by Monte Carlo simulation under $\widehat{\theta}_n^{CPL}$.
3. By parametric bootstrap, where asymptotic arguments are not necessary.

5.4.3 The coding method

Suppose that X is a Markov random field on $(S, \mathcal{G})$. Recall that C is a *coding subset* of $(S, \mathcal{G})$ if pairs of points of C are never neighbors. In this case, the *conditional variables* $\{(X_i | x_{\partial i}), i \in C\}$ *are independent*. The coding log-pseudo-likelihood l_n^C on $C_n = C \cap D_n$ and the associated coding estimator are respectively,

$$l_n^C(\theta) = \sum_{C_n} \log \pi_i(x_i \mid x_{\partial i}, \theta) \quad \text{and} \quad \widehat{\theta}_n^C = \operatorname*{argmax}_{\theta \in \Theta} l_n^C(\theta).$$

Conditional on a random field x^C observed outside of C, l_n^C is the log-likelihood of non-identically distributed independent variables. However, compared with CPL estimation we lose information found at sites $i \in D_n \backslash C_n$. We nevertheless retain the expression for a global likelihood (conditional on x^C). This means that under a certain subergodicity condition we can hold on to the asymptotic properties of ML for non-identically distributed independent variables.

Also, if the specification (π_i) follows a generalized linear model (GLM), we are able to use GLM software to calculate coding estimations. While the estimation $\widehat{\theta}_n^C$ provided by GLM software of the calculated variance is indeed correct, the difference with CPL is that here, variables $\{(X_i|x_{\partial i}), i \in C\}$ are, as for standard GLMs, independent.

Consider now the context in §5.4.2 and suppose that X satisfies conditions (C1–3). Denote by $\pi_i^{(k)}$, $k = 1,2,3$ the kth-order derivatives at α of $\pi_{i,\alpha}$ and for $i \in C$, set:

$$Z_i = -\frac{\partial}{\partial\theta}\{\log \pi_i(X_i, x_{\partial i};\theta)\}_\theta^{(1)},$$

$$I_i(x_{\partial i};\theta) = Var_{\theta,x_{\partial i}}(Z_i) \text{ and } I_n^C(\theta) = \frac{1}{c_n}\sum_{C_n} I_i(x_{\partial i};\theta).$$

Define the conditions:

(N1) $\forall i \in S$, $\pi_{i,\alpha}$ is $\mathscr{C}^3(V)$ at α in a neighborhood of θ and $\pi_i^{-1}, \pi_i^{(k)}$, $k = 1,2,3$ are uniformly bounded in $i, x_i, x_{\partial i}$ and $\alpha \in V$.

(N2) $\exists$ a positive-definite symmetric non-random $p \times p$ matrix $I(\theta)$ such that:

$$\liminf_n \frac{I_n^C(\theta)}{c_n} \geq I(\theta). \tag{5.29}$$

(5.29) is a *subergodicity condition* on X. We have:

Theorem 5.6. *Normality of $\widehat{\theta}_n^C$ and coding test (97; 96)*

Suppose that X is a Markov random field on $(S,\mathscr{G})$ that satisfies conditions (C1–3) and (N1-2). Then:

1. $\{I_n^C(\theta)\}^{1/2}(\widehat{\theta}_n^C - \theta) \xrightarrow{d} \mathscr{N}_p(0, Id_p)$.
2. *If $\theta = r(\varphi)$, $\varphi \in \mathbb{R}^q$ is a subhypothesis (H_0) of rank q and class $\mathscr{C}^2(V)$, we have:*

$$2\{l_n^C(\widehat{\theta}_n^C) - l_n^C(\widehat{\varphi}_n^C)\} \xrightarrow{d} \chi^2_{p-q} \text{ under } (H_0).$$

Comments:

1. The first result says that under subergodicity condition (5.29), ML results for i.i.d. observations are preserved by taking the Fisher information $I_n^C(\theta)$ conditional on x^C. $I_n^C(\theta)$ can be estimated using $I_n(\widehat{\theta}_n^C)$ or Monte Carlo empirical variances under $\widehat{\theta}_n^C$.
2. To show (5.29), we could use the result giving lower bounds for empirical frequencies in Lemma 5.1: (5.29) is satisfied if there exists $V : \chi \to \mathbb{R}^p$ such that for $i \in C$,

$$E_{\theta,x_{\partial i}}(\{\log \pi_i\}_\theta^{(1)}\, {}^t\{\log \pi_i\}_\theta^{(1)}) \geq V^t V(x_{\partial i}) \text{ and } \int_\chi V^t V(y)\lambda_\chi(dy) \text{ is p.d.}$$

3. The proof is similar to that of Th. C.2 in Appendix C (see (108; 96) for further details). We have:

$$0 = \sqrt{c_n}U_n^{(1)}(\theta) + \Delta_n(\theta, \widehat{\theta}_n^C)\sqrt{c_n}(\widehat{\theta}_n^C - \theta), \text{ with } \sqrt{c_n}U_n^{(1)}(\theta) = \frac{1}{\sqrt{c_n}} \sum_{i \in C_n} Z_i,$$

where $\Delta_n(\theta, \widehat{\theta}_n^C) = \int_0^1 U_n^{(2)}(\theta + t(\widehat{\theta}_n^C - \theta))dt$. As the variables $\{Z_i, i \in C\}$ are centered, bounded and independent conditionally on x^C, we apply under (5.29) the CLT for independent non-identically distributed bounded variables (32, Th. 9.2). It is straightforward to show that $\Delta_n(\theta, \widehat{\theta}_n^C) + I_n^C(\theta) \xrightarrow{P_\theta} 0$ and under (N1), $I_n^C(\theta) \equiv J_n^C(\theta)$. Combining asymptotic normality of $\sqrt{c_n}U_n^{(1)}(\theta)$ with this result gives the required result for the coding likelihood ratio test.
4. Several choices of coding C of S are possible: for example, for $S = \mathbb{Z}^2$ and the 4-NN graph, $C^+ = \{(i,j) : i+j \text{ even}\}$ and $C^- = \{(i,j) : i+j \text{ odd}\}$ are two maximal codings. To each coding C corresponds an estimator $\widehat{\theta}_n^C$, but estimators associated with different codings are not independent.

Example 5.11. Subergodicity for inhomogeneous Ising random fields

Let us go back to the example (5.9) of an inhomogeneous Ising model. The conditional energy at $i \in C$ is ${}^t\theta h_i(x_i, x_{\partial i})$, where

$$h_i(x_i, x_{\partial i}) = x_i \begin{pmatrix} u_i \\ v_i \end{pmatrix} \quad \text{and} \quad v_i = \sum_{j \in \partial i} w_{ij} x_j.$$

First we show that $I_n^C(\theta) = c_n^{-1} \sum_{i \in C_n} Var_{\theta, x_{\partial i}}(X_i \binom{u_i}{v_i})$. We begin by finding a lower bound on the conditional variance of X_i. Noting $p_i(x_{\partial i}, \theta) = P_\theta(X_i = -1 \mid x_{\partial i})$, it is easy to see that there exists $0 < \delta < 1/2$ such that, uniformly in $i \in C$, $x_{\partial i}$ and θ, $\delta \leq p_i(x_{\partial i}, \theta) \leq 1 - \delta$ and thus $Var_{\theta, x_{\partial i}}(X_i) \geq \eta = 4\delta(1-\delta) > 0$. Therefore, let us consider a vector ${}^t(c,d) \neq 0$, where c and d have the same sign. Even if it means removing the contribution of certain sites, we have:

$$(c,d) I_n^C(\theta)\, {}^t(c,d) \geq \eta \frac{\sum_{i \in C_n : v_i = +1} (c a_i + d)^2}{c_n}. \tag{5.30}$$

As the empirical frequency over C_n of the constant configuration $x_{\partial i} = \mathbf{1}$ (which gives $v_i = +1$) has a positive lower bound for large n (cf. Lemma 5.1), we deduce that $\liminf_n \{(c,d) I_n^C(\theta)\, {}^t(c,d)\} > 0$; if c and d have opposite signs, we can get the same lower bound by replacing $\sum_{i \in C_n : v_i = +1} (c a_i + d)^2$ by $\sum_{i \in C_n : v_i = -1} (c a_i - d)^2$ in (5.30). Thus, (5.29) holds.

Example 5.12. Tests for isotropy for Ising models

Let X be the Ising model observed on $\{1, 2, \ldots, n\}^2$ with conditional energy at (i,j) of

$$(H) : h(x_{i,j}, x_{\partial(i,j)}; \theta) = x_{i,j}(h + \beta_1(x_{i-1,j} + x_{i+1,j}) + \beta_2(x_{i,j-1} + x_{i,j+1})).$$

Table 5.9 Mercer and Hall data: coding and ML estimation.

	Coding (even)			Coding (odd)			ML		
	$\widehat{c}_{10}$	$\widehat{c}_{01}$	l_n^C	$\widehat{c}_{10}$	$\widehat{c}_{01}$	l_n^C	$\widehat{c}_{10}$	$\widehat{c}_{01}$	l_n
$L(0)$	–	–	−104.092	–	–	−116.998	–	–	488.046
$L_{iso}(1)$	0.196	–	−69.352	0.204	–	−88.209	0.205	–	525.872
$L(1)$	0.212	0.181	−69.172	0.260	0.149	−86.874	0.233	0.177	526.351

To test for isotropy $(H_0) : \beta_1 = \beta_2$, we use the statistic for coding difference on $C = \{(i,j) : i+j \text{ is even}\}$: asymptotically, under (H_0) this statistic has a χ^2_1 distribution.
Further examples of coding tests are given in Exercise 5.8.

Example 5.13. Modeling blank experiments (continuation of Example 1.12)

Let us take another look at Mercer and Hall's data giving harvested wheat data from a blank experiment in a rectangular field divided into 20×25 equally sized parcels. As the initial graphical analysis (cf. Fig. 1.10-b) suggested there was no effect with respect to rows, we propose modeling the quantity of harvested wheat using a spatial regression with only a mean effect with respect to columns and a stationary $CAR(L)$ error model:

$$X_t = \beta_j + \varepsilon_t, \qquad t = (i,j),$$

$$\varepsilon_t = \sum_{s\in L(h)} c_s \varepsilon_{t+s} + e_t, \qquad Var(e_t) = \sigma^2,$$

this for neighborhoods $L(h) = \{(k,l) : 0 < |k| + |l| \le h\}$ with range $h = 0$ (independence) and $h = 1$ (4-NN model). We also estimate the 4-NN isotropic $L_{iso}(1)$ model. Table 5.9 gives coding and ML estimations for each of the three models.

Next, we test $L(0)$ against $L_{iso}(1)$. The two tests (coding and ML) with statistic $T_n^a = 2\{l_n^a(\widehat{\theta}_{n,a}) - l_n^a(\widehat{\varphi}_{n,C})\}$ reject independence (H_0) at 5%. The two tests of $L_{iso}(1)$ against $L(1)$ accept isotropy.

5.4.4 Comparing asymptotic variance of estimators

While it seems intuitive that we lose information on passing from ML to MCPL and from MCPL to Coding, this can only be justified if we know how to calculate the asymptotic variance of each of the three methods. This is indeed possible when X is an ergodic Ising model (99) or if X is the $2d$-NN isotropic Markov Gaussian random field on $\mathbb{Z}^d$,

$$X_t = \beta \sum_{s:\|s-t\|_1=1} X_s + e_t\ , \ |\beta| < \frac{1}{2d}. \tag{5.31}$$

Let us consider further this second example. This specification is associated with a *unique* exponentially-mixing ergodic Gaussian random field X (cf. §B.2). If ρ_1 denotes the correlation at distance 1 and if X is observed on the cube with sides of n points, the asymptotic variance of ML, MCPL and coding estimators of β are respectively (26):

$$\lim_n (n \times Var\widehat{\beta}_{ML}) = \frac{1}{2(2\pi)^d} \int_{T^d} \left(\frac{\sum_{i=1}^d \cos \lambda_i}{1 - 2\beta \sum_{i=1}^d \cos \lambda_i} \right)^2 d\lambda_1 \ldots d\lambda_d,$$

$$\lim_n (n \times Var\widehat{\beta}_{MCPL}) = \begin{cases} \frac{2\beta^2(1-2d\rho_1)^2}{2d\rho_1^2} & \text{if } \beta \neq 0 \\ 1/d & \text{otherwise} \end{cases},$$

$$\lim_n (n \times Var\widehat{\beta}_C) = \begin{cases} \frac{\beta(1-2d\beta\rho_1)}{d\rho_1} & \text{if } \beta \neq 0 \\ 1/d & \text{otherwise} \end{cases},$$

the retained coding set being $C = \{i = (i_1, i_2, \ldots, i_d) \in \mathbb{Z}^d$ such that $i_1 + i_2 + \ldots + i_d$ is even$\}$. Table 5.10 gives for $d = 2$ the relative efficiencies $e_1(\beta) = ML/coding$ and $e_2(\beta) = ML/MCPL$ as well as the NN correlation $\rho_1(\beta)$ for $0 \leq \beta < 1/2d$. We see that the loss of efficiency of MCPL compared with ML is small if $\beta < 0.15$ (remember that for $d = 2$, we must impose $|\beta| < 0.25$).

Table 5.10 Relative efficiencies $e_1 = ML/Coding$, $e_2 = ML/MCPL$ and NN correlation ρ_1 for the 4-NN isotropic Gaussian random field as a function of β.

4β	0.0	0.1	0.2	0.3	0.4	0.6	0.8	0.9	0.95	0.99
ρ_1	0.0	0.03	0.05	0.08	0.11	0.17	0.27	0.35	0.60	0.73
e_1	1.00	0.99	0.97	0.92	0.86	0.68	0.42	0.25	0.15	0.04
e_2	1.00	1.00	0.99	0.97	0.95	0.87	0.71	0.56	0.42	0.19

We also see that $\beta \mapsto \rho_1(\beta)$ initially increases slowly, then rapidly when $\beta \uparrow .25$, ρ_1 equaling 0.85 when $(1 - 4\beta) = .32 \times 10^{-8}$ (20; 21). An explanation for this behavior is given in Exercise 1.17.

Under independence ($\beta = 0$), ML, MCPL and coding have the same efficiency. While this is easy to explain between ML and MCPL, the coding efficiency is more

Table 5.11 Efficiency $e_1(0) = ML/coding$ for a ν-NN isotropic CAR model in the independent case ($\beta = 0$) and for various regular networks of nodes with ν neighbors.

Regular network S	ν	τ^{-1}	e_1
linear	2	2	1
square	4	2	1
square+diagonals	8	4	1/2
triangular	6	3	2/3
hexagonal	3	2	1
Body-centered cubic	8	2	1
Face-centered cubic	12	4	1/2
tetrahedral	4	2	1
d-dimensional cubic	$2d$	2	1

surprising. In fact, $e_1(0) = 1$ is unique to regular lattices whose geometry allows asymptotic rates of optimal coding $\tau = \lim_n(\sharp C_n/\sharp D_n) = 1/2$. This is the case in $\mathbb{Z}^d$. Table 5.11 (27) shows that for isotropic ν-NN CAR models $X_i = \beta \sum_{\partial i} X_j + e_i$, $i \in S$, $|\beta| < 1/\nu$ on regular networks, but that this is no longer true for graphs with optimal coding rates $\tau < 1/2$.

5.4.5 Identification of the neighborhood structure of a Markov random field

The goal here is to identify the neighborhood structure L of a Markov random field with local specification $\pi_i(x_i \mid x^i) = \pi_i(x_i \mid x_{i+L})$ using a $\tau(n)$-penalized contrast (cf. Appendix C, §C.3). If $L_{\max}$ denotes an upper bounding neighborhood, we select the $\widehat{L}_n$ minimizing:

$$\widehat{L}_n = \underset{L \subseteq L_{\max}}{\operatorname{argmin}} \left\{ U_n(\widehat{\theta}_L) + \frac{\tau(n)}{n} \sharp L \right\}.$$

We now give two results.

Identification of a Gaussian CAR model on $\mathbb{Z}^d$

Let $L_{\max}$ be a finite subset of $(\mathbb{Z}^d)^+$, the positive half-space of $\mathbb{Z}^d$ with respect to the lexicographic order and, for $m = \sharp L_{\max}$,

$$\Theta = \{c \in \mathbb{R}^m : [1 - 2 \sum_{l \in L_{\max}} c_l \cos({}^t\lambda\, l)] > 0 \text{ for all } \lambda \in \mathbb{T}^d\}.$$

If $\theta \in \Theta$, the equations $X_i = \sum_{l \in L_{\max}} c_l (X_{i-l} + X_{i+l}) + e_i$ with $E(e_i X_j) = 0$ when $i \neq j$ define a $CAR(L_{\max})$ model. Suppose that the true model is a Gaussian $CAR(L_0)$, $L_0 \subseteq L_{\max}$. If X is observed on $D_n = \{1, 2, \ldots, n\}^d$, then using for U_n Whittle's Gaussian contrast (cf. (5.13)), we have the following result: if

$$T \log\log(n) < \tau(n) < \tau \times n,$$

then $\widehat{L}_n = L_0$ for large n if $\tau > 0$ is small enough and if T is large enough, for example if $\tau(n) = \log(n)$ or $\tau(n) = \sqrt{n}$ (102, Prop. 8). More precisely, Guyon and Yao (102) give bounds on the probability of two sets related to an incorrect parametrization of the model, these being the overfitted set $M_n^+ = \{\widehat{P}_n \supsetneq P_0\}$ and the underfitted one $M_n^- = \{\widehat{P}_n \not\supseteq P_0\}$.

Neighborhood of a Markov random field with a finite number of states

For a finite and symmetric $L_{\max} \subset \mathbb{Z}^d$, $0 \notin L_{\max}$, consider an $L_{\max}$-Markov random field X with finite state space E and translation-invariant specification belonging to the exponential family (5.23). For $(x,v) \in E \times E^{L_{\max}}$, note by $\langle \theta, h(x,v) \rangle$ the conditional energy at 0, where $\theta \in \Theta \subset \mathbb{R}^m$ and $h = {}^t(h_l(x,v), l = 1,\ldots,m) \in \mathbb{R}^m$. We know (cf. (5.24)) that if the matrix $H = (h(x,v))$ with columns $h(x,v)$, $(x,v) \in E \times E^{L_{\max}}$ has full rank m, then the parametrization by θ is identifiable. Suppose that the neighborhood support of X is $L_0 \subseteq L_{\max}$. By considering the penalized conditional log-pseudo-likelihood contrast, it is possible to show that if the rate of penalization satisfies:

$$\tau(n) \to \infty \text{ and } \tau(n) < \tau \times n$$

for a small enough $\tau > 0$, then $\widehat{L_n} = L_0$ for large n (102, Prop. 9). We remark that the search for the support of Markov random fields is also studied in (124; 52); in particular, (52) do not suppose known an upper bound for $L_{\max}$.

Example 5.14. Modeling spatial distribution of agricultural land use

The data studied here (5) come from satellite images of the Nebrodi region of Sicily, Italy (cf. http://stat.uibk.ac.at/smij). The image (cf. Fig. 5.14) covers a surface of 2.016 km^2 and is divided into $40 \times 56 = 2240$ pixels, each representing a parcel of land of 30×30 m^2. The output variable considered is binary, $X_s = 1$ (resp. $X_s = 0$) representing cultivated (resp. uncultivated) land. The cultivated parts can be divided into arable land, citrus, olive and almond plantations whereas the uncultivated parts include fields, forests, scrub and all other land types.

Also available to us are covariates describing certain characteristics of each parcel: soil density (DS) in grams per cubic centimeter, thickness of the layer of topsoil (HT) in centimeters, annual rainfall (AR), soil permeability under saturated condi-

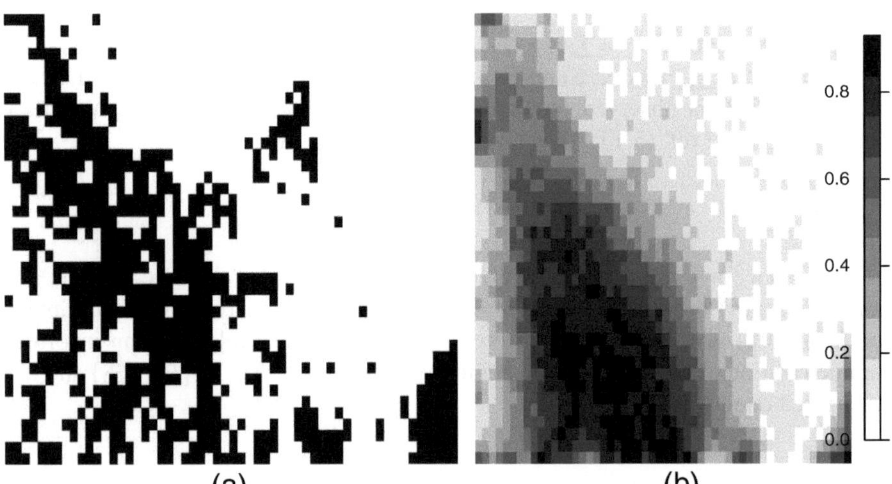

Fig. 5.14 (a) satellite image X with 40×56 pixels of the Nebrodi region (Sicily, Italy): cultivated (■) and uncultivated (□) land; (b) prediction of X using land covariates, including the neighborhood auto-covariate and altitude DTM (model (5.34)) using the estimated ML parameter.

tions (SP) and numerical data giving the altitude in meters (DTM). The goal of the analysis is to find out whether these land characteristics influence land use.

To do this, we are going to work with two auto-logistic models, the first with 4-NN and the second with 8-NN where, noting $v_1(x_{\partial i}) = \sum_{j:\|i-j\|=1} x_j$ and $v_2(x_{\partial i}) = \sum_{j:\|i-j\|=\sqrt{2}} x_j$,

$$P(X_i = 1|x_j, j \neq i) = \exp(\eta_i + \delta_i)/\{1+\exp(\eta_i + \delta_i)\},$$

with:

$$\eta_i = \beta_0 + \beta_1 DS_i + \beta_2 HT_i + \beta_3 PM_i + \beta_4 PC_i + \beta_5 DTM_i$$

and

$$\delta_i = \beta_6 v_1(x_{\partial i}) \tag{5.32}$$

or

$$\delta_i = \beta_6 v_1(x_{\partial i}) + \beta_7 v_2(x_{\partial i}). \tag{5.33}$$

Model (5.32) is one of those studied by Alfó and Postiglione (5). We are also going to consider the model where, apart from the two neighbor covariates, only altitude (DTM) is significant among the land covariates:

$$\eta_i = \beta_0 + \beta_5 TA_i + \beta_6 v_1(x_{\partial i}) + \beta_7 v_2(x_{\partial i}). \tag{5.34}$$

As for maximum CPL, coding estimates of β (cf. Table 5.12) are found using the function `glm` in `R`: in effect, the minimum contrast associated with each method corresponds to maximization of the likelihood function of a generalized linear model, the logistic model for which we had added one (resp. two) neighbor covariates.

For the (5.32) (resp. (5.33)) model, there are two (resp. four) possible codings.

Table 5.12 Coding estimates of the 4- and 8-NN auto-logistic models characterizing spatial distribution of land use.

	Model 5.32		Model 5.33				Model 5.34			
	C_0	C_1	C_0^*	C_1^*	C_2^*	C_3^*	C_0^*	C_1^*	C_2^*	C_3^*
$\widehat{\beta_0}$	0.361	7.029	0.829	−0.888	1.508	13.413	1.998	5.221	0.223	5.643
s.e.	10.351	10.197	14.317	14.807	15.239	14.797	3.080	3.123	3.245	3.009
$\widehat{\beta_1}$	−1.975	−1.625	−1.640	−2.645	−0.168	−1.171	–	–	–	–
s.e.	1.691	1.647	2.419	2.213	2.742	2.301	–	–	–	–
$\widehat{\beta_2}$	0.069	0.045	0.039	0.077	0.020	0.038	–	–	–	–
s.e.	0.051	0.050	0.073	0.067	0.083	0.069	–	–	–	–
$\widehat{\beta_3}$	0.008	−0.001	0.006	0.015	−0.006	−0.012	–	–	–	–
s.e.	0.018	0.018	0.026	0.026	0.027	0.026	–	–	–	–
$\widehat{\beta_4}$	1.360	1.418	1.658	2.461	−0.165	0.969	–	–	–	–
s.e.	1.298	1.273	1.859	1.734	2.069	1.778	–	–	–	–
$\widehat{\beta_5}$	−0.010	−0.011	−0.009	−0.014	−0.002	−0.010	−0.006	−0.010	−0.004	−0.010
s.e.	0.004	0.004	0.005	0.006	0.006	0.005	0.004	0.004	0.004	0.003
$\widehat{\beta_6}$	1.303	1.203	1.150	0.851	1.254	1.015	1.139	0.873	1.248	1.024
s.e.	0.095	0.089	0.156	0.161	0.166	0.158	0.155	0.160	0.165	0.156
$\widehat{\beta_7}$	–	–	0.194	0.515	0.359	0.203	0.177	0.489	0.378	0.206
s.e.	–	–	0.155	0.166	0.154	0.159	0.153	0.159	0.152	0.157

Estimated variances $\widehat{Var}(\hat{\beta}_i^{(k)})$ for each coding are those given in the output of the R function `glm`: in effect, coding contrasts are associated with exact (conditional) likelihoods which are in turn associated with (conditional) Fisher information. For stationary models, estimated variances $\widehat{Var}(\widehat{\beta}_i^{(k)})$ converge to the true values.

Though we can summarize the coding estimation by taking the mean $\widehat{\beta}_i$ of estimates $(\widehat{\beta}_i^{(k)}, k = 1, \ldots, m)$ over the different codings, it is important to note that these different coding estimates are not independent; therefore $\sqrt{\sum_{k=1}^m \widehat{Var}(\widehat{\beta}_i^{(k)})/m}$ *is not* the standard error of $\widehat{\beta}_i$.

There are three ways to calculate standard errors of maximum CPL estimators: we use here parametric *bootstrap* by simulating $m = 100$ times $X^{(i)}$ from the model $\widehat{\beta}_{CPL}$, estimating parameters $\beta^{(i)}$ for each $X^{(i)}$ and calculating the square root of the empirical variance over these 100 estimations.

For ML, we use a Newton-Raphson Monte Carlo algorithm (cf. §5.5.6) with $N = 1000$ simulations $X^{(i)}$ of a model under the initial estimator $\psi = \widehat{\beta}^{CPL}$ to calculate a value close to the actual likelihood. This choice should be examined more closely if the simulated exhaustive statistics $T^{(1)}, \ldots, T^{(1000)}$ do not contain the observed statistic T^{obs} in their convex envelope (in such cases, there is no ML solution). One possible choice is to take the output of the first iteration of the Newton-Raphson algorithm as the new initial value ψ and simulate again the random field $X^{(i)}$ under ψ. The estimator obtained in this way was suggested by Huang and Ogata (115).

Table 5.13 Estimates (coding, CPL and ML) of 4- and 8-NN auto-logistic models describing the spatial distribution of land use and their standard errors (s.e.). The bottom row of the table gives the residual sum of squares (RSS) of predictions of X by each of the three methods.

	Model 5.32			Model 5.33			Model 5.34		
	C	CPL	ML	C	CPL	ML	C	CPL	ML
$\widehat{\beta}_0$	3.695	4.115	-5.140	3.715	3.086	-4.764	3.271	3.394	-0.614
s.e.	7.319	5.618	4.510	7.397	5.896	3.066	1.558	1.936	1.124
$\widehat{\beta}_1$	-1.800	-1.810	-1.230	-1.406	-1.495	-0.950	–	–	–
s.e.	1.196	0.972	0.670	1.213	1.048	0.721	–	–	–
$\widehat{\beta}_2$	0.057	0.057	0.039	0.043	0.047	0.029	–	–	–
s.e.	0.036	0.030	0.020	0.037	0.032	0.021	–	–	–
$\widehat{\beta}_3$	0.003	0.003	0.009	0.001	0.002	0.008	–	–	–
s.e.	0.013	0.009	0.007	0.013	0.010	0.005	–	–	–
$\widehat{\beta}_4$	1.389	1.387	0.897	1.231	1.279	0.660	–	–	–
s.e.	0.918	0.745	0.455	0.932	0.798	0.520	–	–	–
$\widehat{\beta}_5$	-0.010	-0.010	-0.005	-0.009	-0.009	-0.004	-0.007	-0.008	-0.003
s.e.	0.003	0.002	0.002	0.003	0.003	0.002	0.002	0.002	0.001
$\widehat{\beta}_6$	1.253	1.240	1.409	1.068	1.055	1.109	1.071	1.061	1.097
s.e.	0.067	0.092	0.090	0.080	0.142	0.128	0.079	0.099	0.123
$\widehat{\beta}_7$	–	–	–	0.318	0.303	0.294	0.313	0.308	0.319
s.e.	–	–	–	0.079	0.132	0.118	0.078	0.059	0.113
RSS	–	–	–	–	–	–	350.111	347.815	308.118

Table 5.13 gives estimations for the three models using each of the three methods, coding (C), CPL and ML. These results show that the two autoregressive parameters of the 8-NN model (5.33) are significant and that among the land covariates, only altitude is significant. Model selection based on the values of statistics $\widehat{\beta}^a/\sqrt{\widehat{Var}(\widehat{\beta}_k^a)}$, $a = C, CPL, ML$ chooses model (5.34) for each method.

For this model, let us compare the three estimation methods with respect to their predictive qualities (226). Using Gibbs sampling, we simulate $m = 100$ times a random field $X^{(i)}$ under $\beta = \widehat{\beta}$ and note by m_k the number of times we observe $X_k^{(i)} = 1$ at site k. An empirical estimation of the probability $P(X_k = 1)$ is $\widehat{p}_k = m_k/m$ and the sum of the squares of residuals $RSS = \sum_k (X_k - \widehat{p}_k)^2$ gives a measure of the prediction error for the estimation $\widehat{\beta}$. We notice that for this criteria, ML is better than MCPL which is in turn better than coding.

5.5 Statistics for spatial point processes

5.5.1 *Testing spatial homogeneity using quadrat counts*

Suppose that $x = \{x_i, i = 1, \dots, n\}$ is generated by a PP over a set A, itself inside an larger observation window in order to remove boundary effects. The first modeling step is to test the CSR hypothesis:

$$(H_0)\ X \text{ is a homogeneous Poisson PP on } A.$$

There are many techniques for testing (H_0), including: (i) statistics based on quadrat counts, elements of a partition of A (62; 48; 194) and (ii) statistics based on distances between pairs of points of x (cf. §5.5.3).

We divide the observation window A into m disjoint quadrats A_i, $i = 1, \dots, m$ having the same measure $\nu(A_i) = \overline{\nu}$ and count the number of points $N_i = n(A_i)$ of x in each A_i. We then calculate the statistic:

$$Q = \sum_{i=1}^{m} \frac{(N_i - \overline{N})^2}{\overline{N}}$$

with $\overline{N} = \sum_{i=1}^{m} N_i/m$. If X is a homogeneous Poisson PP with intensity τ, the N_i are independent Poisson variables with mean $\tau\overline{\nu}$ and the distribution of Q can be approximated by a χ^2 with $m-1$ degrees of freedom. This approximation is judged to be reasonable if $E(N_i) \geq 1$ and $m > 6$ (Diggle (62)).

Fig. 5.15 gives an illustration of this method on the `finpines` dataset. We find $Q_{obs} = 46.70$ and as $P(\chi^2_{15} > Q_{obs}) \simeq 0$, we reject the homogeneous Poisson hypothesis for the pine distribution.

This method can be extended to testing the inhomogeneous case. If X is a Poisson PP with intensity $\rho(\cdot, \theta)$, we can use ML to estimate θ (cf. §5.5.2), then estimate the expected totals per block by $N_i(\widehat{\theta}) = \int_{A_i} \rho(u, \widehat{\theta})du$. The test statistic for spatial

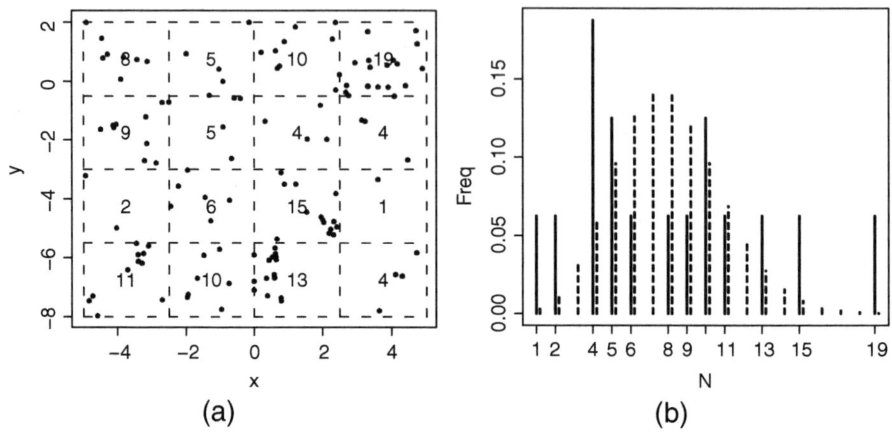

Fig. 5.15 Statistics of the 4×4 quadrats for the Finnish pines data (`finpines`). To the left, the 16 quadrats and the associated counts N_i; to the right, the empirical distribution of the N_i (solid line) and the fitted theoretical Poisson distribution (dotted line).

independence is now

$$Q = \sum_{i=1}^{m} \frac{(N_i - N_i(\widehat{\theta}))^2}{N_i(\widehat{\theta})}.$$

For large enough m and if each $N_i(\widehat{\theta}) \geq 1$, the distribution of $(Q-m)/2m$ is close to a Gaussian random variable with variance 1 if there is spatial independence.

5.5.2 Estimating point process intensity

5.5.2.1 Estimating parametric models for a Poisson PP

Let X be a PP observed on a measure positive compact subset A of $\mathbb{R}^d$ and let $x = \{x_1, \ldots, x_n\}$ be $n = n(A)$ points generated by this PP. If X is stationary, the intensity τ can be estimated by:

$$\widehat{\tau} = \frac{n(A)}{\nu(A)}.$$

$\widehat{\tau}$ is an unbiased estimator of τ. Furthermore, if X is a homogeneous Poisson PP, $\widehat{\tau}$ is in fact the maximum likelihood estimator of τ.

If X is ergodic (cf. §B.1) and if (A_n) is an increasing sequence of bounded convex sets such that $d(A_n) \longrightarrow \infty$, where $d(A) = \sup\{r : B(\xi, r) \subseteq A\}$ denotes the interior diameter of A, then

$$\widehat{\tau}_n = \frac{n(A_n)}{\nu(A_n)} \to \tau \qquad a.s.$$

Homogeneous Poisson PPs as well as Neyman-Scott PPs derived from homogeneous Poisson PPs are themselves homogeneous.

If X is an inhomogeneous Poisson PP with intensity $\rho(\cdot;\theta)$ parametrized by θ, the ML estimator of θ can be found by maximizing the log-density of X on A (cf. (3.5)):

$$l_A(\theta) = \sum_{\xi \in x \cap A} \log \rho(\xi;\theta) + \int_A \{1 - \rho(\eta;\theta)\} d\eta. \tag{5.35}$$

If $\rho(\cdot;\theta)$ follows a log-linear model, maximizing $l_A(\theta)$ is done in the same way as for the log-likelihood of generalized linear models (cf. also §5.5.5.1). Baddeley and Turner (14) suggest approximating $\int_A \rho(\eta;\theta)d\eta$ by $\sum_{j=1}^m \rho(\eta_j;\theta)w_j$ for $\eta_j \in A$ and suitable integration weights w_i, leading to the log-likelihood:

$$l_A(\theta) \simeq \sum_{\xi \in x \cap A} \log \rho(\xi;\theta) - \sum_{j=1}^m \rho(\eta_j;\theta)w_j. \tag{5.36}$$

If the set of η_j contains $x \cap A$, (5.36) can be rewritten

$$l_A(\theta) \simeq \sum_{j=1}^m \{y_j \log \rho(\eta_j;\theta) - \rho(\eta_j;\theta)\} w_j, \tag{5.37}$$

where $y_j = \mathbf{1}[\eta_j \in x \cap A]/w_j$. The term on the right hand side of (5.37) is formally equivalent to a log-likelihood reweighted by weights w_j for independent Poisson variables with mean $\rho(\eta_j;\theta)$ and (5.37) can be maximized using GLM software (cf. the `spatstat` package in R).

Consistency and asymptotic normality of this ML estimator are studied in Rathbun and Cressie (179) and Kutoyants (135).

We present briefly Rathbun and Cressie's result. Suppose X is an inhomogeneous Poisson PP with log-linear intensity

$$\log \rho(\xi;\theta) = {}^t z(\xi)\theta, \ \theta \in \mathbb{R}^p,$$

where the $z(\xi)$ are observable covariates. Suppose furthermore that the sequence of bounded domains of observation (A_n) is increasing in that $d(A_n) \longrightarrow \infty$ and:

(L1) $\int_A \rho(\xi;\theta)d\xi < \infty$ for every Borel set A satisfying $\nu(A) < \infty$.
(L2) There exists $K < \infty$ such that $\nu(\xi \in S : \max_{1 \le i \le p} |z_i(\xi)| > K) < \infty$.
(L3) The matrix $M_n = \int_{A_n} z(\xi)z(\xi)^t d\xi$ is p.d. and satisfies $\lim_n M_n^{-1} = 0$.

Then, $\widehat{\theta}_n = \underset{\theta \in \Theta}{\operatorname{argmax}}\, l_{A_n}(\theta)$ is consistent and for large n:

$$(\widehat{\theta}_n - \theta) \sim \mathcal{N}_p(0, I_n(\theta)^{-1}), \ \text{where } I_n = \int_{A_n} z(\xi)\, {}^t z(\xi) \exp\{{}^t z(\xi)\theta\} d\xi.$$

Estimating PP intensity

When the Poisson hypothesis is no longer valid, Møller and Waagepetersen (160) still use the functional $l_A(\theta)$ (5.35) to estimate the parameter θ of the density $\rho(\cdot;\theta)$. In this case, $l_A(\theta)$ is known as a "Poisson pseudo-likelihood" or "composite likelihood" and is useful for evaluating intensity parameters.

5.5.2.2 Nonparametric intensity estimation

Nonparametric estimation of intensity functions $x \mapsto \rho(x)$ is a similar task as that for probability densities. Diggle (62) suggests the estimator

$$\widehat{\rho}_\sigma(x) = \frac{1}{K_\sigma(x)} \sum_{i=1}^{n} \frac{1}{\sigma^d} k\left(\frac{x - x_i}{\sigma}\right),$$

with $k : \mathbb{R}^d \to \mathbb{R}^+$ a symmetric kernel integrating to 1, $\sigma > 0$ a smoothing parameter and

$$K_\sigma(x) = \int_A \frac{1}{\sigma^d} k\left(\frac{x - \xi}{\sigma}\right) \nu(d\xi).$$

The choice of σ is important and we suggest trying several values. On the other hand, the choice of $k(\cdot)$ is less so, some classical choices being the Gaussian kernel $k(x) = (2\pi)^{-d/2} e^{-\|x\|^2/2}$ and the Epanechnikov kernel $k(x) = c(d)\mathbf{1}\,(\|x\| < 1)(1 - \|x\|^2)$.

Fig. 5.16 gives the location of 514 maple trees in a Michigan forest (`lansing` data from the `spatstat` package). These data, scaled to fit in a unit square, exhibit a non-constant intensity. This intensity has been estimated with a Gaussian kernel for three values of the smoothing parameter, $\sigma = 0.01$, 0.05 and 0.1.

5.5.3 Estimation of second-order characteristics

5.5.3.1 Empirically estimating NN distributions

If X is a stationary PP, the distribution $G(\cdot)$ of the distance of the point $\xi \in X$ to its nearest neighbor in X is (cf. §3.5.2):

$$G(h) = P_\xi(d(\xi, X\backslash\{\xi\}) \leq h),\ r \geq 0.$$

Let n be the number of points of X in $A_{\ominus h} = \{\xi \in A : B(\xi,h) \subseteq A\}$, the h-interior of A, where $B(\xi,h)$ is the ball of center ξ and radius h, and h_i the distance from an observation $x_i \in x$, $i = 1, \dots, n$, to its nearest neighbor in x. A nonparametric estimator of G is the empirical cumulative distribution function

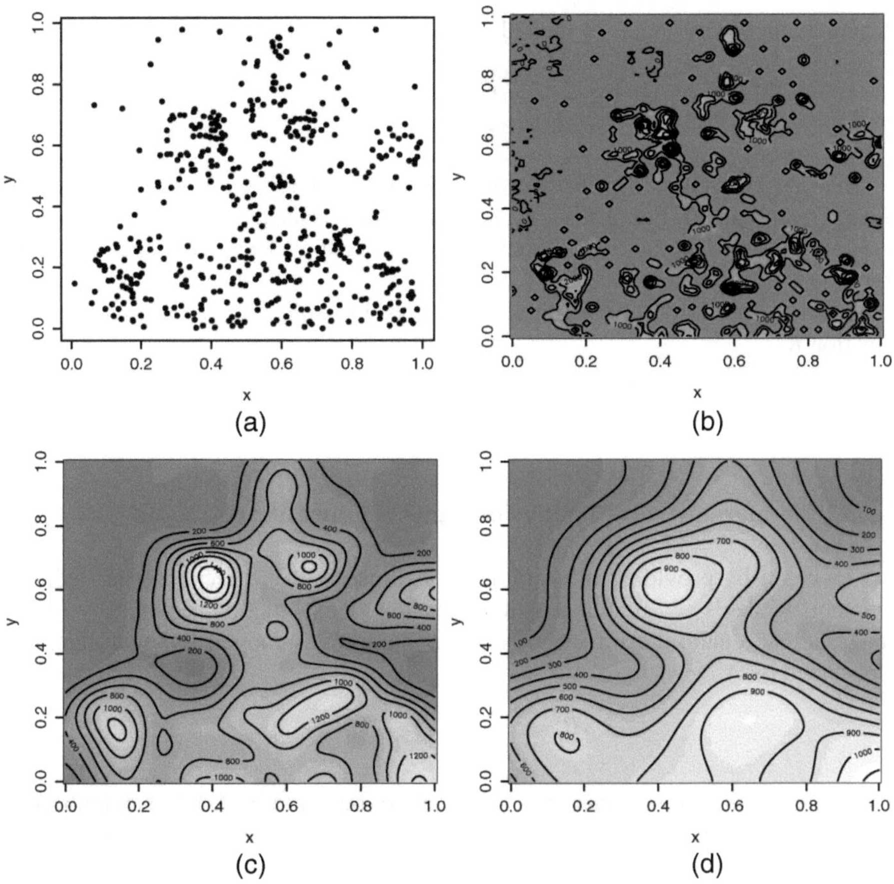

Fig. 5.16 (a) Locations of 514 maple trees in a Michigan forest. Nonparametric estimates of intensity, with (b) $\sigma = 0.01$, (c) $\sigma = 0.05$ and (d) $\sigma = 0.1$.

$$\widehat{G}(h) = \frac{1}{n} \sum_{i=1}^{n} \mathbf{1}_{(0,h_i]}(h).$$

The distribution of the distance of a point $\xi \in \mathbb{R}^d$ to its NN in X is

$$F(h) = P(d(\xi, X \backslash \{\xi\}) \leq h),\ h \geq 0.$$

To estimate F, we start by choosing independently of X a regular grid of points over $A_{\ominus h}$. Let m be the number of points of the grid and h_i^* the distance from a point on the grid to its nearest neighbor in X_A, $i = 1, \ldots, m$. A nonparametric estimator of F is the empirical cumulative distribution function

$$\widehat{F}(h) = \frac{1}{m} \sum_{i=1}^{n} \mathbf{1}_{(0,h_i^*]}(h).$$

A nonparametric estimator of the index J of a Poisson PP is thus:

$$\widehat{J}(h) = \frac{1-\widehat{G}(h)}{1-\widehat{F}(h)}, \quad \text{if } \widehat{F}(h) < 1.$$

The first column in Fig. 5.17 gives estimates of J for the `ants`, `cells` and `finpines` data (cf. Fig. 3.1-a–c). The null hypothesis is accepted for the ants data and rejected for the other two, the cells data having higher regularity and the pines lower.

Another useful tool for testing if a PP is Poisson uses an estimate of the function K.

5.5.3.2 Nonparametric estimation of Ripley's K function

Let us begin by providing nonparametric estimators of the second-order reduced moment, Ripley's $\mathscr{K}$ function or its extension. This function, defined by (3.12) suggests that a natural estimator of $\tau^2\mathscr{K}(B)$ is, for X homogeneous with intensity τ observed on A,

$$\tau^2\widehat{\mathscr{K}}(B) = \frac{1}{\nu(A)} \sum_{\xi,\eta\in X\cap A}^{\neq} \mathbf{1}_B(\xi-\eta).$$

This estimator underestimates $\tau^2\mathscr{K}$ as boundary points of A have less neighbors in A than interior points. We can limit this bias by "enlarging" A but this is not always possible. A possible way to decrease the bias of this estimator is to consider (160):

$$\widehat{\mathscr{K}_A}(B) = \sum_{\xi,\eta\in X\cap A}^{\neq} \frac{\mathbf{1}_B(\xi-\eta)}{\nu(W_\xi \cap W_\eta)}$$

if $\nu(W_\xi \cap W_\eta) > 0$, where $W_\xi = \xi + W$ is the ξ-translation of W. In effect:

$$\begin{aligned}
\mathbb{E}\left(\widehat{\mathscr{K}_A}(B)\right) &= \mathbb{E}\left(\sum_{\xi,\eta\in X}^{\neq} \frac{\mathbf{1}_B(\xi-\eta)\mathbf{1}_A(\xi)\mathbf{1}_A(\eta)}{\nu(W_\xi \cap W_\eta)}\right)\\
&= \int \frac{\mathbf{1}_B(\xi-\eta)\mathbf{1}_A(\xi)\mathbf{1}_A(\eta)}{\nu(W_\xi \cap W_{\eta-\xi})}\alpha_2(d\xi \times d\eta)\\
&= \tau^2 \int\int \frac{\mathbf{1}_B(u)\mathbf{1}_A(\xi)\mathbf{1}_A(\xi+\zeta)}{\nu(W_\xi \cap W_{\xi+\zeta})} d\xi\, \mathscr{K}(d\zeta)\\
&= \tau^2 \int \mathbf{1}_B(\zeta)\mathscr{K}(d\zeta) = \tau^2\mathscr{K}(B).
\end{aligned}$$

For X an isotropic PP on $\mathbb{R}^2$ and $B = B(0,h)$ the ball with center at 0 and radius h, Ripley (184) proposes the estimator:

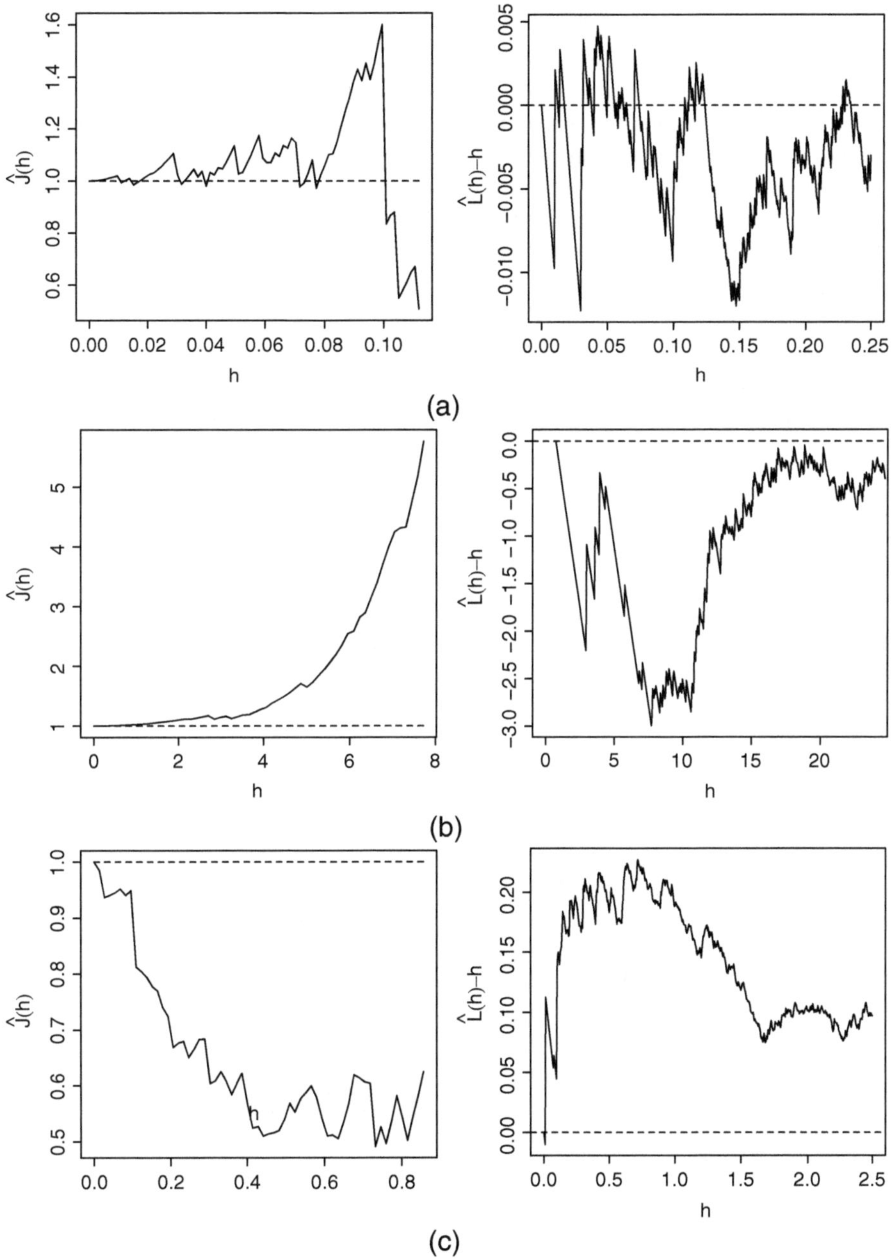

Fig. 5.17 Two second-order summary statistics: left, $h \mapsto \widehat{J}(h)$ and right, $h \mapsto (\widehat{L}(h) - h)$, for the `ants`, `cells` and `finpines` spatial distributions (cf. Fig. 3.1-a–c). When X is a homogeneous Poisson PP, $L(h) - h \equiv 0$ and $J \equiv 1$.

$$\widehat{K}_A(h) = \frac{\nu(A)}{n(A)} \sum_{\xi,\eta \in X\cap A}^{\neq} w(\xi,\eta)\mathbf{1}[0 < \|\xi-\eta\| < h],$$

where $w(\xi,\eta)^{-1}$ is the proportion of the perimeter of the circle $C(\xi;\|\xi-\eta\|)$ with center at ξ and with the point η on its boundary that is found in A. This estimator is unbiased if $h < h^*$, where h^* is the largest radius r for a circle $C(\xi;r)$ centered at some $\xi \in A$ with a boundary cutting A. For example, if $A = [0,1]^2$, $h^* = \sqrt{2}$. In $d = 2$ dimensions, we can deduce an estimator of L:

$$\widehat{L}(h) = \sqrt{\widehat{K}(h)/\pi}.$$

Theorem 5.7. *(Heinrich (111)) Suppose that X is an ergodic PP and (A_n) an increasing sequence of bounded convex sets with interior diameters satisfying $d(A_n) \longrightarrow \infty$. Denote $\widehat{k}_n(h) = \widehat{\mathscr{K}_{A_n}}(B(0,h))$ and $k(h) = \mathscr{K}(B(0,h))$. Then, for any fixed h_0,*

$$\lim_n \sup_{0\le h\le h_0} \left|\widehat{k}_n(h) - k(h)\right| = 0 \qquad a.s.$$

Proof: By the ergodic theorem (cf. §B.1), for all h,

$$\lim_n \widehat{k}_n(h) = k(h) \qquad \text{a.s.}$$

An argument in the style of Glivenko-Cantelli (cf. (55)) then gives the required result.

□

In the same way as homogeneous PPs, inhomogeneous PPs may exhibit aggregation, regularity and independence tendencies. In order to identify this behavior, Baddeley et al. (13) extended Ripley's K function to second-order stationary inhomogeneous PPs with respect to the reweighted correlation g (cf. (3.1)) by introducing the function (cf. (3.14)):

$$K_{BMW}(h) = \frac{1}{\nu(A)}\mathbb{E}\left[\sum_{\xi,\eta\in X}^{\neq} \frac{\mathbf{1}_{\{\|\xi-\eta\|\le h\}}}{\rho(\xi)\rho(\eta)}\right],$$

whose natural empirical estimator is:

$$\widehat{K}_{BMW}(h) = \frac{1}{\nu(A)} \sum_{\xi,\eta\in X\cap A}^{\neq} \frac{\mathbf{1}_{\{\|\xi-\eta\|\le h\}}}{\widehat{\rho}(\xi)\widehat{\rho}(\eta)}.$$

Here again, it is possible to correct for boundary effects.

For *inhomogeneous Poisson PPs* in 2 dimensions, we still have: $K_{BMW}(h) = \pi h^2$. Thus, by looking at $\widehat{K}_{BMW}(h) - \pi h^2$, we could test the hypothesis that a process is an inhomogeneous Poisson PP, just as Ripley's K function allows us to test whether a process is a homogeneous Poisson PP.

5.5.3.3 Monte Carlo tests

In general, the distribution of the previous estimators, useful for constructing tests and confidence intervals, is unknown. Monte Carlo approximation of the null distribution of the test provides a convenient tool for testing model fit. Here are the details of the method.

Suppose that x_A represents the configuration in window A of a homogeneous Poisson PP X on $\mathbb{R}^2$ with intensity τ (hypothesis (H_0)). We want to construct a confidence interval for $G(h)$, the cumulative distribution function of the distance of a point in X to its NN as a function of h. Conditional on $n = n(x)$, we generate m independent data points $\mathbf{x}_A^{(i)}$ of X on A, a PPP($\widehat{\lambda}$) with intensity $\widehat{\lambda} = n/\nu(A)$ and for each $\mathbf{x}_A^{(i)}$ we calculate $\widehat{G}_i(h)$, $i = 1,\dots,m$. We then approximate the quantiles of $\widehat{G}(h)$ using the empirical distributions of the $\{\widehat{G}_i(h),\ i = 1,\dots,m\}$. To test (H_0), it remains to compare these quantiles with the statistic $\widehat{G}_0(h)$ calculated using the data x_A. If calculating $\widehat{G}$ is not quick and simple, we can choose a smaller m so that the bounds $\widehat{G}_{inf}(h) = \min_i \widehat{G}_i(h)$ and $\widehat{G}_{sup}(h) = \max_i \widehat{G}_i(h)$ define under (H_0) the confidence interval of level $1 - 2/(m+1)$,

$$P(\widehat{G}_0(h) < \widehat{G}_{inf}(h)) = P(\widehat{G}_0(h) > \widehat{G}_{sup}(h)) \leq \frac{1}{m+1},$$

with equality if the values $\widehat{G}_i$ are all different. We could also calculate the functions $\widehat{G}_{inf}$, $\widehat{G}_0$, $\widehat{G}_{sup}$ and compare them with πh^2. If $\widehat{G}_0$ is in the confidence band $\{[\widehat{G}_{inf}(h), \widehat{G}_{sup}(h)] : h \geq 0\}$, we can conclude that X is a homogeneous Poisson PP.

Other Poisson-type test functions can also be used, for example:

$$T = \sup_{h_1 \leq h \leq h_2} |\widehat{K}(h) - \pi h^2| \quad \text{or} \quad T = \int_0^{h_3} (\widehat{J}(h) - 1)^2 dh,$$

where the h_i, $i = 1,2,3$ are certain chosen radii. We could then compare the value $T_0 = T(x_A)$ obtained for observation x_A with the ordered sample $T^{(1)} \leq T^{(2)} \ldots \leq T^{(m)}$ obtained by generating m independent values $x_A^{(i)}$ of X. A bilateral rejection region at level $\alpha = 2k/(m+1)$ is

$$\mathscr{R} = \{T_0 \leq T^{(k)}\} \cup \{T_0 \geq T^{(m-k+1)}\}.$$

Analyzing the same data as in Fig. 3.1 using $\widehat{L}$ leads again to the previous conclusions (cf. Fig. 5.17, second column).

These types of Monte Carlo procedures can be easily extended to other situations including validation of general models.

Example 5.15. Biodiversity in tropical rainforests

We turn to a dataset analyzed by Waagepetersen (220) giving the spatial distribution of 3605 trees of the species *Beilschmiedia pendula Lauraceae* in the tropical rainforests of Barro Colorado Island (cf. Fig. 5.18-a).

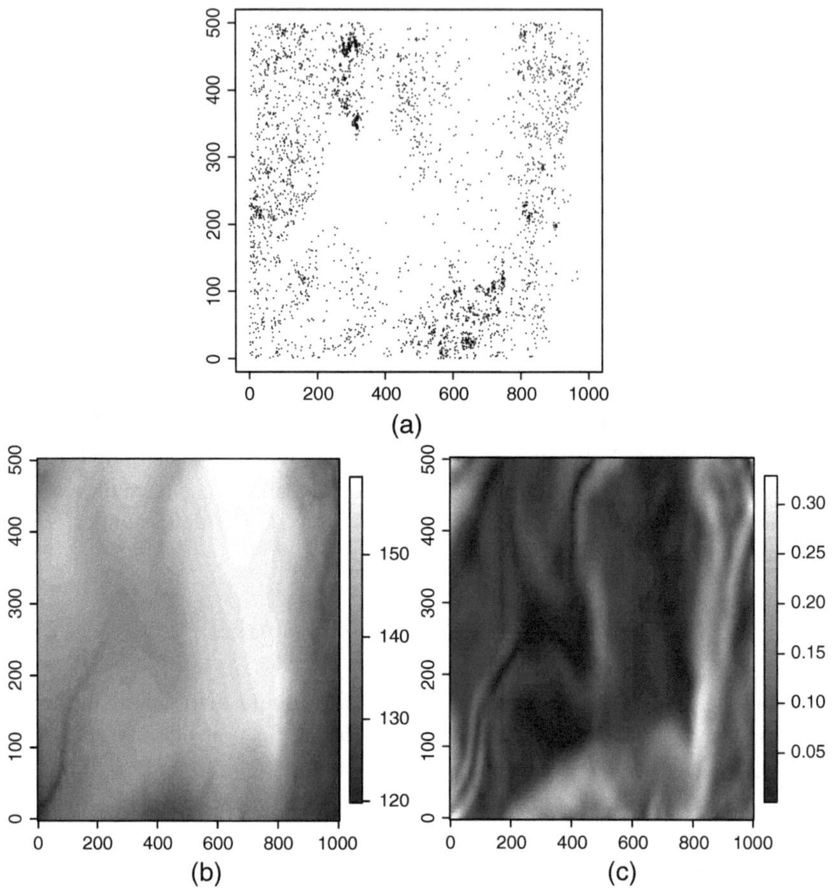

Fig. 5.18 (a) Locations of 3605 trees in a part of the Barro Colorado forest; covariates altitude (`elev`, (b)) and slope (`grad` (c)).

Tropical rainforests are characterized by tall and densely distributed vegetation in a warm and humid climate. One question is to know whether the spatial distribution is linked to altitude (`elev`) and land gradient (`grad`). An initial model to consider is a Poisson PP with log-linear intensity $\log \rho(\xi;\theta) = {}^t z(\xi)\theta$, where $z(\xi) = {}^t(1, z_{\texttt{elev}}(\xi), z_{\texttt{grad}}(\xi))$. The ML of the parameters and their standard errors are, in order, -8.559 (0.341), 0.021 (0.002) and 5.841 (0.256). Thus, elevation and slope are significant as the forest density increases with both. As for the nonparametric estimation of K_{BMW} based on $\rho(\xi;\widehat{\theta})$ (cf. Fig. 5.19):

$$\widehat{K}_{BMW}(h) = \frac{1}{\nu(A)} \sum_{\xi,\eta \in X \cap A}^{\neq} \frac{\mathbf{1}_{\{\|\xi-\eta\| \leq h\}}}{\rho(\xi;\widehat{\theta})\rho(\eta;\widehat{\theta})},$$

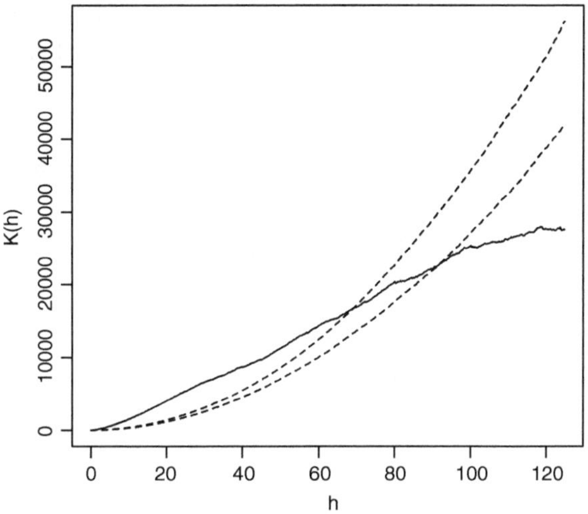

Fig. 5.19 Estimated function $\widehat{K}_{BMW}$ (solid line) and lower and upper confidence bands (95%) after 40 trials of a homogeneous Poisson PP with the same estimated intensity.

it is significantly different to πh^2, that of a Poisson PP with intensity $\rho(\cdot;\widehat{\theta})$. This leads us to doubts about using a Poisson PP model, either for modeling intensity or in terms of spatial independence of the tree distribution. More precisely, Waagepetersen (220) proposed modeling these data using an inhomogeneous Neyman-Scott PP.

Example 5.16. Differences in spatial distribution depending on plant species

This example comes from a study of Shea et al. (198) on the spatial distribution of the aquatic tulip species *Nyssa aquatica* with respect to three characteristics (male, female, not mature) in three swamps of size 50×50 m^2 in South Carolina (cf. `nyssa` dataset on the website). We take a look here at data for only one swamp and the trait "sex" (male or female, cf. Fig. 5.20). The random distribution of males and females is *a priori* judged optimal as it facilitates reproduction. Aggregation of males or females can be explained by different resource needs. The distribution of each sex can be seen as a MPP (Marked Point Process) $Y = (x_i, m_i)$, where m_i is the binary variable "sex."

We choose to evaluate spatial aggregation via Ripley's K function. If the male (1) and female (0) populations have the same spatial aggregation, then $D(h) = K_1(h) - K_0(h)$ will be exactly zero. To test this hypothesis, we consider the statistic $\widehat{D}(h) = \widehat{K}_1(h) - \widehat{K}_0(h)$. As the distribution of $\widehat{D}$ under the null hypothesis is difficult to obtain, we take a Monte Carlo approach conditional on the superposition of male and female locations. To this end, we generate k MPPs $Y^{(j)} = (x_i, m_i^{(j)})$, $j = 1, \ldots, k$, where $m^{(j)} = (m_i^{(j)})$ is a random permutation of marks $m = (m_i)$, followed by calculating $\widehat{D}^{(j)}(h) = \widehat{K}_1^{(j)}(h) - \widehat{K}_0^{(j)}(h)$. For $k = 40$ trials, the obtained confidence envelopes (cf. Fig. 5.20-b) show that, without considering population density, differences between the sexes is not seen in the spatial pattern.

5.5.4 Estimation of a parametric model for a point process

If X follows a parametric model $\mathscr{M}(\theta)$ that allows us to calculate K_θ, we can estimate θ with OLS by minimizing

$$D(\theta) = \int_0^{h_0} \{\widehat{K}(h)^c - K(h;\theta)^c\}^2 \, dh. \tag{5.38}$$

for some chosen power $c > 0$ and range h_0. Diggle (62, p. 87) recommends $h_0 = 0.25$ if $A = [0,1]^2$, $c = 0.5$ for regular PPs and $c = 0.25$ for PPs with aggregation. For calculations, $D(\theta)$ must be approximated by the sum

$$D^*(\theta) = \sum_{i=1}^{k} w_i \{\widehat{K}(h_i)^c - K(h_i;\theta)^c\}^2$$

for suitable weights w_i. If the analytic form of K_θ is unknown, we can use a Monte Carlo approximation $K^{MC}(\theta) = \sum_{i=1}^{m} \widehat{K}_i(\theta)/m$ by performing m independent trials $x_A^{(i)}$ for X under the model $\mathscr{M}(\theta)$ and then calculating $\widehat{K}_i(\theta)$ for each i.

Consistency of this OLS estimator was studied by Heinrich (111) for ergodic X. Guan and Sherman (95) established asymptotic normality of estimators under a mixing condition that is satisfied if X is a Neyman-Scott PP or log-Gaussian Cox PP.

Example 5.17. Estimation and validation of a parametric model

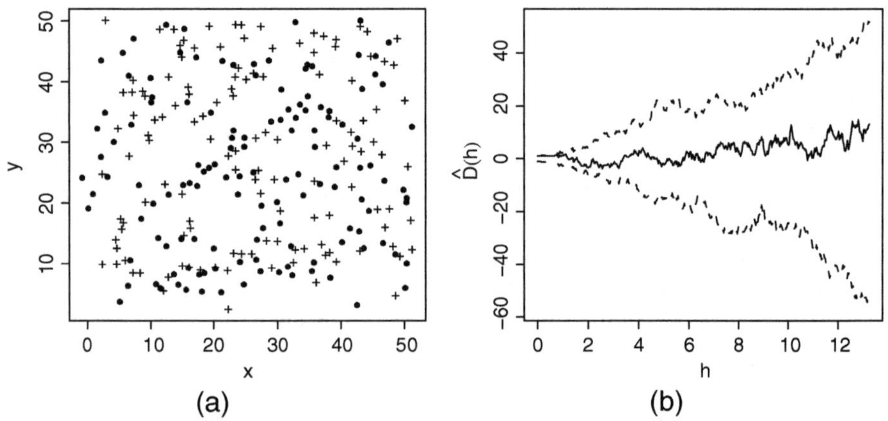

Fig. 5.20 (a) Spatial pattern of the species *Nyssa aquatica* in terms of male (•) and female (+); (b) (continuous) estimate of $D(h) = K_1(h) - K_2(h)$ and upper and lower confidence bands (dashed lines) at 95% obtained from 40 simulations.

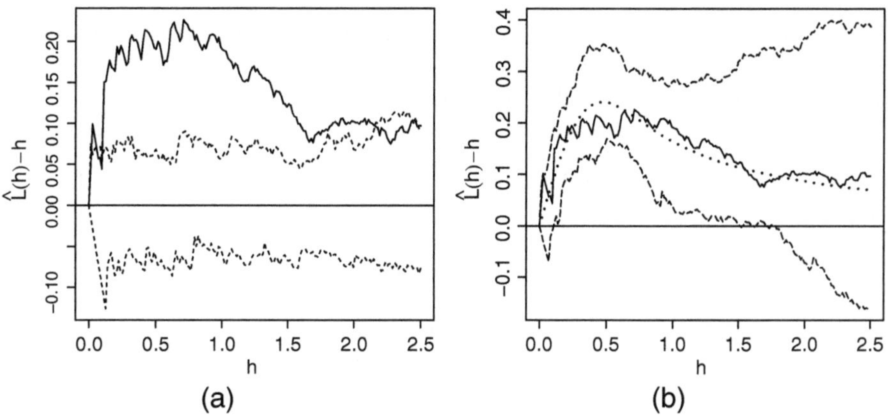

Fig. 5.21 Estimation of $D(h) = L(h) - h$ for the finpines data (cf. Fig. 3.1-c): solid line (a and b), nonparametric estimation of D; in (b) the dotted line is a parametric estimation using the Neyman-Scott model. These estimates are then compared with upper and lower confidence bands (95%) obtained after 40 simulations: (a) of a homogeneous Poisson PP of the same intensity; (b) of a Neyman-Scott process with the given estimated parameters.

We remark that Fig. 5.17-c, representing the finpines data seems to show that the spatial pattern has aggregates. We therefore propose modeling X using a Neyman-Scott process with location of parents given by a homogeneous $PPP(\lambda)$, with the number of descendants following a Poisson distribution with mean μ and position of descendants around a parent a $\mathcal{N}_2(0,\sigma^2 I_2)$. We estimate parameters by minimizing (5.38) with $h_0 = 2.5$, $c = 0.25$, performing the integration over a regular grid with 180 points. Under this model, the function K is:

$$K(h;\theta) = \pi h^2 + \theta_1^{-1}\{1 - \exp(-h^2/4\theta_2)\},$$

with $\theta_1 = \mu$ and $\theta_2 = \sigma^2$. We find $\widehat{\mu} = 1.875$, $\widehat{\sigma}^2 = 0.00944$ and $\widehat{\lambda} = n(\mathbf{x})/(\widehat{\mu}\nu(A)) = 0.672$. We then validate the model using parametric bootstrap by generating $m = 40$ values from a Neyman-Scott PP with parameters $(\widehat{\mu}, \widehat{\sigma^2}, \widehat{\lambda})$. The confidence region (cf. Fig. 5.21) shows that this model is a reasonable choice as $\widehat{K}_0(h)$ is inside the region for $h > 0.2$.

5.5.5 Conditional pseudo-likelihood of a point process

5.5.5.1 Definition and evaluation of the conditional pseudo-likelihood function

As the notion of conditional density of PPs at individual sites makes no sense, we have to create a new notion of conditional pseudo-likelihood. Intuitively, this can be developed based on the one for random fields on a network in the following way (184):

1. For a finely-spaced partition of X, we associate a counting process with the PP on each node of the partition.
2. We then define the CPL of this counting process.
3. We study the limit of this CPL as the area of each partition element goes to 0.

This limit is identifiable if the density f_θ of X is hereditary (cf. (3.6)) and if f_θ is *stable*, i.e., if:

$$\exists c_\theta \text{ and } K_\theta > 0 \text{ finite such that} : \forall x,\ f_\theta(x) \leq c_\theta K_\theta^{n(x)}.$$

We thus consider a sequence of nested partitions of S, $A_{i,j} \subseteq A_{i-1,j}$, $(S = \bigcup_{j=1}^{m_i} A_{i,j}, i = 1,2,\ldots)$ satisfying, if $\delta_i = \max\{\nu(A_{i,j}), j = 1, m_i\}$,

$$m_i \longrightarrow \infty, \qquad m_i \delta_i^2 \longrightarrow 0. \tag{5.39}$$

Theorem 5.8. *Pseudo-likelihood of a point process (Jensen-Møller (121))*
Note μ_S the measure of the PPP(1). If the density f_θ of X is hereditary and stable, then, under (5.39):

$$\lim_{i\to\infty} \prod_{j=1}^{m_i} f_\theta(x_{A_{i,j}} | x_{S\backslash A_{i,j}}) = \exp\{\lambda \nu(S) - \Lambda_\theta(S,x)\} \prod_{\xi \in x} \lambda_\theta(\xi, x\backslash\{\xi\}), \quad \mu_S \text{ a.s.},$$

where

$$\lambda_\theta(\xi,x) = \frac{f_\theta(x \cup \{\xi\})}{f_\theta(x)} \mathbf{1}\{f_\theta(x) > 0\} \quad \textit{and} \quad \Lambda_\theta(A,x) = \int_A \lambda_\theta(\eta,x)d\eta.$$

If $A \in \mathscr{B}(\mathbb{R}^d)$, the CPL of X on A is defined as:

$$pl_A(x,\theta) = \exp\{-\Lambda_\theta(A,x)\} \prod_{\xi \in x} \lambda_\theta(\xi, x\backslash\{\xi\}). \tag{5.40}$$

The CPL of a PP is proportional to the product across sites $x_i \in X$ of the Papangélou conditional intensities. If $\xi \in x_A$, the definition of $\lambda_\theta(\xi,x)$ does not change if the joint density on S is replaced by the conditional density $f_\theta(x_A | x_{S\backslash A})$. This is important to note as it means we can decide to model a set of points observed in an observation window A either without needing to take into account what happens outside or, instead, conditionally on $S\backslash A$. The second approach avoids boundary effects.

The maximum CPL estimator on S is

$$\widehat{\theta}_S = \underset{\theta \in \Theta}{\operatorname{argmax}}\, pl_S(x,\theta).$$

If $f_\theta(x) = Z(\theta)^{-1} h(x) \exp\{{}^t\theta v(x)\}$ belongs to an exponential family, the CPL is concave and strictly so if the model is identifiable at θ:

$$\text{if } \theta \neq \theta_0,\ \mu_S\{(\xi,x) : \lambda_\theta(\xi,x) \neq \lambda_{\theta_0}(\xi,x)\} > 0. \tag{5.41}$$

Example 5.18. Strauss process

For a Strauss PP (3.9) with density $f_\theta(x) = \alpha(\theta)\beta^{n(x)}\gamma^{s(x)}$, $\beta > 0$, $\gamma \in [0,1]$, the conditional intensity and CPL are respectively,

$$\lambda_\theta(\xi, x) = \beta\gamma^{s(\xi;x)}$$

and

$$pl_S(x,\theta) = \beta^{n(x)}\gamma^{s(x)} \exp\left(-\beta \int_S \gamma^{s(\xi,x)} d\xi\right),$$

where $s(\xi,x) = \sum_{i=1}^{n} \mathbf{1}\{\|\xi - x_i\| \leq r\}$.

If $\min_{i\neq j} |x_i - x_j| > r$, then $s(x) = 0$ and the CPL is maximal when $\gamma = 0$. Also, $\widehat{\gamma} > 0$ and we get $(\widehat{\beta}, \widehat{\gamma})$ by solving $pl_S^{(1)}(x,\widehat{\theta}) = 0$, i.e.,

$$n(x) = \beta \int_S \gamma^{s(\xi,x)} d\xi \quad \text{and} \quad \sum_{\xi \in x} s(\xi, x\backslash\{\xi\}) = \beta \int_S s(\eta,x)\gamma^{s(\eta,x)} d\eta.$$

If $\widehat{\gamma} > 1$, we take $\widehat{\gamma} = 1$ and $\widehat{\beta} = n(x)/\nu(S)$.

The advantage of CPLs is that they avoid having to calculate the normalizing constant of joint densities. It nevertheless remains necessary to calculate the factor $\Lambda_\theta(S,x) = \int_S \lambda_\theta(\xi,x)d\xi$. Baddeley and Turner (14) propose approximating this integral by:

$$\log pl_S(x,\theta) \simeq \sum_{i=1}^{n(x)} \log \lambda_\theta(x_i, x\backslash\{x_i\}) - \sum_{i=1}^{m} \lambda_\theta(u_j,x) w_j, \tag{5.42}$$

where u_j, $j = 1,\ldots,m$ are points of S and w_i the weights associated with the integration formula. If the set of u_j contains x, (5.42) can be rewritten

$$\log pl_S(x,\theta) \simeq \sum_{j=1}^{m} (y_j \log \lambda_j^* - \lambda_j^*) w_j, \tag{5.43}$$

where $\lambda_j^* = \lambda_\theta(u_j, x\backslash\{u_j\})$ if $u_j \in x$ and $\lambda_j^* = \lambda_\theta(u_j,x)$ otherwise, and $y_j = \mathbb{1}[u_j \in x]/w_j$. The expression on the right hand side of (5.43) is analogous to (5.37) and can therefore be maximized using GLM estimation software.

5.5.5.2 Asymptotic properties of the conditional pseudo-likelihood estimator for Gibbs point processes

Consistency of the conditional pseudo-likelihood estimator

For X a PP observed in a sequence of windows $(S(n))$, Jensen and Møller (121) establish consistency of estimation by maximum CPL when X is a Markov PP with bounded range R (the range of a potential φ is R if $\varphi(x)=0$ as soon as two points in x are more than R apart) with a translation-invariant conditional specification belonging to the exponential family:

$$f_\theta(x_S|x_{\partial S}) = \frac{1}{Z(\theta;x_{\partial S})} \prod_{\emptyset\neq y\subseteq x_S} \prod_{z\subseteq x_{\partial S}} \psi(y\cup z)\exp\{{}^t\theta\phi(y\cup z)\}.$$

Here, ∂S is the R-neighborhood of S, $\psi(x) \geq 0$, $\phi(x) \in \mathbb{R}^k$, with $\psi(x) = 1$ and $\phi(x) = 0$ if x is not a clique. Furthermore, one of the two following conditions has to hold:

(P1) $\forall\, x$ such that $n(x) \geq 2$, $\psi(x) \leq 1$ and ${}^t\alpha\phi(x) \geq 0$ for α in the neighborhood of the true value θ.
(P2) $\exists N$ and $K < \infty$ such that if $n(x) \geq 2$, $\psi(x) \leq K$, $\|\phi(x)\| \leq K$ and $n(x_{A_i}) \leq N$, where $\{A_i\}$ is a partition of bounded Borel sets that cover $S(n)$.

(P1) means that interactions are repulsive; (P2) allows consideration of attractive potentials in the case of Markov processes with hard-core potentials.

5.5.5.3 Asymptotic normality of the conditional pseudo-likelihood estimator

Exploiting a property that resembles that of a martingale difference sequence, Jensen and Künsch (123) proved that even under phase transition we have asymptotic normality of the CPL estimator. We now present their framework. Let X be a Gibbs PP with pair interactions, finite range R and density

$$f_\theta(x_S|x_{\partial S}) = \frac{1}{Z(\theta;x_{\partial S})} \exp\{-\theta_1 n(x_S) - \theta_2 \sum_{x_i,x_j\in x_S\cup x_{\partial S}} \phi(x_i - x_j)\}.$$

We associate with X a lattice process $X^* = (X_i^*)$, where $X_i^* = X \cap S_i$ is a partition $\mathbb{R}^d = \bigcup_{i\in\mathbb{Z}^d} S_i$, with $S_i = \tilde{R}\times(i+]-1/2,1/2]^d)$, $\tilde{R} > R$. If $D_n \subseteq \mathbb{Z}^d$ is an increasing sequence such that $\sharp\partial D_n/\sharp D_n \to 0$ with $\partial D_n = \{i\in D_n | \exists\, j\notin D_n : |j-i| = 1\}$, and if X is observed in the window $\bigcup_{i\in D_n\cup\partial D_n} S_i$, the estimated value is the one maximizing the pseudo-likelihood $pl_{S(n)}(x,\theta)$ of X calculated over $S(n) = \bigcup_{i\in D_n} S_i$.

We suppose that $\theta_2 > 0$ and that one of the two following conditions on ϕ holds:

(J1) $0 \leq \phi(\xi) < \infty$.
(J2) $\phi(\xi) = \kappa(\|\xi\|)$, where $\kappa : \mathbb{R}^+ \to \mathbb{R}$ satisfies:

1. $\kappa(r) \geq -K$ for $K < \infty$, $\kappa(r) = \infty$ if $0 \leq r < r_1$ for some $r_1 > 0$ and $\kappa(\cdot)$ is a $\mathscr{C}^1$ function except at a finite number of points.
2. For all $\theta > 0$, the function $\kappa'(r)\exp\{-\theta\kappa(r)\}$ is bounded.
3. $\kappa(r) \to k_0 \neq 0$ if $r \to R$ and its derivative $\kappa'(r) \to k_1 \neq 0$ if $r \to R$.

If furthermore X is stationary, Jensen and Künsch (123) showed that:

$$J_n^{-1/2}(\theta)I_n(\theta)(\widehat{\theta}_n - \theta) \xrightarrow{d} \mathscr{N}(0,I_2),$$

where the pseudo-information matrices are given by:

$$I_n(\theta) = -\frac{\partial^2 pl_{S(n)}(\theta)}{\partial^t\theta\partial\theta},$$

$$J_n(\theta) = \sum_{i\in D_n}\sum_{|j-i|\le 1,\, j\in D_n} \frac{\partial pl_{S_i}(\theta)}{\partial\theta}\, {}^t\left(\frac{\partial pl_{S_j}(\theta)}{\partial\theta}\right).$$

Mase (151) extended these results to second-order pseudo-likelihoods and to marked point processes.

5.5.6 Monte Carlo approximation of Gibbs likelihood

Let X be a Gibbs random field (*pointwise or lattice*) with distribution P_θ and density

$$\pi(x;\theta) = Z^{-1}(\theta)g(x;\theta),$$

where $Z(\theta) = \int g(x;\theta)\mu(dx) < \infty$. Using ML estimation is problematic if it is hard to calculate $Z(\theta)$. This is the case for Gibbs random fields. We now describe a Monte Carlo method to asymptotically calculate $Z(\theta)$. Let θ be the value of the parameter and $\psi \in \Theta$ some other fixed value of the parameter. We then estimate the ratio:

$$\frac{Z(\theta)}{Z(\psi)} = E_\psi\left[\frac{g(X;\theta)}{g(X;\psi)}\right]$$

using the strong law of large numbers under the distribution $\pi(\cdot;\psi)$. If we do not have access to an exact simulation method, we can use an MCMC method which like Gibbs sampling and the Metropolis-Hastings algorithm does not require knowledge of $Z(\psi)$. If x is generated from X, a Monte Carlo approximation of the logarithm of the likelihood ratio

$$l(\theta) = \frac{f(x;\theta)}{f(x;\psi)} = \frac{Z(\psi)g(x;\theta)}{Z(\theta)g(x;\psi)}$$

can be obtained by simulating N samples from X under ψ,

$$l_N(\theta) = \log\frac{g(x;\theta)}{g(x;\psi)} - \log\frac{1}{N}\sum_{j=1}^{N}\frac{g(X_j;\theta)}{g(X_j;\psi)}. \qquad (5.44)$$

This approximation converges to $l(\theta)$ when $N \longrightarrow \infty$. Importance sampling theory shows that the approximation improves with proximity of ψ to θ. We thus obtain $l_N(\theta)$ and $\widehat{\theta}_N$, the Monte Carlo approximations of $l(\theta)$ and $\widehat{\theta}$, the ML estimator. If $\pi(x;\theta) = Z(\theta)^{-1}\exp\{{}^t\theta v(x)\}$, (5.44) and its derivative are given by:

$$l_N(\theta) = {}^t(\theta-\psi)v(x) - \log N - \log \sum_{j=1}^{N} \exp\left[{}^t(\theta-\psi)v(X_j)\right],$$

$$l_N^{(1)} = v(x) - \sum_{j=1}^{N} v(X_i) w_{N,\psi,\theta}(X_i),$$

where $w_{N,\psi,\theta}(X_i) = \exp\{{}^t(\theta-\psi)v(X_i)\}\{\sum_{j=1}^{N} \exp[{}^t(\theta-\psi)v(X_j)]^{-1}$.

$\widehat{\theta}_N$, the value maximizing $l_N(\theta)$ depends on observation x and simulated $X_1,\ldots,X_N$ under ψ. However, it is possible to show that for large N, if $\widehat{\theta}$ is the exact ML estimator, the Monte Carlo error $e_N = \sqrt{N}(\widehat{\theta}_N - \widehat{\theta})$ is approximately normal (86). Thus, for a large number N of simulations, $\widehat{\theta}_N$ tends to $\widehat{\theta}$. One step in the Newton-Raphson algorithm is given by:

$$\theta_{k+1} = \theta_k - [l_N^{(2)}(\theta_k)]^{-1} \left\{ v(x) - \sum_{j=1}^{N} \frac{v(X_i)}{N} \right\},$$

where

$$-l_N^{(2)}(\theta_k) = \sum_{j=1}^{N} \frac{v(X_i)\,{}^t v(X_i)}{N} - \sum_{j=1}^{N} \frac{v(X_i)}{N}^{t} \left\{ \sum_{j=1}^{N} \frac{v(X_i)}{N} \right\}.$$

In effect, $w_{N,\psi,\theta}(X_i) = 1/N$ if $\psi = \theta$ and $-[l_N^{(2)}(\theta_k)]^{-1}$ estimates $Cov(\widehat{\theta})$.

As the approximation (5.44) improves the closer ψ is to θ, we can iterate starting from the maximum CPL estimation of θ.

5.5.6.1 Recursive algorithm for calculating the maximum likelihood estimate

Penttinen (169) applied the Newton-Raphson method to the Strauss process case. When there are a fixed number $n = n(x)$ of points of x and the density belongs to an exponential family, Moyeed and Baddeley (162) suggest resolving the ML equation $E_\theta[v(X)] = v(x)$ by a stochastic approximation method,

$$\theta_{k+1} = \theta_k + a_{k+1}\left[v(x) - v(X_{k+1})\right], \tag{5.45}$$

where a_k is a sequence of positive numbers satisfying $a_k \to 0$ and $X_{k+1} \sim f_{\theta_k}$ (cf. Younes (230) for Gibbs random fields on a regular network and Duflo (71) for general properties of such algorithms). If we do not have access to an exact simulation of X_k, we can use MCMC methods, which belong to the class of Markov stochastic algorithms (23).

Example 5.19. The Strauss process

For fixed $n(x) = n$, the density at $x = \{x_1, x_2, \ldots, x_n\}$ is proportional to $\exp\{-\theta s(x)\}$, with $\theta = \log\gamma$. Equation (5.45) becomes

$$\theta_{k+1} = \theta_k + a_{k+1}\left[s(X_{k+1}) - s(x))\right].$$

We can couple this equation with the MCMC algorithm for the purpose of simulating X_k. If $X_k = x$, the transition towards $X_{k+1} = y$ is the following:

1. `Delete one point` $\eta \in x$ `chosen uniformly in` x.
2. `Generate` ξ `from a density conditional on` $(x\backslash\{\eta\})$ `proportional to` $f((x\backslash\{\eta\})\cup\{\xi\})$ `and keep` $y = x\backslash\{\eta\}\cup\{\xi\}$.

This transition is that of Gibbs sampling with random sweeping on indices $\{1,\ldots,n\}$ and density conditional on $(x\backslash\{\eta\})$ proportional to $f(x\backslash\{\eta\})\cup\{\xi\})$.

5.5.6.2 Asymptotic properties of the maximum likelihood estimator for point processes

There are few results dealing with consistency and asymptotic normality of ML estimation. Jensen (122) gives a partial response to the question of asymptotic normality of ML when X is a PP with pair interactions and density

$$f_\theta(x) = \frac{1}{Z(\theta)}\exp\{-\theta_1 n(x) - \theta_2 \sum_{i<j}\phi(x_i - x_j)\},$$

with $-\infty < \theta_1 < \infty$, $\theta_2 > 0$ and when either:

(J1) X is a Markov process whose potential has bounded range.
(J2) X is a hard-core process.

Then if X satisfies a certain weak dependency condition, the ML estimator of θ is asymptotically normal.

Example 5.20. Model characterizing the spatial distribution of pine trees

The `swedishpines` dataset in the `spatstat` package gives the spatial distribution of 71 pine trees in a Swedish forest. Ripley (185) analyzed these using the Strauss model

$$f_\theta(x) = c(\theta)\exp\{\theta_1 n(x) + \theta_2 s(x)\},$$

where $s(x) = \sum_{i<j} 1(\|x_i - x_j\| \le r)$. We conclude the study by estimating the interaction radius r by the value $\widehat{r}$ maximizing the pseudo-likelihood profile

$$pl_A(r) = \max_\theta pl_A(x,\theta;r),$$

(cf. Fig. 5.22-a): we find $\widehat{r} = 7.5$. The maximum conditional pseudo-likelihood (cf. Table 5.14) is then calculated using approximation (5.42). For the maximum likelihood, we perform 1000 trials of $X^{(i)}$ obtained using the Metropolis-Hastings algorithm and stopping after 10000 iterations.

To finish, we validated the model by generating X 40 times from $f_{\widehat{\theta}_{MV}}$ (cf. Fig. 5.22-b).

Table 5.14 Maximum CPL and ML estimates of the Strauss model (with $r = 7.5$) for describing the location of 71 pine trees in a Swedish forest. Values in brackets are the estimated standard deviations. For the maximum CPL, the standard error was calculated using parametric bootstrap with 1000 trials.

	$\widehat{\theta}_1$	$\widehat{\theta}_2$
MCPL	-3.937 (0.211)	-1.245 (0.286)
ML	-3.760 (0.252)	-1.311 (0.321)

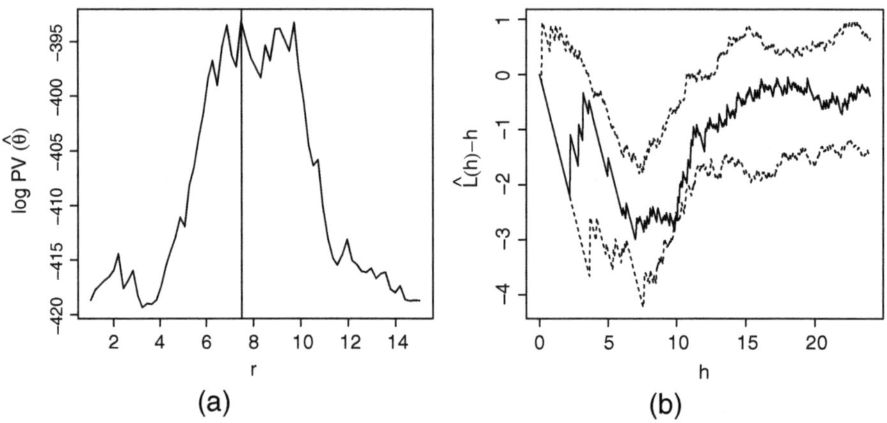

Fig. 5.22 Estimation and validation of the Strauss model for describing the spatial distribution of 71 pine trees: (a) value as a function of r of the profile pseudo-likelihood; (b) nonparametric estimation of $L(h) - h = \sqrt{K(h)/\pi} - h$ (solid line) compared with upper and lower confidence levels (dotted lines) obtained from 40 samples from a Strauss model with parameters $\widehat{\theta}_1 = -3.794$, $\widehat{\theta}_2 = -1.266$ and $r = 7.5$.

5.5.7 Point process residuals

Baddeley et al. (16) introduced h-residuals of Gibbs PPs for any test function $h: E \to \mathbb{R}$ defined on the sample space. These residuals are particularly useful for model validation. We now briefly present their definition and properties and refer the reader to (16) for a more in-depth presentation.

The definition of these residuals relies on the integral representation (3.17) of Gibbs PPs whose Papangélou intensity is $\lambda(\xi, x)$. For a chosen h, we define the *h-innovation* on a set B by:

$$I(B,h,\lambda) = \sum_{\xi \in x \cap B} h(\xi, x \backslash \{\xi\}) - \int_B h(\eta, x)\lambda(\eta, x) d\eta, \; B \subseteq S.$$

If h depends on the choice of model, it must be estimated before calculating residuals.

Baddeley et al. (16) look at the special case of three functions: $h=1$, $h=1/\lambda$ and $h=1/\sqrt{\lambda}$ representing respectively raw, λ-inverse and Pearson innovations:

$$I(B,1,\lambda) = N(B) - \int_B \lambda(\eta,x)d\eta,$$

$$I(B,\frac{1}{\lambda},\lambda) = \sum_{\xi\in x\cap B} \frac{1}{\lambda(\xi,x)} - \int_B 1[\lambda(\eta,x)>0]d\eta,$$

$$I(B,\frac{1}{\sqrt{\lambda}},\lambda) = \sum_{\xi\in x\cap B} \frac{1}{\sqrt{\lambda(\xi,x)}} - \int_B \sqrt{\lambda(\eta,x)}d\eta.$$

Using (3.17), it can be shown that $I(B,h,\lambda)$ is centered. If X is an inhomogeneous Poisson PP with intensity $\rho(\eta)$, then $\lambda(\eta,x)=\rho(\eta)$ and the variances of these innovations are:

$$Var(I(B,1,\rho)) = \int_B \rho(\eta)d\eta,$$

$$Var\left(I(B,\frac{1}{\rho},\rho)\right) = \int_B \frac{1}{\rho(\eta)}d\eta \text{ and } Var\left(I(B,\frac{1}{\sqrt{\rho}},\rho)\right) = |B|.$$

Note that the first equation equates the mean with the variance of $N(B)$.

For $\widehat{h}$ and $\widehat{\lambda}$ estimators of h and λ, the h-residuals are defined by:

$$R(B,\widehat{h},\widehat{\lambda}) = \sum_{\xi\in x\cap B} \widehat{h}(\xi,x\backslash\{\xi\}) - \int_B \widehat{h}(\eta,x)\widehat{\lambda}(\eta,x)d\eta, \qquad B\subseteq S.$$

As for standard linear models (with intercepts) where the sum of residuals is zero, we have the same "centered" property for the raw residuals $R(B,1,\widehat{\rho}) = N(x\cap B) - |B|\widehat{\rho}$ of a homogeneous Poisson PP with intensity ρ: if ρ is being estimated by ML, $\widehat{\rho} = N(x)/|S|$ and $R(S,1,\widehat{\rho}) = 0$.

The use of λ-inverse residuals was proposed by Stoyan and Grabarnik (203) while Pearson residuals are defined as in log-linear Poisson regression models.

Similarly, we can define innovations and residuals by replacing Papangélou's intensity λ by the PP intensity ρ for any test function $h(\cdot)$ such that

$$\int_S h(\eta)\rho(\eta)d\eta < \infty.$$

The following example gives some graphical tools useful for model validation. The heuristic behind these tools is based on the similarity of the logarithm of the likelihood of log-linear regression models for Poisson variables and the discretized version of the logarithm of the pseudo-likelihood of Gibbs PPs (cf. (5.43) and (5.37)).

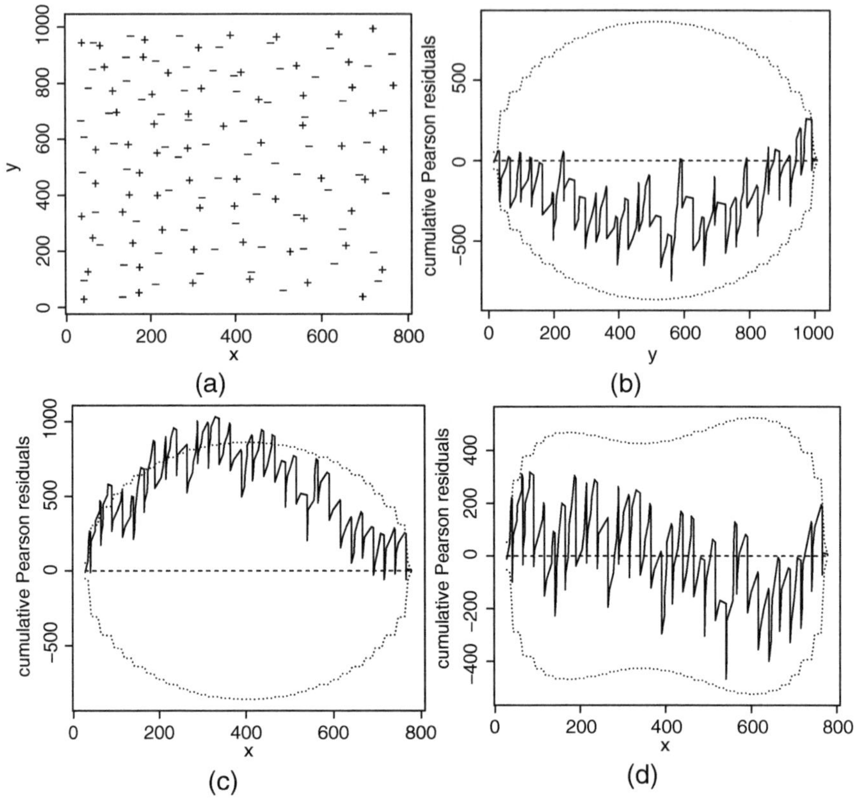

Fig. 5.23 (a) Spatial distribution of *on* (+) and *off* (−) ganglionic cells in the retina of a cat's eye; *diagnostic graphs*: (b) in y and (c) in x for a homogeneous Poisson PP model; (d) for inhomogeneous Poisson PP model (5.46). Dotted lines indicate the confidence band $C(v) \pm 2\sqrt{Var(C(v))}$.

Example 5.21. Spatial pattern of ganglionic cells in a cat's retina

Fig. 5.23-a (`betacells` dataset in `spatstat`) shows locations of beta-type ganglionic cells of a cat's retina. Ganglionic cells are sensitive to contrasts of light, some react to a thin beam of light surrounded by darkness (cell *on*), others to the opposite (cell *off*). The observation window is a $[0,1000] \times [0,753.3]$ μm^2 rectangle.

A study by van Lieshout and Baddeley (218) showed that there is repulsion between cells of the same type and that the *on* and *off* configurations are independent. Suppose therefore that both sets of locations are results of the same process to be modeled. As for standard linear models, *diagnostic tools* use the graph of residuals (y-axis) as a function of an observable spatial covariate or spatial coordinate of observation sites (x-axis). If in the graph there appears some pattern, this indicates inadequacy of the model. Associated with a spatial covariate $u(\eta)$, $\eta \in B$, let us define the subset of level $B(v) = \{\eta \in B : u(\eta) \leq v\}$ and the *cumulative residual function*:

$$C(v) = R(B(v), \widehat{h}, \widehat{\lambda}), \qquad v \in \mathbb{R}.$$

If the model is correct, $C(v)$ will be approximately equal to 0. Figures 5.23-b and c give *diagnostic curves* of Pearson residuals obtained from an estimated homogeneous Poisson PP, respectively as a function of y and x. Confidence bands $C(v) \pm 2\sqrt{Var(C(v))}$ can be calculated using the approximation $Var(C(v)) \approx Var(I(B(v),\widehat{h},\widehat{\lambda})$, where $\widehat{h}$ and $\widehat{\lambda}$ have been estimated under the hypothesis that X is a Poisson PP. Values outside of these bands for the curve relative to x suggest a spatial trend in x. Estimation of a second inhomogeneous Poisson PP model with intensity

$$\rho(\eta;\theta) = \exp\{\theta_1 + \theta_2 x_\eta\}, \qquad \eta = {}^t(x_\eta, y_\eta), \tag{5.46}$$

improves the results (cf. Fig. 5.23-d).

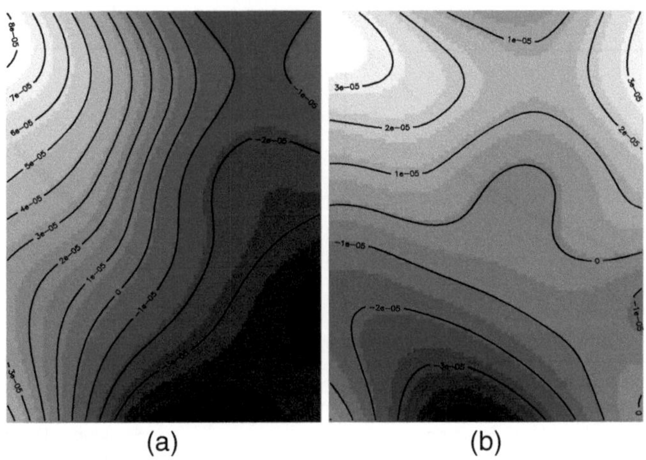

Fig. 5.24 Graphical representation of smoothed raw residuals for: (a) the homogeneous Poisson PP model; (b) the inhomogeneous Poisson PP model (5.46).

A second diagnostic tool is the *smoothed residuals field* (cf. Fig. 5.24):

$$l(\eta,x) = \frac{\sum_{x_i \in x} k(\eta - x_i)\widehat{h}(x_i, x\backslash\{x_i\}) - \int_S k(\eta-\xi)\widehat{\lambda}(\xi,x)\widehat{h}(\xi,x)d\xi}{\int_S k(\eta-\xi)d\xi},$$

where k is a smoothing kernel and the denominator a normalizing factor. In the case of raw residuals, the expectation of $l(\eta,x)$ is proportional to $\int_S k(\eta-\xi)E[\lambda(\xi,X) - \widehat{\lambda}(\xi,X)]d\xi$; positive values (resp. negative) suggest that the model has underestimated (resp. overestimated) the intensity function. Fig. 5.24-a shows that the homogeneous Poisson PP is inadequate and that the inhomogeneous Poisson PP model is better (cf. Fig. 5.24-b).

A *Q-Q* plot is a tool introduced by Baddeley et al. (16) for validating the interaction component of PPs. It works as follows: we simulate m examples $x^{(i)}$, $i = 1,\dots,m$ from the estimated model. For a grid of points η_j, $j = 1,\dots,J$ and for each trial $x^{(i)}$, $i = 0,\dots,m$ (with $x^{(0)} = x$), we calculate the smoothed residuals $l_j^{(i)} = l(\eta_j, x^{(i)})$ and the ordered sample $l_{[1]}^{(i)} \leq \ldots \leq l_{[J]}^{(i)}$. The observed quantiles $l_{[j]}^{(0)}$ are then compared with the means of the simulated quantiles $\sum_{i=1}^{m} l_{[j]}^{(i)}/m$. If the model is well adapted, the two quantiles should be almost equal. Fig. 5.25 gives a *Q-Q* plot for two different models: the previous inhomogeneous Poisson PP model and the Strauss model with conditional intensity:

$$\lambda(\eta, x; \theta) = \exp\{\theta_1 + \theta_2 x_\eta + \theta_3 \sum_{x_i \in x} 1(\|\eta - x_i\| \leq r)\}, \quad \theta = {}^t(\theta_1, \theta_2, \theta_3). \quad (5.47)$$

The intersection radius $\widehat{r} = 74$ is identified as the point at which the profile of the CPL is maximal. Fig. 5.25 invalidates the inhomogeneous Poisson PP model and suggests that the Strauss model is more appropriate.

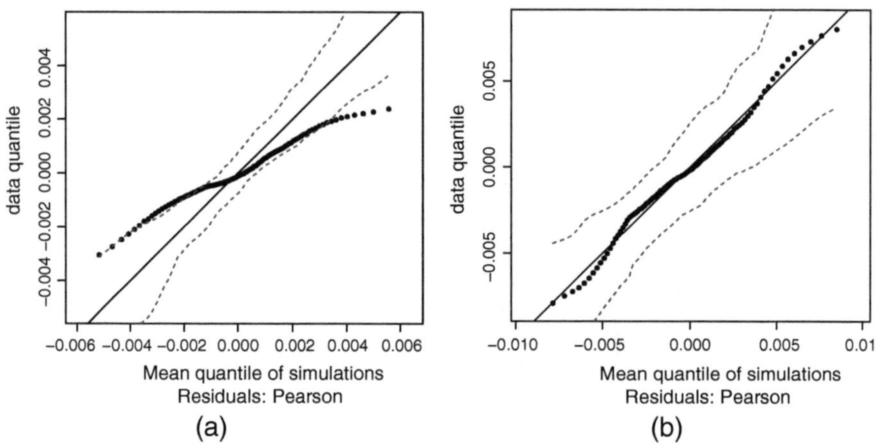

Fig. 5.25 *Q-Q* plot of Pearson residuals for: (a) the inhomogeneous Poisson PP model (5.46); (b) the Strauss model (5.47). Dotted lines show the 2.5 and 97.5 percentiles of the simulated quantiles, over one hundred simulations.

5.6 Hierarchical spatial models and Bayesian statistics

Given three random variables U, V and W, we can always decompose the joint distribution of the triplet (U, V, W) by successive conditioning,

$$[U, V, W] = [W|U, V][V|U][U],$$

where $[a]$ denotes the distribution of the variable a. This decomposition forms the basis of hierarchical modeling.

If the process of interest X is *not observed* and if observations Y are generated conditional on X, we can consider the hierarchical model:

$$[Y,X,\theta_Y,\theta_X] = [Y|X,\theta_Y,\theta_X][X|\theta_X][\theta_Y,\theta_X].$$

In this decomposition there are three levels of hierarchy. At level 1, we specify how observations Y are generated from X by giving the distribution of Y conditional on the process X and model parameters (θ_Y,θ_X). For level 2 we give the distribution of the process of interest X conditional on the parameters (θ_X). As for level 3, it specifies uncertainty we have on the parameters (θ_Y,θ_X). This methodology gives a certain liberty to the modeling, allowing us to incorporate simultaneously uncertainties and *a priori* knowledge on the observed phenomenon. When combined with developments in numerical MCMC methods we can understand the use and popularity of spatial Bayesian models.

The most common approach, which we follow here, supposes that observations $Y = {}^t(Y_{s_1},\dots,Y_{s_n})$ are, conditional on X, independent:

$$[Y|X,\theta_Y,\theta_X] = \prod_{i=1}^{n}[Y_{s_i}|X,\theta_Y,\theta_X],$$

where the distribution of X is that of a *spatial process*. X is usually some unobserved signal (image, shape, covariates) that we would like to extract (reconstruct) from observations Y. As in Bayesian image reconstruction (cf. (82)), we endow X with information in the form of a prior distribution.

For this type of model, Bayesian inference is oriented towards obtaining various *posterior distributions* $[X|Y]$, $[\theta_Y|Y]$ and *predictive distributions* $[Y_s|Y]$ when no observation has been made at s. In many models, analytic forms of these distributions are impossible to obtain and it is the conditional structure of the hierarchical model which allows them to be empirically evaluated using Monte Carlo methods and the MCMC algorithm (cf. Ch. 4).

Our goal here is not to give a general overview of such models but rather to sketch a description based on two examples: Bayesian kriging for spatial regressions and Bayesian analysis of GLMs with random spatial effects. We point the reader to the book of Banerjee et al. (18) for a complete treatment of spatial hierarchical models and (187) or (68) for Bayesian statistics in general.

5.6.1 Spatial regression and Bayesian kriging

Consider the regression with random spatial component X:

$$Y_{s_i} = {}^t z_{s_i}\delta + X_{s_i} + \varepsilon_{s_i}, \qquad i=1,\dots,n, \qquad \delta \in \mathbb{R}^p, \tag{5.48}$$

with $X = \{X_s\}$ an unobserved, centered Gaussian random field. Suppose for simplicity that X is stationary with covariance $c(h) = \sigma^2\rho(h,\phi)$ and that $\varepsilon = \{\varepsilon_s\}$ is a Gaussian WN with variance τ^2 representing small-scale local variability (cf. (1.31) for other possible choices of ε).

The first two levels of model (5.48) are specified by:

$$[Y|X,\theta_Y,\theta_X] = \mathcal{N}_n(Z\delta + X,\tau^2 I) \text{ and } [X|\theta_X] = \mathcal{N}_n(0,\sigma^2 R(\phi)),$$

where $R(\phi) = (\rho(s_i - s_j, \phi))_{i,j=1,n}$ is a correlation matrix.

Choosing Gaussian prior distributions on parameters allows explicit analytic calculations of posterior distributions and predictive distributions without requiring MCMC simulations (in this case we say we have conjugate distributions, cf. (187; 68)). Suppose for example that σ^2, ϕ and τ^2 are known and $\delta \sim \mathcal{N}_p(\mu_\delta, \Sigma_\delta)$. In this case, the distribution of δ conditional on Y is:

$$[\delta|Y, \sigma^2, \phi, \tau^2] = \mathcal{N}_p(\tilde{\delta}, \Sigma_{\tilde{\delta}}),$$

where $\tilde{\delta} = (\Sigma_\delta^{-1} + {}^tZ\Sigma^{-1}Z)^{-1}(\Sigma_\delta^{-1}\mu_\delta + {}^tZ\Sigma^{-1}Y)$, $\Sigma_{\tilde{\delta}} = (\Sigma_\delta^{-1} + {}^tZ\Sigma^{-1}Z)^{-1}$ and $\Sigma = \sigma^2 R(\phi) + \tau^2 I$. Furthermore, the predictive distribution of Y_s at an unobserved site s is given by:

$$[Y_s|Y, \sigma^2, \phi, \tau^2] = \int [Y_s|Y, \delta, \sigma^2, \phi, \tau^2][\delta|Y, \sigma^2, \phi, \tau^2] d\delta.$$

If $\{(Y_s, X_s)|\delta, \sigma^2, \phi, \tau^2\}$ is a Gaussian process, the universal kriging formula (1.36) gives:

$$[Y_s|Y, \delta, \sigma^2, \phi, \tau^2] = \mathcal{N}(\widehat{Y}_s, \sigma^2_{\widehat{Y}_s}), \text{ with:}$$

$$\begin{aligned}\widehat{Y}_s &= {}^tz_s\widehat{\delta} + {}^tc\Sigma^{-1}(Y - Z\widehat{\delta}), \quad c = Cov(X_s, X) \text{ and} \\ \sigma^2_{\widehat{Y}_s} &= \sigma^2 - {}^tc\Sigma^{-1}c + {}^t(z_s - {}^tZ\Sigma^{-1}c)({}^tZ\Sigma^{-1}Z)^{-1}(z_s - {}^tZ\Sigma^{-1}c).\end{aligned}$$

We deduce that $[Y_s|Y, \sigma^2, \phi, \tau^2]$ is a Gaussian distribution with mean and variance:

$$\begin{aligned}\mu^* &= ({}^tz_s - {}^tc\Sigma^{-1}Z)(\Sigma_\delta^{-1} + {}^tZ\Sigma^{-1}Z)^{-1}\Sigma_\delta^{-1}\mu_\delta + \\ &\quad \left[{}^tc\Sigma^{-1} + ({}^tz_s - {}^tc\Sigma^{-1}Z)(\Sigma_\delta^{-1} + {}^tZ\Sigma^{-1}Z)^{-1}\,{}^tZ\Sigma^{-1}\right] Y, \\ \sigma^{*2} &= \sigma^2 - {}^tc\Sigma^{-1}c \\ &\quad + ({}^tz_s - {}^tc\Sigma^{-1}Z)(\Sigma_\delta^{-1} + {}^tZ\Sigma^{-1}Z)^{-1}\,{}^t({}^tz_s - {}^tc\Sigma^{-1}Z).\end{aligned}$$

We remark that if we choose a relatively uninformative prior distribution on the parameter δ ($\Sigma_\delta \geq kI$ for large k), these formulae are merely those of universal kriging (set $\Sigma_\delta^{-1} = 0$).

5.6.2 Hierarchical spatial generalized linear models

The model (5.48) described in the previous paragraph is not useful for non-continuous data (number of ill people per region, reaching a certain level of pollution or not, binary presence/absence variables), nor for continuous data with strong asymmetry (maximum rainfall data). The model can be naturally extended by considering generalized linear models for the likelihoods (cf. (155)).

When studying geostatistical data, Diggle et al. (64) suggest examining the GLM

$$f(y_{s_i}|X_{s_i},\delta,\psi)=\exp\left\{\frac{y_{s_i}\eta_{s_i}-b(\eta_{s_i})}{\psi}+c(y_{s_i},\psi)\right\},\qquad i=1,\dots,n,\tag{5.49}$$

with $b'(\eta_{s_i})=\mathbb{E}(Y_{s_i}|X_{s_i},\delta,\psi)$ and for a link function g (to be chosen),

$$g(\mathbb{E}(Y_{s_i}|X_{s_i},\delta,\psi))={}^{t}z_{s_i}\delta+X_{s_i}.\tag{5.50}$$

Other specifications are possible (cf. (38)). Equations (5.49) and (5.50) characterize the first level of the hierarchical model. For the second level, we suppose that $\{X_s\}$ is a centered stationary Gaussian process with covariance $c(h)=\sigma^2\rho(h,\phi)$.

By taking a look at linear (5.48) and nonlinear models, we can see how introducing random spatial effects X on the mean (the transformed mean) induces a relationship between the conditional means of $(Y_{s_i}|X_{s_i})$ at neighboring sites without inducing correlations between these variables. In this sense, spatial hierarchical models are significantly different to spatial auto-models that do indeed induce spatial correlation.

In general, posterior and predictive distributions are not analytically tractable. We must therefore use MCMC algorithms that exploit the hierarchical conditional structure of models in order to evaluate these distributions. Without going too much into the details of the choice of proposed change distributions (cf. (64) and (44) for modifications to create specific models), let us describe a Metropolis algorithm for dealing with model (5.49). With the symbol $\propto$ signifying "proportional to," we will use the results:

$$\begin{aligned}
[\sigma^2,\phi|Y,X,\delta] &\propto [X|\sigma^2,\phi][\sigma^2,\phi],\\
[X_{s_i}|Y,X_{s_j},\delta,\sigma^2,\phi;s_j\neq s_i] &\propto [Y|X,\delta][X_{s_i}|X_{s_j};\sigma^2,\phi,s_j\neq s_i]=\\
&\prod_{i=1}^{n}[Y_{s_i}|X_{s_i},\delta][X_{s_i}|X_{s_j};\sigma^2,\phi,s_j\neq s_i],\\
[\delta|Y,X,\beta,s_j\neq s_i] &\propto [Y|X,\delta][\delta]=\prod_{i=1}^{n}[Y_{s_i}|X_{s_i},\delta][\delta].
\end{aligned}$$

One iteration of the Metropolis algorithm for simulating $((\sigma^2,\phi),X,\delta\mid Y)$ is given by the following three steps:

1. For (σ^2,ϕ) with current values $\sigma^{2\prime},\phi'$:
   (a) Propose $\sigma^{2\prime\prime},\phi''$ each from independent uniform distributions.
   (b) Accept $\sigma^{2\prime\prime},\phi''$ with probability

$$r(X,(\sigma^{2\prime},\phi'),(\sigma^{2\prime\prime},\phi'')) = \min\left\{1, \frac{[X|\sigma^{2\prime\prime},\phi'']}{[X|\sigma^{2\prime},\phi']}\right\}.$$

2. For X_{S_i}, for $i = 1,\ldots,n$,

   (a) Propose X''_{s_i} sampled from the distribution $[X_{s_i}|X'_{s_j};\sigma^2,\phi,s_j \neq s_i]$ (this is the same as simulating a conditional Gaussian random variable with simple kriging (cf. (1.35)).
   (b) Accept X''_{s_i} with probability

$$r(X_{s_i}{}',X_{s_i}{}'',Y;\delta) = \min\left\{1, \frac{[Y_{s_i}|X_{s_i}{}'',\delta]}{[Y_{s_i}|X_{s_i}{}',\delta]}\right\}.$$

3. For δ, with current value δ',

   (a) Propose δ'' following the distribution $[\delta''|\delta']$.
   (b) Accept δ'' with probability

$$r(\delta',\delta'') = \min\left\{1, \frac{\prod_{i=1}^n [Y_{s_i}|X_{s_i},\delta''][\delta'|\delta'']}{\prod_{i=1}^n [Y_{s_i}|X_{s_i},\delta'][\delta''|\delta']}\right\}.$$

Initial values of σ^2,ϕ and δ are chosen to be compatible with the prior distributions. As for the initial values $\{X_{s_i}, i = 1,\ldots,n\}$, we could choose $X_{s_i} = g(Y_{s_i}) - {}^t z_{s_i}\delta$ for $i = 1,\ldots,n$.

Evaluating the predictive distribution of X_s at an unobserved site s necessitates an extra step: as $[X_s|Y,X,\delta,\sigma^2,\phi] = [X_s|X,\sigma^2,\phi]$, we proceed by simulating a conditional Gaussian variable whose mean is calculated by simple kriging at s, this for values $X^{(k)},\sigma^{2(k)}$ and $\phi^{(k)}$ once the algorithm has entered its stationary regime.

Example 5.22. Spatial distribution of animal species

Studies of spatial distribution of animal species can be used to measure the impact of human activity on the environment. The data investigated here (cf. indigo data on the website) come from a study by Strathford and Robinson (205) whose goal was to determine what were the "environmental parameters" able to predict occupancy across a region by a migratory bird species, *Indigo Buntings*. Four land use factors, measured in the neighborhood of sites where birds were counted were kept: natural woodlands, denoted M, open parks, hayfields and pasture, denoted G, early successional forests, denoted T and impervious surfaces (roads, houses, carparks), denoted U. The sampling map is shown in Fig. 5.26.

For a hierarchical model, we consider that the number of birds Y_s at s is a Poisson random variable conditional on the observed factors (G,M,T,U) and on an unobserved spatial process X:

$$[Y_s|\delta,X_s] = \mathscr{P}(\mu_s),$$

where

$$\mu_s = \delta_1 + \delta_2 G_s + \delta_3 M_s + \delta_4 T_s + \delta_5 U_s + X_s,$$

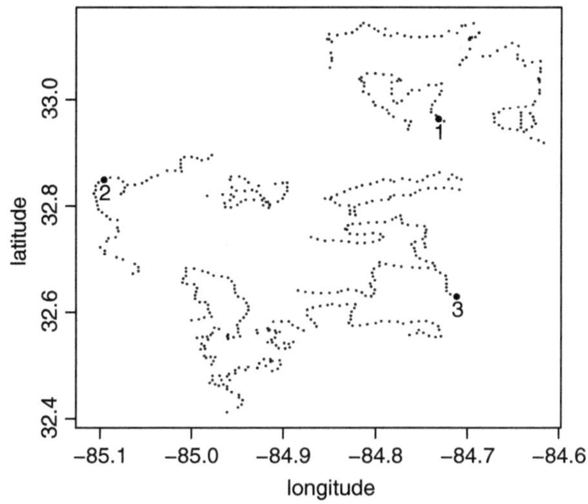

Fig. 5.26 Bird observation sites. The 3 sites • are the ones kept for prediction, with site 1 being where the largest number of birds was seen.

with X an isotropic centered Gaussian process with covariance

$$c(h) = \sigma^2 \exp\{-\|h\|/\phi\},$$

with unknown $\sigma^2 > 0$ and $\phi > 0$. The prior distributions on δ and σ^2 are chosen to be uninformative, i.e., with large variance, and independent. A preliminary analysis led to the choice of a uniform distribution ϕ on $[0.03, 0.20]$. To obtain the provisional posterior distributions, we use the MCMC algorithm from the `geoRglm` package with a "time to stationarity" of 1000 iterations for each simulation, of which there are 50000 with subsampling at every 50 steps. Estimations displayed in Fig. 5.27 show that all the "soil" factors as well as the spatial factor are significant, though the spatial effect is not huge.

We also give the predicted distributions at sites 1, 2 and 3, with site 1 being that with the largest observed number of birds. We see in Fig. 5.28 that the model characterizes well possible absence of birds but has problems where there are large numbers (as at site 1).

GLM (5.49) has been successfully applied in a variety of situations: epidemiology (spatial distribution of disease risk), image analysis (reconstruction) and ecology (spatial distribution of species).

We now take a look at an epidemiological example, that of the study of spatial variations in the risk of rare diseases or the risk of a disease in a small geographic area. This is a realistic example as neighboring zones tend to have similar risks as they are likely to share common risk factors.

For level 1, we model the number of cases (Y_i) of the disease in region i by independent Poisson variables

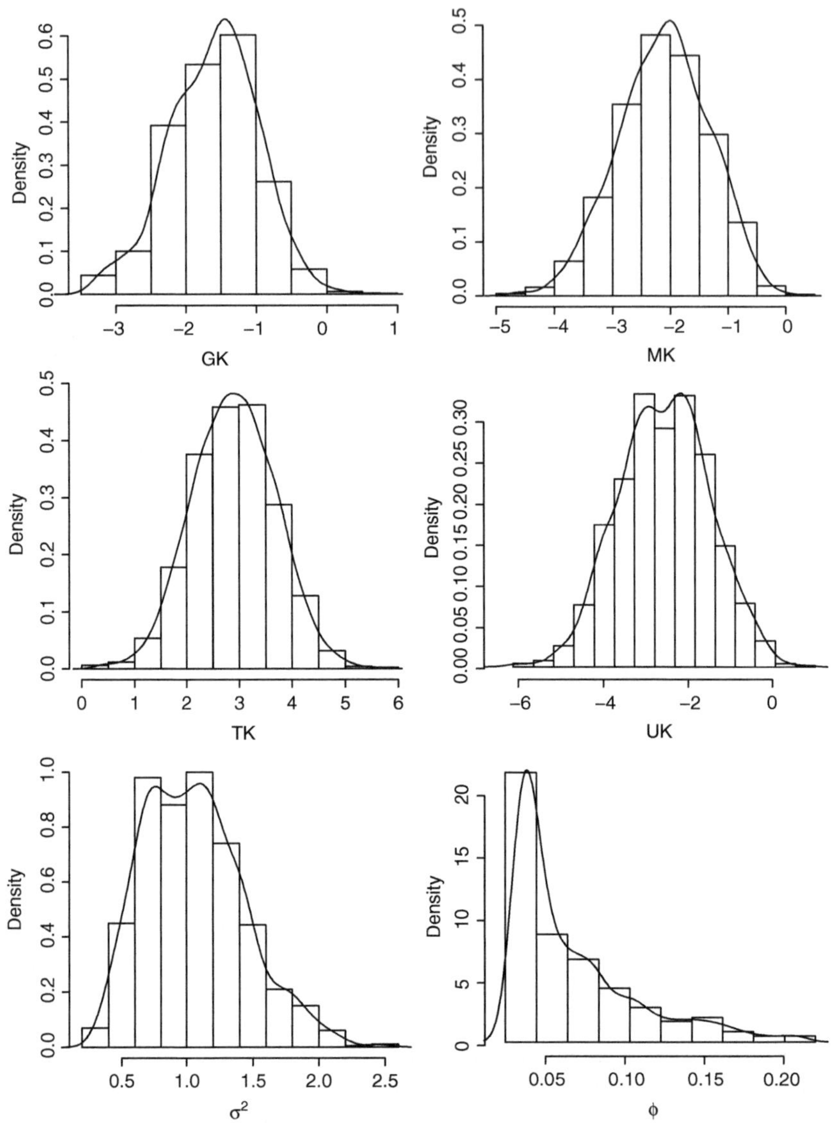

Fig. 5.27 Estimation of posterior distributions of parameters δ_G, δ_M, δ_T, δ_U as well as σ^2 and ϕ.

$$Y_i \sim \mathscr{P}(\theta_i E_i),$$

with E_i representing the expected number of cases expected in region i, precalculated using known socio-demographic covariates for each region. We could equally use a binomial model for Y_i with the maximum possible value being the population

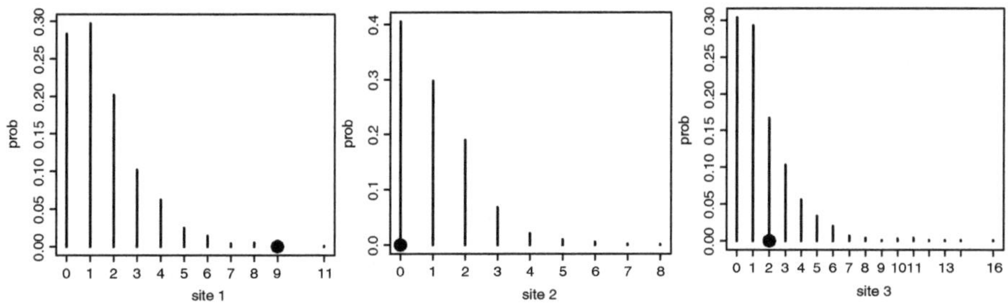

Fig. 5.28 Estimation of the predictive distributions of the number of birds at sites 1, 2 and 3. • represents the observed value.

size u_i in the region. The parameter θ_i, representing the *relative risk* specific to zone i is unknown and is the subject of the study. The maximum likelihood estimator of θ_i, $\widehat{\theta_i} = Y_i/E_i$, which does not take into account spatial structure, has variability proportional to $1/E_i$.

At level 2, the parameter θ_i incorporates either spatial heterogeneity or spatial dependency. Let us take a look at the model proposed by Besag et al. (28): the parameter θ_i of the Poisson regression is considered in a log-linear model

$$\log(\theta_i) = \alpha_i + \sum_{k=1}^{p} \beta_k x_{ik} + \gamma_i, \tag{5.51}$$

with random effects α_i and γ_i representing certain interpretations of the observable covariates (x_k):

- $\alpha_i \sim \mathcal{N}(0, 1/\tau^2)$ i.i.d. represent an unstructured spatial heterogeneity component.
- $(\gamma_i, i \in S)$ follow a structured spatial model, for example an intrinsic Gaussian CAR model:

$$\mathcal{L}(\gamma_i|\gamma^i) \sim \mathcal{N}\{(\sharp\partial i)^{-1} \sum_{j\in\partial i} \gamma_j, (\kappa^2 \sharp\partial i)^{-1}\}.$$

In this formulation, $\kappa^2 > 0$ is a parameter controlling the spatial smoothness of the γ_i and thus the smoothness of the θ_i too. This model is identifiable if the exogenous x have no constant term. If they do, we add the constraint $\sum_{i\in S} \gamma_i = 0$.

At level 3, we model τ^2 and κ^2 with prior distributions that are either Gamma or χ^2 distributions with fixed hyperparameters. Using this conditional Markov formulation then allows us to work with MCMC algorithms.

Under this model, the parameters τ^2 and κ^2 characterizing spatial dependency have a global effect that can potentially hide possible discontinuities due to overly strong “spatial smoothing.” For this reason, Green and Richardson (93) developed a model that replaces the CAR model with a Potts model (2.5), thus allowing inclusion of spatial discontinuity.

Example 5.23. Lung cancer in Tuscany (Italy)

The data presented here (cf. the Tuscany cancer data on the website) come from an epidemiological study of Catelan et al. (39) on lung cancer in men living in the 287 municipalities of Tuscany (Italy) born between 1925 and 1935 and dying between 1971 and 1999. A goal of the study was to see whether risk of cancer is linked to environmental effects and/or lifestyle. We denote by Y_i the observed number of deaths in municipality i and E_i the expected number of deaths. The spatial pattern of risk is estimated by $\widehat{R_i} = Y_i/E_i$, the standardized mortality ratio. This estimate is the maximum likelihood estimate for a Poisson log-linear model for independent random variables with unknown parameters R_i,

$$Y_i \sim \mathscr{P}(\theta_i), \tag{5.52}$$

$$\log(\theta_i) = \log(R_i) + \log(E_i). \tag{5.53}$$

Fig. 5.29-a shows higher estimates in the north and south-west but there are nevertheless large fluctuations. These may be due to local heterogeneity linked to environment and lifestyle. We therefore use a random effects model:

$$\log(\theta_i) = \alpha_i + \beta_1 + \gamma_i + \log(E_i), \tag{5.54}$$

with i.i.d. $\alpha_i \sim \mathscr{N}(0, 1/\tau^2)$ representing unstructured spatial heterogeneity components and the $(\gamma_i, i \in S)$ following an intrinsic Gaussian CAR model (5.51) with constraint $\sum_{i \in S} \gamma_i = 0$. Lastly, we model β_1 as a centered Gaussian random variable with variance 100000^2 and τ^2 and κ^2 by Gamma prior distributions with parameters 0.5 and 0.0005 corresponding to uninformative distributions.

In this model, the marginal posterior distributions of parameters are approximated using MCMC. To do this, we use Gibbs sampling in OpenBUGS (212), the open source version of WinBUGS (146) as well as the `R2WinBUGS` interface (209), allowing the MCMC algorithm to run in `R`. To have some control over convergence of Gibbs sampling, we use two independent chains and the procedure suggested by Gelman and Rubin (80) as described in §4.5.2. The ‘burn-in’ period used here was 2000 iterations for each simulation with each estimation made up of 8000 simulations. Fig. 5.29-b shows clearly the smoothing effect of model (5.54).

Models with covariates: among lifestyle indicators, that which is retained here is the education score ED_i, representing the quotient of the number of illiterate people divided by the number of people knowing how to read but not having successfully finished school (no school certificate received). The larger this is, the lower the level

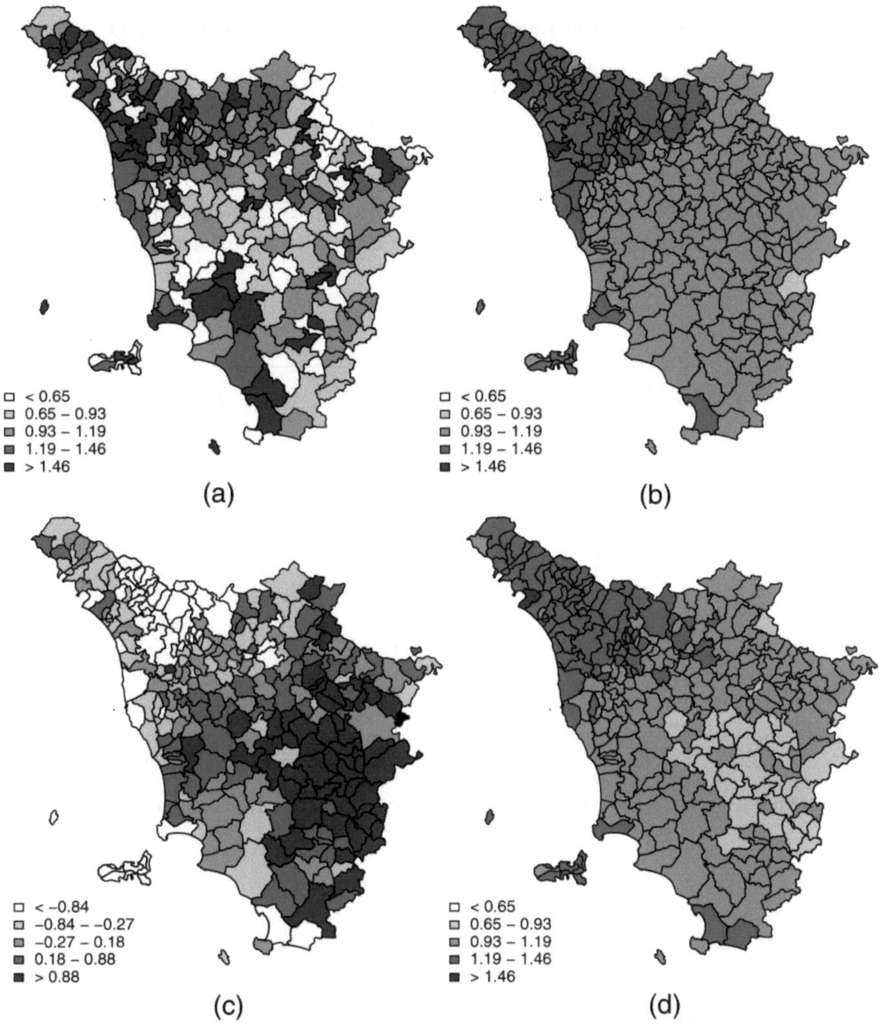

Fig. 5.29 (a) Spatial distribution of standardized mortality rates from lung cancer in the 287 municipalities of Tuscany (Italy); (b) median posterior risks estimated using model (5.53); (c) spatial distribution of the education score in 1951; (d) median posterior risks estimated by model (5.54) that includes the education score.

of school achievement. We also suppose that there is no exposure risk to cancer before the age of 20. Here, mortality is associated with the education score observed 20 years later and derived from the 1951 census.

Fig. 5.29-c shows the spatial distribution of the *ED* score. It is obvious that places with higher levels of school achievement in the northwest correspond to places where industrialization took place earlier and where the mortality rate is higher. However, the empirical correlation between $\widehat{R_i}$ and ED_i of -0.20 is relatively weak.

We therefore consider a log-linear model integrating the covariate ED,

$$\log(\theta_i) = \alpha_i + \beta_1 + \beta_2 ED_i + \gamma_i + \log(E_i), \tag{5.55}$$

where β_2 is a centered Gaussian random variable with variance 100000. Results obtained using MCMC (cf. Fig. 5.29-d) show median posterior risks that are less smooth but more realistic than those given in model (5.53); estimation of β_2 (cf. Fig. 5.30) shows a negative link between lung cancer mortality rate and education level.

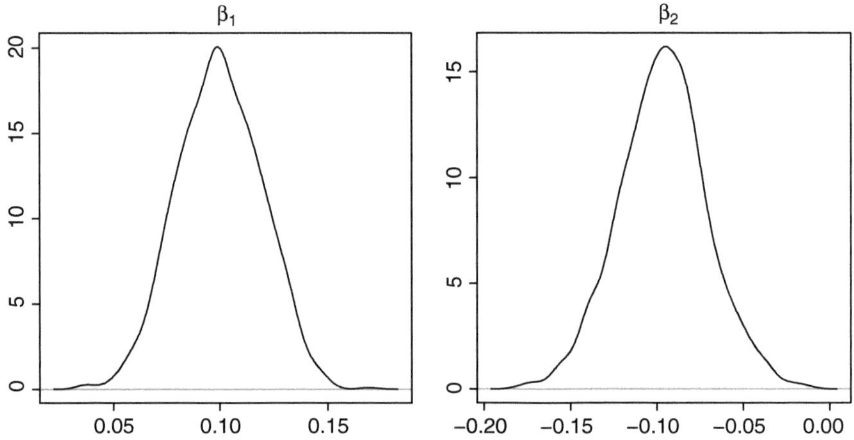

Fig. 5.30 MCMC estimates of posterior distributions of parameters β_1 and β_2 in model (5.55).

Exercises

5.1. Soil retention potentials.

The `potentials` dataset (see website) comes from an experiment in which soil samples from zone S are saturated with water at three different pressures and measure the retention potentials (`W5`, `W200` and `W1500`). We also have data on each sample corresponding to particle size fractions of clay (`ARG`) and four types of silt going from finest to coarsest (`LF`, `F3`, `F4.5`, `F6.7`). We aim to link the retention potential variables with the easily measured porosity variables (granulometry). We first consider the variable `W1500`.

1. Empirically estimate the variogram for various directions.
2. Correct any discovered anisotropy by introducing granulometric covariates.
3. Suggest variogram models and estimate them. Choose one model using the AIC criterion.
4. Propose methods for drawing a map of retention potential.
5. Are there differences between the potentials measured at each of the 3 pressures?

5.2. Moran's index.

The sample $(4,-2,-4,0,-1,3)$ is generated from X on the set $S=\{1,2,\ldots,6\}$ endowed with the symmetric graph with 10 edges $\mathcal{G}=\{\langle 1,2\rangle,\langle 1,3\rangle,\langle 1,4\rangle,\langle 1,6\rangle,\langle 2,3\rangle,\langle 3,6\rangle,\langle 2,6\rangle,\langle 3,4\rangle,\langle 4,5\rangle,\langle 5,6\rangle\}$. Calculate Moran's index I^M when we have weights $w_{i,j}=1$ if $\langle i,j\rangle$, $w_{i,j}=0$ otherwise. Calculate the mean and variance of I^M under the permutation distribution. Compare with results from a normal approximation.

5.3. Calculating $E(I^M)$ and $Var(I^M)$ under Gaussian hypotheses.

1. Let $X=(X_1,X_2,\ldots,X_n)$ be n samples from $\mathcal{N}(\mu,\sigma^2)$ and $Z_i=X_i-\overline{X}$ for $i=1,\ldots,n$. Show that for distinct indices i, j, k and l:

$$E(Z_i)=0,\quad E(Z_i^2)=\left(1-\frac{1}{n}\right)\sigma^2,\quad E(Z_iZ_j)=-\frac{\sigma^2}{n},$$

$$E(Z_i^2Z_j^2)=\frac{n^2-2n+3}{n^2}\sigma^2,\ E(Z_i^2Z_jZ_k)=-\frac{n-3}{n^2}\sigma^4,\ E(Z_iZ_jZ_kZ_l)=\frac{3}{n^2}\sigma^4.$$

2. Use Proposition 5.5 to deduce the expectation and variance of Moran's index under independence hypotheses.
3. Suppose that X is a sample but without hypotheses supposing that the X_i follow the same distribution. Denoting $\mathbb{E}_P$ the expectation of the permutation distribution, show that:

$$\mathbb{E}_P(Z_i)=0,\ \mathbb{E}_P(Z_i^2)=m_2=\frac{1}{n}\sum_{i=1}^n z_i^2,\ \mathbb{E}_P(Z_iZ_j)=-\frac{m_2}{n-1},\quad \text{if } i\neq j.$$

Deduce that

$$\mathbb{E}_P(I_n^M)=-\frac{1}{n-1}.$$

5.4. Limit distribution of Moran's index.

Find the asymptotic distribution of Moran's index under independence hypotheses and with weights $w_{i,j}=1$ if $\langle i,j\rangle$, $w_{i,j}=0$ otherwise for the following graphs:

1. $S=\mathbb{Z}^2$ and the 4-NN relation (resp. 8-NN).
2. S the regular triangular network in $\mathbb{R}^2$ with the 6-NN relation.
3. $S=\mathbb{Z}^d$ and the $2d$-NN relation.

5.5. Test for factorizing covariances.

Suppose X is an 8-NN stationary centered Gaussian CAR model on $\mathbb{Z}^2$.

1. Characterize this model and the factorizing covariance submodel denoted (F).
2. Suppose that X is observed over the square with side n. Give ML and MCPL equations for each model. Test (F). Determine under (F) the asymptotic distribution of

$$\Delta=n\{(\widehat{r}_{00}\widehat{r}_{11}-\widehat{r}_{10}\widehat{r}_{01})^2+(\widehat{r}_{00}\widehat{r}_{1,-1}-\widehat{r}_{-1,0}\widehat{r}_{01})^2\}.$$

5.6. Estimating 4-NN CAR models.

Suppose that X is a 4-NN stationary centered Gaussian CAR model with parameter $\theta = (\alpha, \beta, \sigma^2)$, where σ^2 is the conditional residual variance and $|\alpha| + |\beta| < 1/2$. Suppose that X is observed on the square of side n.

1. Find the asymptotic distribution of the ML (resp. MCPL) estimator of θ if the data are sufficiently tapered at the boundary. Test for isotropy.
2. Same question if $Y = X + \varepsilon$, with ε a Gaussian WN(σ^2) independent of X.

5.7. Auto-logistic modeling of the spatial distribution of a plant species.

A We measure presence ($X_s = 1$) or absence ($X_s = 0$) of the sedge (plant species) in a swamp subdivided into a regular 24×24 grid (cf. `laiche` dataset on the website). We would like to fit the following translation-invariant auto-logistic models: (i) 8-NN (5 parameters); 4-NN (3 parameters); (iii) isotropic 4-NN (2 parameters).

1. Estimate these models (parameters and variance) using ML, MCPL and coding estimators. Test for 4-NN isotropy in the 8-NN model using deviation of the likelihood and a χ^2 coding test. Test for this same isotropy in the 4-NN model.
2. Which model would you choose after using the CPL penalized contrast at rate $\sqrt{n}$ (in the case of n observations)?

B Wu and Huffer (226) studied the spatial distribution of presence/absence of plant species as a function of climate covariates on a domain S with $n = 1845$ sites, a subset of a regular 68×60 network (cf. `castanea` dataset on the website). The 9 covariates are: TMM (minimum annual temperature, in Celsius), TM (mean temperature for the coldest month), TAV (mean annual temperature), LT (lowest temperature in the period 1931–1990), FZF (number of days with frost), $PRCP$ (mean annual precipitation in mm), MI (annual index of mean humidity), $PMIN$ (mean precipitation in the driest months) and ELV (altitude in feet).

1. Plot the network S and show presence/absence of *Castanea pumila*.
   We now look at regression/autoregression models with the climate covariates x_i and the 4-NN covariate $v_i = \sum_{j \sim i} y_j$:
$$P(Y_i = 1 \mid y^i, x) = \frac{\exp \eta_i}{1 + \exp \eta_i},$$
   where $\eta_i = a + {}^t x_i b + c v_i$.
2. Fit a logistic regression using the climate covariates x_i. Using AIC, which ones would you keep?
3. Fit a complete logistic regression/autoregression model using ML, CPL and coding estimators. Is the neighbor covariate v significant? Using penalized log-CPL, which climate covariates would you keep? (hint: notice that introduction of the neighbor variable simplifies the model).

4. Suppose we keep model 3, the one with only the covariates TAM, PR, MI and v. In order to compare ML, CPL and coding estimates, Wu and Huffer suggest measuring the distance $DMA = \sum_i |y_i - \widehat{y}_i|$ between observations y and their predictions $\widehat{y}$ obtained by each estimation method ($\widehat{y}_i$ is the empirical mean at i of Gibbs sampling simulations at the estimated parameter value). Compare the quality of prediction of the three methods.

5.8. Several χ^2 coding tests.
Suppose that X is a Markov random field with translation-invariant specification on $S = \{0,1,\dots,n-1\}^2 \subset \mathbb{Z}^2$.

1. Give the test for isotropy if X is the 4-NN Ising model.
2. Suppose that a 4-NN isotropic model with K states $E = \{1,2,\dots,K\}$ has the following conditional energy at i: if $x_i = k$,

$$h_i(k, x_{\partial i}) = \alpha_k + \sum_{l:l\neq k} \beta_{kl} n_i(l),$$

where $n_i(l) = \sum_{j\in\partial i} \mathbf{1}(X_j = l)$.
We impose the following identifiability conditions: $\alpha_K = 0$, for $k \neq l$, $\beta_{kl} = \beta_{lk}$ and for all l, $\beta_{Kl} = 0$. Test the interchangeability hypothesis (E) : β_{kl} is constant at all $k \neq l$.
3. Suppose X is a V-Markov random field with K states and translation-invariant specification, where $V = \partial 0 \subset \mathbb{Z}^2$ is finite and symmetric, $0 \notin V$.
   (a) Give a general model for X (make some choice of V and K).
   (b) Characterize the following submodels: (a) cliques have at most 2 points; (b) the model is isotropic; (c): (a)∩(b); (d) the model is auto-binomial on $E = \{0,1,\dots,K-1\}$.
   (c) Propose tests of these submodels with respect to the general model.

5.9. Estimating Markov random field dynamics.
Suppose we have $X = (X(t), t \in \mathbb{N})$, $X(t) = (X_i(t), i \in S) \in \{0,1\}^S$ some dynamic on $S = \{1,2,\dots,n\}$ modeled by a homogeneous and ergodic Markov chain. The transition $x \mapsto y$ of these dynamics is defined by:

$$p(x,y;\theta) = Z^{-1}(x,\theta)\exp U(x,y,\theta), \text{ with}$$

$$U(x,y;\theta) = \sum_{i=1}^{n} y_i\{\alpha x_i + \beta(x_{i-1} + x_{i+1}) + \gamma(y_{i-1} + y_{i+1})\}.$$

1. Calculate the conditional distributions $\mathscr{L}(X_i(t) \mid x^i(t), x(t-1))$. Deduce the associated CPL.
2. For some choice of network S and some θ, simulate these dynamics at T successive instants of time, then estimate θ by CPL.

5.10. A Markov growing stain model.
A growing stain model involves defining an increasing sequence $A = \{A(t), t = 0,1,2,\dots\}$ of finite subsets of $\mathbb{Z}^2$. Markov dynamics can be characterized by the

distribution of the initial stain $A(0) \neq \emptyset$ and the Markov transitions $P(a(t-1), a(t))$ for $t \geq 1$. These dynamics are of the NN kind if $a(t-1) \subseteq a(t) \subseteq a(t) \cup \partial a(t-1)$, where ∂B is the NN boundary of B. These NN dynamics are local and independent if the transition satisfies:

$$P(a(t-1), a(t)) = \prod_{i \in \partial a(t-1)} P(a_{\partial i}(t-1), a_i(t)).$$

1. Propose a NN Markov dynamic. Simulate it for a given stain $a(0)$. Give the CPL of this model.
2. Same question for local and independent dynamics. Show convergence of the CPL estimation and give the rate of convergence.
3. Test whether a NN Markov dynamic is local and independent.

5.11. Independence of bivariate spatial random variables.
Suppose that (U, V) is a 4-NN isotropic Markov random field on $S = \{1, 2, \ldots, n\}$ with states $(U_i, V_i) \in \{0, 1\}^2$. This model is translation-invariant.

1. Show that the general model has 12 parameters, with singleton and pair potentials:

$$\phi_1(u, v) = \alpha u + \beta v + \gamma uv,$$
$$\phi_2((u, v), (w, t)) = \delta_1 uw + \delta_2 vt + \delta_3 ut + \delta_4 vw + \delta_5 uwt + \delta_6 vwt$$
$$+ \delta_7 uvw + \delta_8 uvt + \delta_9 uvwt.$$

2. Present and then test the submodel for independence of U and V.
3. Let (ω) be the submodel where, in ϕ_2, only δ_1 and δ_2 are non-zero. Find the conditional distributions $\pi_i(u_i, v_i \mid \cdot)$ and $\nu_i(u_i \mid \cdot)$ with conditioning on all the other observations. Construct a χ^2 coding test of the independence subhypothesis for U and V based on the conditional distributions $\nu_i(\cdot \mid \cdot)$.

5.12. Gaussian spatial regression.
Consider the Gaussian spatial regression model $Z = X\beta + \delta$, Z and $\delta \in \mathbb{R}^n$, $\beta \in \mathbb{R}^p$, where δ is the SAR model

$$(I - \varphi W)\delta = \varepsilon,$$

with ε a Gaussian WN with variance σ^2.

Give the information matrix $J_{p+2}(\theta)$ of $\theta = (\beta, \sigma^2, \delta)$. Same question if δ is a CAR model.

5.13. Experimental design on random fields.
Suppose that p treatments with mean effects $\mu = {}^t(\mu_1, \mu_2, \ldots, \mu_p)$ are applied, each r times, thus covering $n = r \times p$ zones of a field $S = \{1, 2, \ldots, n\}$. We observe Y with mean $D\mu$, where D is the $n \times p$ matrix defining the coordinates of applied treatments (D satisfies: ${}^tDD = rI_p$). Suppose that the error process $X = Y - D\mu$ is a spatial CAR model:

$$X = \beta W X + e,$$

where $W_{ij} = 1$ if $|i-j| = 1$ and $W_{ij} = 0$ otherwise.

1. Estimate μ by OLS. Deduce an estimation $\widehat{X}$ of the residuals.
2. Estimate β by *LS* conditional on the basis $\widehat{X}$.
3. Deduce the $GLS(\widehat{\beta})$ estimation of μ and interpret the result.

5.14. Pointwise spatial distribution of child leukemia.
An important question in spatial epidemiology is to know whether or not the spatial distribution of sick individuals (*cases*) is the same as that of the general population (*controls*). The `humberside` dataset in the `spatstat` package gives the location of 62 cases of child leukemia and 141 homes of healthy children randomly chosen from the birth register in the study zone for a fixed time period (North-Humberside, G.B., 1974–1982). These data, representing a MPP with two marks (1 ill and 0 healthy) were first studied by Cuzick and Edwards (53) (cf. Diggle (62)). Using the Monte Carlo method, test whether the function $D(h) = K_1(h) - K_0(h)$ equals zero, i.e., that except for a spatial intensity factor of location of children, there is no spatial risk influencing leukemia cases.

5.15. CPL for soft-core (interaction) point processes.
Suppose that X is a PP with density:

$$f_\omega(\mathbf{x}) = \alpha(\theta)\beta^{n(\mathbf{x})} \prod_{1\le i<j\le n} \exp - \left(\frac{\sigma}{\|x_i - x_j\|}\right)^{2/\kappa}$$

on $S = [0,1]^2$, where $\theta = (\beta, \sigma)'$ with $\beta > 0$ and $0 \le \sigma < \infty$ unknown, $0 < \kappa < 1$ known and $\alpha(\theta)$ the normalization constant. Analyze the influence of σ and κ on the distribution of the spatial configuration. Give the model in an exponential form and calculate its pseudo-likelihood.

5.16. ML for hard-core processes.
Let X be the hard-core PP on $[0,1]^2$ with density

$$f_\theta(\mathbf{x}) = \alpha(\theta) h(\mathbf{x}) \beta^{n(\mathbf{x})},$$

where $h(\mathbf{x}) = \prod_{1\le i<j\le n} \mathbb{1}\{\|x_i - x_j\| \ge \gamma\}$, $\theta = (\beta, \gamma)'$, $\beta > 0$, $\gamma > 0$ and $\alpha(\theta)$ is the normalizing constant. Show that it is possible to find an explicit expression for the ML estimate of γ but not of β.

5.17. An analysis of the `lansing` and `spruce` datasets.
The `lansing` dataset in the `spatstat` package gives spatial location of three types of oak: red, white and black.

1. Estimate the second-order characteristics K and J for the three sets of locations.
2. Test the CSR hypothesis using a Monte Carlo method.

The `spruce` dataset, also in the `spatstat` package gives location and diameters of trees in a forest in Saxony. In the present question we will only consider the location variable.

1. Estimate the Strauss model using MCPL by making a preliminary estimation of the interaction radius r.
2. Validate this choice of model by performing 40 simulations of the estimated model and using the K function.
3. Would working with a model having a second-order potential modeled by a step function be better?

The following exercises look at asymptotic behavior of estimators in a framework that is not necessarily spatial. They require properties of the minimum contrast estimation method given in Appendix C.

5.18. Convergence of OLS for regression models.
Consider the following regression model:

$$y_i = f(x_i, \theta) + \varepsilon_i, \qquad i = 1, \ldots, n$$

with i.i.d. errors (ε_i) from $\mathcal{N}(0, \sigma^2)$, where f is continuous at (x, θ) and $\mathscr{X} = (x_i)$ are i.i.d. data generated from a distribution g on $\mathbb{R}^k$ with compact support. Suppose also that θ is an interior point of a compact Θ in $\mathbb{R}^p$. Give a condition that ensures convergence of the OLS estimator. Can we weaken the hypothesis on $\mathscr{X}$?

5.19. CPL for bivariate Gaussian distributions.
Let $Z = (X, Y)$ be a centered bivariate Gaussian distribution where X and Y have variance 1 and correlation ρ. Suppose we have a sample $Z(n) = (Z_1, Z_2, \ldots, Z_n)$ drawn from Z. Give the CPL of $Z(n)$. Show that the maximum CPL estimator of ρ converges and is asymptotically normal. What is its efficiency compared to the ML estimator?

5.20. Estimating inhomogeneous Markov chains.
Suppose that $Y = \{Y_i, i \in \mathbb{N}\}$ is a Markov chain with finite state space E and transition $P(Y_{i+1} = z \mid Y_i = y) = p(y, z; \theta, x_i)$, where $x_i \in X$ is a covariate and X some measurable compact space $(X, \mathscr{X})$. Suppose that p is a $\mathscr{C}^2$ function with respect to θ. Let μ be a positive measure on $(X, \mathscr{X})$ such that:

(C1) $\alpha \mapsto \sum_{y \in E} \int_X p(y, \cdot; \alpha, x) \mu(dx)$ is one-to-one.

(C2) $\forall A \in \mathscr{X}$, $\liminf_n (n^{-1} \sum_{i=1}^n \mathbf{1}(x_i \in A)) \geq \mu(A)$.

1. Show that under (C), the ML estimator converges.

2. Find the asymptotic distribution of the likelihood ratio test.
3. *Example:* Suppose that $X \subset \mathbb{R}$, $E = \{0,1\}$ and F is a $\mathscr{C}^2$ cumulative distribution function such that $f = F' > 0$ and the transition $y \mapsto z = y$ is given by $p(y,y;x,\alpha,\beta) = F(\alpha xy + \beta y)$. Check conditions (C) and test the hypothesis $\alpha = 0$.

5.21. Marginal pseudo-likelihood estimation.
Suppose that X is an exponentially-mixing ergodic real-valued random field on $S = \mathbb{Z}^d$ whose distribution depends on $\theta \in \Theta$, a compact in $\mathbb{R}^p$. Suppose we have a finite subset M of S, $M_i = M + i$ for $i \in S$, $g : \mathbb{R}^{|M|} \times \Theta \mapsto \mathbb{R}$ continuous, $D_n = \{1,2,\dots,n\}^d$ and the marginal contrast:

$$U_n(\alpha) = d_n^{-1} \sum_{i \in D_n} g(X(A_i,\alpha)).$$

1. Give conditions on (X,g) ensuring convergence of the minimum contrast estimator.
2. Examine asymptotic normality of this estimator.
3. *Example:* consider a Markov chain Y with states $\{-1,+1\}$ and transition $p = P(Y_i \neq Y_{i-1} \mid Y_{i-1})$. We add i.i.d. noise so that the noisy response variable X_i satisfies: $P(X_i = Y_i) = 1 - \varepsilon = 1 - P(X_i \neq Y_i)$. Is it possible to identify $\theta = (p,\varepsilon)$ using the distribution of pairs (X_0,X_1)? Estimate θ using the marginal contrast of triplets (X_i,X_{i+1},X_{i+2}). Test for independence of the chain Y.

5.22. Estimating Markov random field dynamics with CPL.
Let $X = (X(t), t \in \mathbb{N})$, $X(t) = (X_i(t), i \in S) \in \{0,1\}^S$, $S = \{1,2,\dots,n\}$ be modeled using an ergodic homogeneous Markov chain. We define the transition $x \mapsto y$ of these dynamics as:

$$P(X(t+1) = y \mid X(t) = x) = p(x,y;\theta) = Z^{-1}(x,\theta)\exp U(x,y,\theta),$$

where

$$U(x,y;\theta) = \sum_{i=1}^{n} y_i\{\alpha x_i + \beta(x_{i-1} + x_{i+1}) + \gamma(y_{i-1} + y_{i+1})\}.$$

1. What difficulty do we encounter when calculating the likelihood?
2. Calculate the conditional distributions $X_i(t) \mid x^i(t), x(t-1)$. Deduce the associated CPL.
3. Study the asymptotic properties of the maximum CPL estimator (resp. coding). Test for temporal independence ($\alpha = \beta = 0$). Test for spatial independence ($\gamma = 0$).

5.23. Consistency of ML estimation of the parametric intensity of a Poisson point process.
Let X be a Poisson PP with intensity $\rho(\cdot,\alpha)$ on $\mathbb{R}^d$, $\alpha \in \Theta$ a compact in $\mathbb{R}^p$ and suppose that the unknown true value θ of the parameter is an interior point of Θ. Suppose that X is observed in a window $D_n = [0,n]^d$ of measure d_n.

1. Using the result

$$E\sum_{\xi\in X\cap S} h(\xi,X\backslash\{\xi\}) = \int_S Eh(\xi,X)\rho(\xi)d\xi,$$

true when X is a PPP(ρ) (160), deduce that the log-likelihood $l_n(\alpha)$ of X observed on D_n is:

$$E[l_n(\theta)-l_n(\alpha)]=K_n(\alpha,\theta)=\int_{D_n}\left\{\left[\frac{\rho(\eta,\alpha)}{\rho(\eta,\theta)}-1\right]-\log\frac{\rho(\eta,\alpha)}{\rho(\eta,\theta)}\right\}\rho(\eta,\alpha)d\eta.$$

2. Suppose that $\rho(\xi,\alpha)$ is uniformly bounded in (ξ,α). Deduce that for some constant $M<\infty$, we have uniformly at α:

$$Var_\theta(l_n(\alpha))\leq Md_n.$$

Noting $U_n(\alpha)=-l_n(\alpha)/d_n$, show that:

$$\liminf_n\,[U_n(\alpha)-U_n(\theta)]\geq K(\alpha,\theta)=\liminf_n K_n(\alpha,\theta)\ \text{ in probability.}$$

3. Give an identifiability condition on the representation $\alpha\mapsto\rho(\cdot,\alpha)$ ensuring that $K(\alpha,\theta)\neq 0$ if $\alpha\neq\theta$. Deduce that the ML estimation of θ is consistent if $n\to\infty$.

Appendix A
Simulation of random variables

We present several well-known methods for simulating random variables. For supplementary details, we suggest the book by Devroye (59).

In the following we suppose that we have a random generator U of the uniform distribution $\mathscr{U}([0,1])$ on $[0,1]$ and that, on a loop, this generator outputs i.i.d. values (U_n) following $\mathscr{U}([0,1])$.

A.1 The inversion method

Let X be a real-valued random variable with cumulative distribution function $F : \mathbb{R} \to [0,1]$ defined for $x \in \mathbb{R}$ by $F(x) = P(X \le x)$. F is increasing, with limits 0 at $-\infty$ and 1 at $+\infty$. Furthermore, F is everywhere continuous if X has a density g. In such cases, $F(x) = \int_{-\infty}^{x} g(u)du$.

Define $F^{-1} : [0,1] \to \mathbb{R}$ to be the *pseudo-inverse* of F:

$$F^{-1}(u) = \inf\{x : F(x) \ge u\} \quad \text{when} \quad u \in [0,1].$$

The following property is crucial for the *inversion method*: if U is uniform on $[0,1]$, then $X = F^{-1}(U)$ has the cumulative distribution function F. In effect:

$$P(X \le u) = P(F^{-1}(U) \le x) = P(U \le F(x)) = F(x). \tag{A.1}$$

Each time that F is explicitly given, $(F^{-1}(U_n), n \ge 1)$ is a sequence of i.i.d. random variables from the distribution of X. Let us give some examples.

Simulating Bernoulli variables with parameter p

Suppose $X \in \{0,1\}$ with $p = P(X=1) = 1 - P(X=0)$. We generate a variable $U \sim \mathscr{U}([0,1])$: if $U < 1-p$ we keep $X = 0$; otherwise we take $X = 1$. This methodology

C. Gaetan, X. Guyon, *Spatial Statistics and Modeling*, Springer Series in Statistics,
DOI 10.1007/978-0-387-92257-7,

is valid any time that we are simulating distributions that take two possible values $X \in \{a,b\}$, $a \neq b$.

Simulating variables that take a finite numbers of values

Suppose that X takes K values $\{a_1,a_2,\dots,a_K\}$ each with probability $p_k = P(X = a_k)$, $k = 1,\dots,K$, $\sum_1^K p_k = 1$. We construct a partition of $[0,1[$ into K adjacent intervals $I_k = [c_{k-1},c_k[$, $k = 1,\dots,K$ where $c_0 = 0$, $c_k = \sum_1^k p_i$ for $k = 1,\dots,K$ and we draw $U \sim \mathscr{U}([0,1])$: if $U \in I_k$, we take $X = a_k$. This method simulates X because

$$P(c_{k-1} \leq U < c_k) = p_k = P(X = a_k) \qquad \text{for } k = 1,\dots,K. \tag{A.2}$$

Variables taking a countably infinite number of values

If $X \in E$ takes a countably infinite number of values, we begin by finding, for some small fixed value $\alpha > 0$ (for example, $\alpha = 10^{-3}$) a finite subset $E_\alpha \subset E$ such that $P(X \in E_\alpha) \geq 1 - \alpha$. Then, we simulate X over the finite set of states $E_\alpha \cup \{E \backslash E_\alpha\}$, where $\{E \backslash E_\alpha\}$ regroups all states outside E_α into a single state.

For example, to simulate a Poisson distribution with parameter λ, we first find the value n_0 satisfying $P(X > n_0) < 10^{-3}$, then simulate X over $\{0,1,2,\dots,n_0\} \cup \{larger\}$ as described above, with probabilities:

$$P(X = n) = p_n = e^{-\lambda}\frac{\lambda^n}{n!} \quad \text{if } 0 \leq n \leq n_0 \quad \text{and} \quad P(X \in \{larger\}) = 1 - \sum_0^{n_0} p_n.$$

Note also that the value of X can be qualitative.

Simulating an exponential distribution

Equation (A.1) allows us to simulate distributions whose state space $E \subseteq T$ is continuous whenever F^{-1} is given. If $X \sim \mathscr{E}xp(\lambda)$ is exponential with parameter $\lambda > 0$, then $F(x) = 0$ if $x < 0$ and $F(x) = 1 - e^{-\lambda x}$ if $x \geq 0$ and we deduce that, for any $u \in [0,1]$,

$$F^{-1}(u) = -\frac{\log(1-u)}{\lambda}.$$

As both U and $1-U$ are uniform on $[0,1]$ when $U \sim \mathscr{U}([0,1])$, $X = -\log(U)/\lambda$ simulates an exponential variable with parameter λ. Similarly,

$$X = -\frac{\sum_{n=1}^N \log U_n}{\lambda}$$

simulates a Gamma distribution $\Gamma(\lambda,N)$ for any integer $N \geq 1$. We will show later how to simulate such distributions when this parameter is not an integer.

A.2 Simulation of a Markov chain with a finite number of states

Let $X = (X_0, X_1, X_2, \ldots)$ be a homogeneous Markov chain taking values in a finite state space $E = \{a_1, a_2, \ldots, a_K\}$ with transition $P = (p_{ij})$ (120; 103; 229):

$$p_{ij} = P(X_{n+1} = j \mid X_n = i).$$

With the convention $p_{i,0} = 0$, we define for each $i = 1, \ldots, K$ the partition of $[0,1[$ into K adjacent intervals $I_i(j)$:

$$I_i(j) = [c_{j-1}(i), c_j(i)[, \qquad j = 1, \ldots, K$$

where $c_j(i) = \sum_{l=0}^{j} p_{i,l}$.

Consider next the mapping $\Phi : E \times [0,1] \to E$ defined for any $i = 1, \ldots, K$ and $u \in [0,1]$ by:

$$\Phi(a_i, u) = a_j \qquad \text{if } u \in I_i(j).$$

If $U \sim \mathscr{U}([0,1])$, we thus have for any $i, j = 1, \ldots, K$:

$$P(\Phi(a_i, U) = a_j) = P(U \in I_i(j)) = p_{ij}.$$

The sequence $\{x_0, X_{n+1} = \Phi(X_n, U_n), n \geq 0\}$ simulates a Markov chain with transition P and initial condition $X_0 = x_0$. If $X_0 \sim \nu_0$ is generated using (A.2), the sequence $(U_n, n \geq 0)$ allows us to simulate a Markov chain with initial distribution ν_0 and transition P.

A.3 The acceptance-rejection method

This method simulates a random variable X with density f on $\mathbb{R}^p$ when there exists some easily simulated distribution with density g on $\mathbb{R}^p$ such that, for all x, $f(x) \leq cg(x)$ for some $c < \infty$. In effect, consider a random variable Y with density g and suppose $U \sim \mathscr{U}([0,1])$ is independent of Y. Then, the following conditional variable has the distribution of X:

$$(Y \mid \text{if } cUg(Y) < f(Y)) \sim X.$$

Indeed, as $c = \int cg(y)dy \geq \int f(y)dy > 0$ and $f(y) = 0$ if $g(y) = 0$, we have:

$$\begin{aligned} P(Y \in [x, x+dx] \mid cUg(Y) < f(Y)) &= \frac{g(x)dxP(U < \frac{f(x)}{cg(x)})}{P(U < \frac{f(Y)}{cg(Y)})} \\ &= \frac{g(x)\frac{f(x)}{cg(x)}dx}{\int \frac{f(y)}{cg(y)} g(y)dy} \\ &= f(x)dx = P(X \in [x, x+dx]). \end{aligned}$$

For the method to be practical, Y has to be easily simulated and c not too large (so that rejection is infrequent). When possible, the optimal choice for c is $c = \sup_x f(x)/g(x)$.

Example: simulation of Gamma and Beta distributions

A Gamma distribution $\Gamma(\lambda,a)$, $\lambda > 0$, $a > 0$ has density

$$f(x) = \frac{\lambda^a}{\Gamma(a)} e^{-\lambda x} x^{a-1} \mathbf{1}(x > 0),$$

where $\Gamma(a) = \int_0^{+\infty} e^{-x} x^{a-1} dx$.

If $Y \sim \Gamma(1,a)$, then $X = Y/\lambda \sim \Gamma(\lambda,a)$. Also, the sum $X+Z$ of two independent $\Gamma(\lambda,a)$ and $\Gamma(\lambda,b)$ gives a $\Gamma(\lambda,a+b)$. It therefore suffices to be able to simulate $\Gamma(1,a)$ with $a \in]0,1]$ in order to be able to simulate any $\Gamma(\lambda,a^*)$ with $\lambda > 0$, $a^* > 0$.

Simulating a random variable Y following $\Gamma(1,a)$, $a \in]0,1]$ can be done using the acceptance-rejection method: we remark that for density $f(x) = \Gamma(a)^{-1} e^{-x} x^{a-1} \mathbf{1}(x > 0)$ of Y,

$$f \le \frac{a+e}{ae\Gamma(a)} g\,,$$

where $g(x) = (a+e)^{-1}[e g_1(x) + a g_2(x)]$, with $g_1(x) = a x^{a-1} \mathbf{1}(0 < x < 1)$ and $g_2(x) = e^{-x+1} \mathbf{1}(1 < x < \infty)$. g_1 is a density on $]0,1[$, g_2 on $]1,+\infty[$ with both able to be simulated using the inversion method and g the mixture of these two distributions with weights $(e/(a+e), a/(a+e))$. An initially generated uniform variable lets us choose whether the simulation of g is in $]0,1[$ or $]1,+\infty[$, after which we simulate the retained variable with density g_i using the inversion method, the simulation of Y thus being obtained by acceptance-rejection. Three $\mathscr{U}([0,1])$ are therefore used during this simulation. Other methods can bring this down to two, in particular by directly simulating the distribution with density g using the inversion method.

The Beta distribution with parameters $a,b > 0$ is noted $\beta(a,b)$ and has density

$$f(x) = \frac{\Gamma(a+b)}{\Gamma(a)\Gamma(b)} x^{a-1} (1-x)^{b-1} \mathbf{1}(0 < x < 1).$$

Simulating such distributions can be done using simulated Γ distributions and the following property: if $X \sim \Gamma(\lambda,a)$ and $Y \sim \Gamma(\lambda,b)$ are independent, then $X/(X+Y) \sim \beta(a,b)$.

A.4 Simulating normal distributions

Simulating $\mathscr{N}(0,1)$

Statistics software packages have functions that can generate a $\mathscr{N}(0,1)$. This can be done starting from two independent U_1 and $U_2 \sim \mathscr{U}([0,1])$ in the following way

(Box-Muller method): the two following variables X_1 and X_2 follow a $\mathcal{N}(0,1)$:

$$X_1 = \sqrt{-2\log U_1}\cos(2\pi U_2), \qquad X_2 = \sqrt{-2\log U_2}\cos(2\pi U_1).$$

Simulation of $Y \sim \mathcal{N}(m,\sigma^2)$ can be obtained by calculating $Y = m+\sigma X$ starting from a $X \sim \mathcal{N}(0,1)$. Also, log-normal, χ^2, Student and Fisher distributions are easily simulated starting from a $\mathcal{N}(0,1)$.

Simulating Gaussian random vectors

If Σ has full rank and if p is not too large ($p < 5000$), we can look for a decomposition (for example, the Cholesky decomposition) $\Sigma = A\,{}^tA$ of Σ. Then, if $X = {}^t(X_1, X_2, \dots, X_p)$ is an i.i.d. sample of dimension p from a $\mathcal{N}(0,1)$, $\mu + AX$ simulates Z. If $r = \text{rank}(\Sigma) < p$, we begin by finding the subspace S of dimension r that is the support of Z, then, using an orthonormal basis of S, we simulate the Gaussian vector using the previous method.

Appendix B
Limit theorems for random fields

B.1 Ergodicity and laws of large numbers

We recall here several ergodicity results and laws of large numbers (LLN) for spatial processes: a strong law (SLLN), i.e., almost sure (a.s.) convergence and a weak law, i.e., L^2 convergence.

B.1.1 Ergodicity and the ergodic theorem

Let $X = \{X_s, s \in S\}$ be a real-valued random field on $S = \mathbb{R}^d$ or $S = \mathbb{Z}^d$. Ergodicity is a property that strengthens the idea of stationarity and allows us to obtain *a.s.* convergence of spatial empirical means when the domain of observation "tends to infinity." If we limit ourselves to L^2 convergence, second-order ergodicity suffices. Ergodicity is important in statistics as it allows us to establish *a.s.* consistency of estimators given in the form of spatial means. Nevertheless, it is not always necessary to invoke the (very strong) ergodicity property in order to prove consistency. Subergodicity conditions or even L^2 conditions may be sufficient (consistency of the CPL estimator or Coding estimator, cf. Th. 5.4 and Th. 5.6; consistency of the minimum contrast estimator, cf. Appendix C, Th. C.1).

Noting $\Omega = \mathbb{R}^S$ the state space of X and $\mathscr{E}$ its Borel σ-algebra, we say that an event $A \in \mathscr{E}$ is *translation-invariant* if for any $i \in S$, $\tau_i(A) = A$ where τ_i is the i-translation on Ω defined as: for all $i \in S$ and $\omega \in \Omega$, $(\tau_i(\omega))_j = \omega_{j-i}$. The set of invariant events forms a sub-σ-algebra $\mathscr{I} \subseteq \mathscr{E}$.

Stationary processes X are characterized by the fact that their distribution P is translation-invariant: for any event $A = \{\omega : X_{s_1}(\omega) \in B_1, X_{s_2}(\omega) \in B_2, \dots, X_{s_n}(\omega) \in B_n\}$ generating the Borel σ-algebra, we have:

$$\forall i \in S : P(A) = P(\tau_i(A)).$$

Ergodicity strengthens this property.

Definition B.1. A stationary process X is ergodic if for any invariant event $A \in \mathscr{I}$, $P(A) = 0$ or $P(A) = 1$.

If X is ergodic, the σ-algebra $\mathscr{I}$ of invariants is trivial. The heuristic interpretation of this is that if an invariant event A has probability > 0, then it has probability 1 since the set of translations of A generates the space of all trajectories.

The standard example of an ergodic process is that of a sequence of i.i.d. variables. Ergodicity can be seen as a combined property of stationarity and asymptotic independence.

We now state the ergodic theorem. With $B(x,r)$ the ball centered at x with radius r, define the interior diameter $d(D)$ of $D \subseteq \mathbb{R}^d$ by:

$$d(D) = \sup\{r : B(x,r) \subseteq D\}.$$

Theorem B.1. *(Birkhoff (32) if $d = 1$; Tempelman (211) and Nguyen and Zessin (164) if $d \geq 2$)*

Suppose that X is a stationary real-valued L^p process on $\mathbb{R}^d$ for some $p \geq 1$. Suppose that (D_n) is an increasing sequence of bounded convex sets such that $d(D_n) \longrightarrow \infty$. Then:

1. $\overline{X}_n = |D_n|^{-1} \int_{D_n} X_u du \longrightarrow E(X_0 \mid \mathscr{I})$ *in L^p and a.s. if $S = \mathbb{R}^d$.*
2. *If furthermore X is ergodic, $\overline{X}_n \longrightarrow E(X_0)$ a.s.*

On $S = \mathbb{Z}^d$, we have: $\overline{X}_n = \sharp D_n{}^{-1} \sum_{i \in D_n} X_i \longrightarrow E(X_0 \mid \mathscr{I})$ in L^p and *a.s.* if $S = \mathbb{Z}^d$, where the *a.s.* limit is $E(X_0)$ if X is ergodic.

B.1.2 Examples of ergodic processes

Let us give several examples of ergodic processes:

- $X = \{X_i, i \in S\}$ an i.i.d. sequence of real-valued random variables in L^1 over a countable set S.
- $Y = \{Y_i = g(X \circ \tau_i), i \in S\}$, where X is ergodic on S and $g : E^V \longrightarrow \mathbb{R}$, for finite $V \subset S$, is a measurable mapping.
- X a strongly-mixing stationary random field (116; 67):

$$\lim_{\|h\| \to \infty} P(\tau_h(A) \cap B) = P(A)P(B), \qquad \forall A, B \in \mathscr{E}.$$

- X an m-dependent stationary random field, i.e., $\forall U \subset S$ and $V \subset S$ at least a distance m apart, there is independence between $\{X_u, u \in U\}$ and $\{X_v, v \in V\}$.

- X a stationary Gaussian random field on $\mathbb{R}^d$ ($\mathbb{Z}^d$) whose covariance C tends to 0 at infinity (3):
$$\lim_{\|h\| \longrightarrow \infty} C(h) = 0.$$
- X a Gibbs random field on $\mathbb{Z}^d$ with translation-invariant potential and satisfying Dobrushin's uniqueness condition (85).
- $N = \{N([0,1[^d + i), i \in \mathbb{Z}^d\}$, where $N(A)$ is the number of points in A of an ergodic PP on $\mathbb{R}^d$ (for example a homogeneous Poisson PP, homogeneous Neyman-Scott PP or a Cox PP driven by an ergodic random field Λ (204)).
- Y a random subset of $\mathbb{R}^d$: $Y = \cup_{x_i \in X} B(x_i, r)$ where X is a homogeneous Poisson PP (an ergodic PP) on $\mathbb{R}^d$; Y is a Boolean random field on $\mathbb{R}^d$, $Y_s = 1$ if $s \in Y$ and 0 otherwise (43).

Ergodicity of PPs (and closed random sets of $\mathbb{R}^d$ including the class of PPs) is studied in (125; 204; 56) and (135) (the Poisson PP case). Heinrich (111) and Guan and Sherman (95) use these properties to prove consistency of parametric estimators of PPs. One way to get at the ergodicity of PPs is to associate them with the lattice process $\widetilde{X}$ of their configurations on the partition $\mathbb{R}^d = \cup_{i \in \mathbb{Z}^d} A_i$, where $A_i = [i, i + \mathbf{1}[$ and $\mathbf{1}$ is the vector in $\mathbb{R}^d$ with entries 1, then verify ergodicity of $\widetilde{X}$.

B.1.3 Ergodicity and the weak law of large numbers in L^2

Suppose that X is a second-order stationary random field on $\mathbb{R}^d$ with covariance C. We say that X is ergodic in L^2 if for any sequence (D_n) of bounded convex sets such that $d(D_n) \longrightarrow \infty$, we have

$$\overline{X}_n = \frac{1}{|D_n|} \int_{D_n} X_u du \longrightarrow E(X_0) \text{ in } L^2.$$

Let F be the spectral measure of X. We have the following weak law of large numbers [227, Ch. 3; 96, Ch. 3; 43]:

Theorem B.2. *The following conditions are equivalent:*

(i) *X is ergodic in L^2.*
(ii) *F has no mass at 0: $F(\{0\}) = 0$.*
(iii) $|D_n|^{-1} \int_{D_n} C(u) du \longrightarrow 0$.

In particular, these conditions are satisfied if $\lim_{\|h\| \longrightarrow \infty} C(h) = 0$ or if F is absolutely continuous.

These results can be directly adapted to $\mathbb{Z}^d$: for example, if F is absolutely continuous on T^d,
$$\lim_n \overline{X}_n = E(X_0) \text{ in } L^2.$$
If furthermore the spectral density f is bounded and continuous at 0, then

$$\lim_{n\to\infty} \sharp D_n Var(\overline{X}_n) = (2\pi)^d f(0),$$

with the asymptotic distribution of $\sqrt{\sharp D_n X_n}$ being Gaussian if X is itself a Gaussian random field.

B.1.4 Strong law of large numbers under L^2 conditions

We also have SLLNs under L^2 conditions. A first result (cf. (32), §3.4) deals with sequences of variables that are independent but not necessarily from the same distribution (cf. Lemma 5.1): if $X = \{X_i, i \in \mathbb{N}\}$ are independent centered real-valued random variables in L^2, then $\sum_{i=1}^n X_i \longrightarrow 0$ *a.s.* whenever $\sum_{i=1}^\infty Var(X_i) < \infty$.

Another result deals with empirical estimates of the mean μ and covariance $C_X(k)$ of second-order stationary processes X on $\mathbb{Z}^d$ ((96), §3.2): if $\sum_{h\in\mathbb{Z}^d} |C_X(h)| < \infty$, then $\overline{X}_n \to \mu$ *a.s.* If furthermore X is 4th order stationary, i.e., $Y = \{Y_i = (X_i - \mu)(X_{i+k} - \mu), i \in \mathbb{Z}^d\}$ satisfies $\sum_{h\in\mathbb{Z}^d} |C_Y(h)| < \infty$, then $\overline{Y}_n \to C_X(k)$ *a.s.* If X is a Gaussian random field, this last condition is satisfied if X has a square integrable spectral density.

B.2 Strong mixing coefficients

Suppose that $Z = \{Z_i, i \in S\}$ is a random field on a network S endowed with a metric d and that A and B are two subsets of S. Let $\mathcal{F}(Z,H)$ be the σ-algebra induced by Z on the subset H of S. The *strong mixing* coefficient $\alpha^Z(E,F)$ of Z on E and F, defined by Ibragimov and Rozanov (117) (cf. also Doukhan (67)) is:

$$\alpha^Z(E,F) = \sup\{|P(A\cap B) - P(A)P(B)| : A \in \mathcal{F}(Z,E), B \in \mathcal{F}(Z,F)\}.$$

The random field is α-mixing if $\alpha^Z(E,F) \longrightarrow 0$ as $dist(E,F) \longrightarrow \infty$, where $dist(E,F) = \inf\{d(i,j), i \in E, j \in F\}$.

We can equally use coefficients α parametrized by the sizes k and $l \in \mathbb{N}\cup\{\infty\}$ of E and F: for $n \in \mathbb{R}$,

$$\alpha_{k,l}(n) = \alpha_{k,l}^Z(n) = \sup\{\alpha^Z(E,F) : \sharp E \le k, \sharp F \le l \text{ and } dist(E,F) \ge n\}.$$

Examples of α-mixing random fields

1. If Z is α-mixing, any measurable functional $Z_W = (f_i(Z_{W_i}))$ that is locally dependent on Z is also α-mixing: more precisely, if for any $i \in S$, (W_i) is a family of subsets such that $i \in W_i$ and the diameter $\delta(W_i)$ of W_i is uniformly bounded by Δ, then $Z_W = (f_i(Z_{W_i}))$ is α-mixing, where $\alpha_{k,l}^{Z_W}(n) = \alpha_{k,l}^Z(n - 2\Delta)$ if $n > 2\Delta$.

2. *R-dependent random fields:* Z is an R-dependent random field if, for any pair (i,j) separated by a distance greater than R, Z_i and Z_j are independent. For R-dependent random fields, $\alpha^Z(E,F)=0$ if $d(E,F)\geq R$ and, for any k,l, $\alpha^Z_{k,l}$ is zero if $n\geq 2R$.
3. *Gaussian random fields:* Let Z be a stationary Gaussian random field on $\mathbb{Z}^d$ with spectral density f, $f(\lambda)\geq a>0$ on the torus $\mathbb{T}^d$. Note $D_K(f)$ the distance from f to its best approximation by a trigonometric polynomial with degree $(K-1)$,

$$D_K(f)=\inf\{|f-P|:P(\lambda)=\sum_{|t|<K}c_t\exp\langle\lambda,t\rangle\},\text{ where }|t|=\sum_{i=1}^{d}|t_i|.$$

We have therefore ((117) for $d=1$; (96, §1.7) for random fields) that:

$$\alpha_{\infty,\infty}(k)\leq\frac{1}{a}D_K(f).$$

If f is continuous, this coefficient tends to 0 if $k\longrightarrow\infty$ and at an exponential rate if f is analytic (for example, if Z is an *ARMA* model). If Z can be put into the linear form $Z_t=\sum_{\mathbb{Z}^d}b_{t-s}\varepsilon_t$ where ε is a Gaussian WN, we have more precisely that:

$$\alpha_{\infty,\infty}(k)\leq\frac{2}{a}\|b\|_\infty\{\sum_{|s|\geq k/2}|sb_s|\}.$$

There are also ways to find mixing coefficients for non-Gaussian linear random fields (67; 96).
4. *Gibbs random fields under Dobrushin's uniqueness condition* (Dobrushin (66); Georgii (85); Guyon (96)): Dobrushin's influence measure $\gamma_{a,b}(\pi)$ of site a on site b, $a\neq b$ of a Gibbs specification π (noted $\gamma_{a,b}(\phi)$ if the specification is derived from a potential ϕ) is defined by:

$$\gamma_{a,b}(\pi)=\sup\frac{1}{2}\left\|\pi_b(\cdot\mid\omega)-\pi_b(\cdot\mid\omega')\right\|_{VT},$$

where $\|\cdot\|_{VT}$ is the total variation norm and the sup is taken over configurations ω and ω' that are identical everywhere except at a. If $a=b$, we set $\gamma_{a,b}(\pi)=0$. We say that the Gibbs potential satisfies Dobrushin's condition if

$$(D):\alpha(\phi)=\sup_{a\in S}\sum_{b\in S}\gamma_{a,b}(\phi)<1. \tag{B.1}$$

(D) is a sufficient (but not necessary) condition ensuring that there is no more than one Gibbs measure in $\mathcal{G}(\phi)$. For example, in the 4-NN isotropic Ising model on $\mathbb{Z}^2$ with specification at i:

$$\pi_i(z_i\mid z^i)=\frac{\exp\beta z_i v_i}{\exp-\beta v_i+\exp\beta v_i},\qquad v_i=\sum_{j:\|i-j\|_1=1}z_j,$$

the exact uniqueness condition (Onsager (166); Georgii (85)) is $\beta < \beta_c = \frac{1}{2}\log(1+\sqrt{2}) \simeq 0.441$ whereas Dobrushin's condition, easily expressed in this case, is $\beta < \frac{1}{4}\log 2 \simeq 0.275$.

If furthermore the state space of the Gibbs random field is Polish (for example, finite or compact), endowed with a finite positive reference measure λ and the potential ϕ is summable, then there is *existence* and uniqueness of the Gibbs measure μ associated with π: $\mathscr{G}(\phi) = \{\mu\}$. In this case, if ϕ is bounded and has bounded range, the unique Gibbs measure μ satisfies an exponential uniform mixing condition:

$$\varphi(A,B) \leq C\,(\sharp A)\;\alpha^{d(A,B)},$$

where $\varphi(A,B) = \sup\{|\mu(E\,|F) - \mu(E)|,\, E \in \mathscr{F}(A),\, F \in \mathscr{F}(B),\, \mu(F) > 0\,\}$. μ is also strongly-mixing as we still have $2\alpha(\cdot) \leq \varphi(\cdot)$ (67).

One difficulty in asymptotic statistics for Gibbs random fields μ known in the form of their specification $\pi(\phi)$ is that we do not know in general whether or not μ is unique in $\mathscr{G}(\phi)$. Therefore, we usually do not have a weak dependency property. Similarly, if the potential ϕ is translation-invariant ($S = \mathbb{Z}^d$), we do not know if μ is ergodic or even stationary. Thus, classical tools of asymptotic statistics (ergodicity, weak dependency, CLT) are not always useful.

5. *Mixing property for spatial point processes*: A Poisson PP X, whether homogeneous or not is α-mixing because X exhibits independence. This mixing property extends to Neyman-Scott PPs if the distributions for descendancy are spatially bounded by R. In such cases, PPs are $2R$-dependent.

Other examples of α-mixing random fields are given in Doukhan (67).

B.3 Central limit theorem for mixing random fields

Suppose (D_n) is a strictly increasing sequence of finite subsets of $\mathbb{Z}^d$, Z a real-valued centered random field with finite variance, $S_n = \sum_{D_n} Z_i$ and $\sigma_n^2 = Var(S_n)$. Note by $(\alpha_{k,l}(\cdot))$ the mixing coefficients of Z. We have the following result (Bolthausen (30); Guyon (96) without stationarity):

Proposition B.1. *Central limit theorem for real-valued random fields on* $\mathbb{Z}^d$
Suppose the following conditions are satisfied:

(i) $\sum_{m\geq 1} m^{d-1}\alpha_{k,l}(m) < \infty$ *if* $k+l \leq 4$ *and* $\alpha_{1,\infty}(m) = o(m^{-d})$.
(ii) *There exists* $\delta > 0$ *such that:* $\sup_{i\in S}\mathbb{E}\,|Z_i|^{2+\delta} < \infty$ *and*

$$\sum_{m\geq 1} m^{d-1}\alpha_{1,1}(m)^{\delta/2+\delta} < \infty.$$

(iii) $\liminf_n (\sharp D_n)^{-1}\sigma_n^2 > 0.$

Then $\sigma_n^{-1} S_n \xrightarrow{d} \mathcal{N}(0,1)$.

Comments

1. The mixing conditions are satisfied by stationary Gaussian random fields with sufficiently regular spectral densities and for Gibbs random fields under Dobrushin's uniqueness condition (B.1).
2. If we only want to use one mixing coefficient, we can keep $\alpha_{2,\infty}$ and the condition:
$$\sum_{m\geq 1} m^{d-1}\alpha_{2,\infty}(m) < \infty.$$
3. The conclusion of the theorem still holds if $S \subset \mathbb{R}^d$ is a locally finite countably infinite network: $\exists \delta_0 > 0$ s.t. for any two sites $i \neq j$ of S, $\|i-j\| \geq \delta_0$. In effect, here and for the regular network $\mathbb{Z}^d$, the key property is that for any $i \in S$, the ball centered at i with radius m satisfies, uniformly in i and m: $\sharp\{B(i,m)\cap S\} = O(m^d)$.
4. As for any CLT, the positivity condition (iii) may be difficult to check.
5. If $Z \in \mathbb{R}^k$ is a multidimensional random field, $\Sigma_n = Var(S_n)$ and (iii) is replaced with
$$(iii)' : \liminf_n (\sharp D_n)^{-1}\Sigma_n \geq I_0 > 0$$
for some positive definite matrix I_0, then: $\Sigma^{-1/2}S_n \xrightarrow{d} \mathcal{N}_k(0, I_k)$.
6. The mixing conditions given by Bolthausen seem minor. However, we bring attention to the work of Lahiri (137) who establishes a CLT using only the coefficients $\alpha_k^*(n) \equiv \alpha_{k,k}(n)$, $k < \infty$. The conditions required are of the same type as for Bolthausen but without the need to calculate coefficients $\alpha_{k,\infty}(\cdot)$. Lahiri (137) also gives CLTs when observation sites are randomly chosen with possible infilling of observation sites (*mixed increasing-domain infill asymptotics*).

B.4 Central limit theorem for a functional of a Markov random field

Suppose that Z is a Markov random field on $S = \mathbb{Z}^d$ taking values in E and with *translation-invariant specification* π. Note $V_i = \{Z_j, j \in \partial i\}$ the m values that the local specification depends on, $\pi_i(\cdot \mid Z^{\{i\}}) \equiv \pi_i(\cdot \mid V_i)$ and consider a measurable functional $h : E^{m+1} \longrightarrow \mathbb{R}$ such that $Y_i = h(Z_i, V_i)$ satisfies for all i the *conditional centering* condition:
$$E(Y_i \mid Z_j, j \neq i) = 0. \tag{B.2}$$
The standard example for which this condition is satisfied is the case where Y is the gradient (in θ) of the conditional pseudo-likelihood of a Markov random field.

To keep things simple, suppose that $D_n = [-n,n]^d$ is the sequence of domains of observation of Z and note $S_n = \sum_{D_n} Y_i$.

We present here two CLTs for the functional Y: the first (99; 123) uses ergodicity of the random field Z, replacing the mixing property. The second (Comets and Janzura (47)) is more general, supposing only translation-invariance of the conditional specification π of Z and gives a Studentized version of the CLT. Comets and Janzura's result therefore applies without ergodicity or stationarity hypotheses, whether or not there is a phase transition of the specification π.

Proposition B.2. *(99; 123; 96)*
Suppose that Z is a Markov random field on $\mathbb{Z}^d$ with an ergodic and translation-invariant specification, where the bounded functional h satisfies centering condition (B.2). Then, if $\sigma^2 = \sum_{j\in\partial 0} E(Y_0 Y_j) > 0$, we have:

$$(\sharp D_n)^{-1/2} S_n \xrightarrow{d} \mathcal{N}(0,1).$$

As is usual for CLTs, the condition $\sigma^2 > 0$ may be difficult to check. This condition is shown to hold in (99) for the 4-NN isotropic Ising model on $\mathbb{Z}^d$.

We now come to result (47). Define $A_n = \sum_{i\in D_n}\sum_{j\in\partial i} Y_i Y_j$ and for $\delta > 0$, $A_n^\delta = \max\{A_n, \delta \times \sharp D_n\}$. Notice that A_n is an unbiased estimator of the variance of S_n.

Proposition B.3. *(Comets-Janzura (47))*
Suppose that Z is a Markov random field on $\mathbb{Z}^d$ with translation-invariant specification and that the functional h satisfies: $\sup_i \left\|Y_i^4\right\| < \infty$ as well as centering condition (B.2). Define the Studentized version of S_n:

$$\zeta_n = A_n^{-1/2} S_n \text{ if } A_n > 0 \text{ and } \zeta_n = 0 \text{ otherwise.}$$

Then, under the condition:

$$\exists \delta > 0 \text{ such that } (\sharp D_n)^{-1} E\left|A_n - A_n^\delta\right| \underset{n\to\infty}{\longrightarrow} 0, \tag{B.3}$$

$\zeta_n \xrightarrow{d} \mathcal{N}(0,1)$.

Appendix C
Minimum contrast estimation

We present here the *minimum contrast* estimation method for parametric (or semi-parametric) models (Dacunha-Castelle and Duflo (54)). Some authors still call this *pseudo-likelihood* estimation (Whittle (222); Besag (25)), *quasi-likelihood* estimation (McCullagh and Nelder (155)) or *extremum* estimation (Amemiya (6); Gourieroux and Monfort (92)). However, the underlying principle is the same throughout: the estimated parameter value maximizes some "pseudo-likelihood" functional. This functional replaces the likelihood when it is unavailable, either because the model is semi-parametric and incompletely specified or because the likelihood is impossible to calculate.

Under regularity conditions on the pseudo-likelihood functional and the observation design, the maximum pseudo-likelihood estimation procedure has some good statistical properties: convergence, asymptotic normality and a test for the parameter of interest. After suitable penalization of these functionals, we also obtain model identification criteria.

Two pseudo-likelihood functionals play a central role in statistics for spatial processes:

1. *Gaussian pseudo-likelihood* for second-order models. This is obtained by calculating the likelihood (or approximation of) by supposing that the model is Gaussian. This contrast was introduced by Whittle (222) for time series and for random fields on $\mathbb{Z}^2$ (cf. §5.3.1).
2. *Conditional pseudo-likelihood* (CPL) of Markov random fields on networks (cf. §5.4.2), a product of conditional densities at each site (Besag (25); Guyon (96)). If we restrict this product to a coding subset, we obtain the *coding pseudo-likelihood.* The notion of CPLs also exists for Markov point processes (cf. §5.5.5.1).

The standard example of a contrast is the least squares functional (spatial regression estimation (cf. §5.3.4), variogram model estimation (cf. §5.1.3) and point process estimation (cf. §5.5.4)). Contrasts for spatio-temporal models are given in (41; 98; 100).

Generally speaking, pseudo-likelihood functionals should have the following features:

1. Encode in a simple way the information of interest in the model.
2. Be numerically simple to calculate.
3. Enable the model parameters to be identified.
4. Tentatively allow checking of statistical properties of estimators.

C.1 Definitions and examples

Consider the process $X = \{X_i, i \in S\}$ defined on a finite or countably finite set of sites S. Whether our knowledge of X is partial (semi-parametric models) or complete (parametric models), we denote $\theta \in \Theta \subseteq \mathbb{R}^p$ the parameter of interest.

The goal is to estimate θ using observations $X(n) = \{X_i, i \in D_n\}$, where D_n is a finite subset of sites. Our asymptotic study is associated with a strictly increasing sequence (D_n) of domains of observation. To simplify things, we suppose that θ, the true unknown value of the parameter, is an interior point of a compact Θ of $\mathbb{R}^p$. $\alpha \in \Theta$ denotes a point in Θ.

A *contrast function* for θ is a non-random function

$$K(\cdot,\theta) : \Theta \to \mathbb{R}, \ \ \alpha \mapsto K(\alpha,\theta) \geq 0$$

that has a unique minimum at $\alpha = \theta$. The value of $K(\alpha,\theta)$ can be interpreted as a pseudo-distance between the model under θ and the one under α.

A *contrast process* associated with the contrast function $K(\cdot,\theta)$ and observations $X(n)$ is a sequence of random variables $(U_n(\alpha), n \geq 1)$ related to $X(n)$, $U_n(\alpha) = U_n(\alpha, X(n))$, defined for all $\alpha \in \Theta$ such that:

$$\forall \alpha \in \Theta : \liminf_n [U_n(\alpha) - U_n(\theta)] \geq K(\alpha,\theta) \text{ in } P_\theta\text{-probability.} \tag{C.1}$$

This *subergodicity* condition (C.1) translates the fact that the value $U_n(\alpha) - U_n(\theta)$ estimating the contrast of α on θ on the basis $X(n)$ asymptotically separates the parameters. Condition (C.1) can be strengthened by the "ergodic" condition, giving:

$$\lim_n [U_n(\alpha) - U_n(\theta)] = K(\alpha,\theta) \ \text{ in } P_\theta\text{-probability.} \tag{C.2}$$

Definition C.1. The minimum contrast estimator is the value $\widehat{\theta}_n$ of Θ that minimizes the contrast U_n:

$$\widehat{\theta}_n = \operatorname*{argmin}_{\alpha \in \Theta} U_n(\alpha).$$

Let us give some examples.

Example C.1. Likelihood of a Bernoulli model

Suppose that the X_i are independent Bernoulli random variables with parameters $p_i = p(\alpha, Z_i) = (1 + \exp \alpha Z_i)/(\exp \alpha Z_i)$, where Z_i is a real-valued covariate and $\alpha \in \mathbb{R}$. The contrast of the likelihood is the negative of the likelihood, $U_n(\alpha) = -\sum_1^n \log f_i(X_i, \alpha)$. If the covariate (Z_i) and α are bounded, the contrast function:

$$K(\alpha, \theta) = \liminf_n \frac{1}{n} \sum_1^n \log \frac{f_i(X_i, \theta)}{f_i(X_i, \alpha)}$$

satisfies (C.1) whenever $\liminf_n n^{-1} \sum_{i=1}^n Z_i^2 > 0$.

Example C.2. Least squares contrast for regression

Consider the regression model (linear or otherwise)

$$X_i = m(Z_i, \theta) + \varepsilon_i, \qquad i = 1, \dots, n$$

expressing $X_i \in \mathbb{R}$ as functions of covariate Z_i and error (ε_i) forming a WN with variance $\sigma^2 < \infty$. This model is *semi-parametric* as no hypothesis is made on the distribution of the errors except that they be a WN. The ordinary least squares (OLS) contrast is defined as:

$$U_n(\alpha) = \sum_{i=1}^n (X_i - m(Z_i, \alpha))^2.$$

Now let us define $K(\alpha, \theta) = \liminf_n \sum_{i=1}^n \{m(x_i, \alpha) - m(x_i, \theta)\}^2/n$. If the experimental design $\mathscr{Z} = \{Z_i, i = 1, 2, \dots\}$ is such that $K(\alpha, \theta) > 0$ for $\alpha \neq \theta$, then (U_n) is a contrast process associated with the contrast function $K(\cdot, \theta)$. This condition is satisfied for example when:

1. Z_i are i.i.d. with distribution Z: $\mathscr{Z}$ is ergodic.
2. The model $\theta \mapsto m(\cdot, \theta)$ is identifiable, i.e.,

$$\text{if } \theta \neq \theta', \text{ then } P_Z\{z : m(z, \theta) \neq m(z, \theta')\} > 0.$$

If the errors are Gaussian, $U_n(\alpha)$ is, up to a multiplicative constant, the negative of the log-likelihood.

If errors are correlated and have a known invertible covariance matrix R_n, the *generalized least squares* (GLS) contrast is

$$U_n^{GLS}(\alpha) = \|X(n) - m_n(\alpha)\|^2_{R_n^{-1}},$$

where $m_n(\alpha) = \{m(Z_1, \alpha), \dots, m(Z_n, \alpha)\}$ and $\|u\|^2_\Gamma = {}^t u \Gamma u$ is the norm associated with the positive-definite matrix Γ. The *weighted least squares* contrast corresponds to the norm associated with the diagonal covariance matrix

$$U_n^W(\alpha) = \sum_{i=1}^n \frac{(X_i - m(Z_i, \alpha))^2}{Var(\varepsilon_i)}.$$

Example C.3. The moment method, marginal pseudo-likelihood

Suppose that $X_1, X_2, \dots, X_n$ are real-valued observations each with the same distribution μ_θ that depends on a parameter $\theta \in \mathbb{R}^p$. Note $(\mu_k(\theta), k = 1, \dots, r)$ the first r moments of this shared distribution and $(\widehat{\mu}_{n,k}(\theta), k = 1, \dots, r)$ the empirical estimates of these moments. If D is a metric on $\mathbb{R}^r$, one contrast for estimating θ is

$$U_n(\theta) = D((\mu_k), (\widehat{\mu}_{n,k}(\theta))).$$

A necessary condition allowing this contrast to make θ identifiable is that $r \geq p$. To construct a contrast leading to an identifiable parameter, we may have to consider more than the identifiable marginal distribution μ_θ of X_1, such as for example distributions of pairs, triplets, etc. This method can be extended to cases where X takes values in a general state space E.

If for example pairs (X_i, X_{i+1}) have the same distribution and if such pairs allow us to identify θ, we can use the marginal pseudo-likelihood of pairs:

$$l_n(\theta) = \sum_{i=1}^{n-1} \log f(x_i, x_{i+1}; \theta).$$

Example C.4. Gaussian contrast of second-order processes

Suppose that $X = (X_t, t \in \mathbb{Z})$ is a second-order stationary centered time series with spectral density f_θ. The periodogram associated with the empirical covariances $\widehat{r}_n(k)$ of the observations $X(n) = (X_1, X_2, \dots, X_n)$ is the estimation of the spectral density:

$$I_n(\lambda) = \frac{1}{2\pi} \sum_{k=-n+1}^{n-1} \widehat{r}_n(k) e^{i\lambda k} ,$$

where $\widehat{r}_n(k) = \widehat{r}_n(-k) = n^{-1} \sum_{i=1}^{n-|k|} X_i X_{i+k}$.

The periodogram $I_n(\lambda)$ is a poor estimator of $f(\lambda)$. However, Whittle's contrast, defined by the following regularization of I_n:

$$U_n(\alpha) = \frac{1}{2\pi} \int_0^{2\pi} \left\{ \log f_\alpha(\lambda) + \frac{I_n(\lambda)}{f_\alpha(\lambda)} \right\} d\lambda$$

is a good functional for estimating θ. Under Gaussian hypotheses, $-2U_n(\alpha)$ approximates the log-likelihood. Without Gaussian hypotheses, U_n leads to a good estimation under quite general conditions (Dalhaus and Künsch (57); Guyon (96); cf. §5.3.1). The contrast function associated with U_n is:

$$K(\alpha, \theta) = \frac{1}{2\pi} \int_0^{2\pi} \left\{ \log \frac{f_\alpha(\lambda)}{f_\theta(\lambda)} - 1 + \frac{f_\alpha(\lambda)}{f_\theta(\lambda)} \right\} d\lambda, \ K(\alpha, \theta) > 0 \text{ if } \alpha \neq \theta.$$

The condition $K(\alpha, \theta) \neq 0$ if $\alpha \neq \theta$ is satisfied if the parametrization of f_θ by θ is identifiable.

Example C.5. Conditional pseudo-likelihood of a Markov random field

While the likelihood of Markov chains can be calculated recursively, this is no longer true for spatial Markov random fields, which are fundamentally non-causal models. The reason for this comes partially from difficulty in calculating the normalizing constant. The same difficulty appears in general Gibbs models.

For this reason, Besag proposed in the context of Markov random fields on networks to use the *conditional pseudo-likelihood* (CPL), the product of conditional densities at each i of D_n:

$$l_n^{PV}(\theta) = \prod_{D_n} \pi_i(x_i \mid x_{\partial i}, \theta). \tag{C.3}$$

If we limit this product to a coding set C, we talk of *pseudo-likelihood on the coding* C. For both of these functionals, we require subergodicity condition (C.1) to be satisfied. However, if ergodicity of the random field suffices, this is not necessary. This is an important remark as in general we do not know whether a given Gibbs random field is ergodic.

Example C.6. Least squares estimation of variogram models

If we are modeling geostatistical data using an intrinsic variogram process $\gamma(\cdot,\theta)$, a classical way to estimate θ is to minimize the least squares contrast

$$U_n^{LS}(\theta) = \sum_{i=1}^{k} (\widehat{\gamma}_n(h_i) - \gamma(h_i,\theta))^2.$$

In this expression, $\widehat{\gamma}_n(h_i)$ are empirical estimates of the variogram at h_i for k preselected vectors h_i (cf. §5.1.3). Note that the least squares method is also used in the estimation of parametric models of spatial point processes (cf. §5.5.4).

Example C.7. Marginal pseudo-transitions of a dynamical system of particles

Consider as an example the presence/absence dynamics $X_i(t) \in \{0,1\}$, $i \in \mathbb{Z}^2$, $t = 0,1,\dots$ of a plant species on the spatial network $\mathbb{Z}^2$. Durrett and Levin (74) studied a discrete time contact process with two parameters (γ,λ) characterized as follows (cf. Fig. C.1): let x and y be configurations at successive instants of time t and $(t+1)$ and suppose:

1. The plant at site i survives with probability $(1-\gamma)$.
2. If a plant survives, it seeds independently in each of the 4 neighboring locations with probability λ.
3. $y_i = 1$ if at least one plant is present in position i at time $(t+1)$.

We suppose furthermore that all seeding defined in this dynamical system is independent in space and time.

After the first two steps, note $z_i \in \{0,1,2,3,4,5\}$ the number of plants at i at time $(t+1)$. For example, $z_i = 5$ if $x_i = 1$ and survives and if every neighboring plant

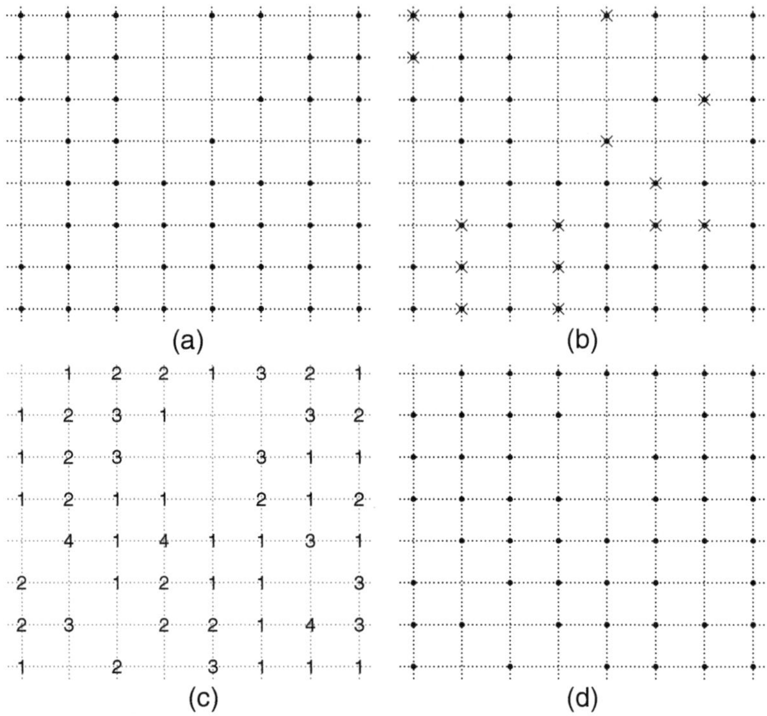

Fig. C.1 An example of the evolution $t \to t+1$ of a contact process on a regular network: (a) configuration $X(t)$ at time t (the $\bullet$ represent living plants); (b) plants ($\times$) that die with probability $\gamma = 0.3$; (c) number of seeds at each location after i.i.d. sowing in the four neighboring locations with probability $\lambda = 0.25$; (d) configuration $X(t+1)$ at time $t+1$.

at sites j survives and spreads to i. The third step tells us that $y_i = \mathbf{1}\{z_i \geq 1\} = 1$. While it is simple to simulate such dynamics, calculating the transition $P_S(x,y)$ is impossible when S is large: in effect, the first thing to notice is that we have to consider every single configuration x in order to calculate the joint probability of the trial $z = \{z_i, i \in S\}$, which has complexity $2^{\sharp(S)}$. Secondly, we are in a missing data situation as only variables $\mathbf{1}\{z_i \geq 1\}$ are observed and calculating the distribution of the observation at time $(t+1)$ is quite complicated.

To get around this, (101) suggest replacing the transition with the *marginal pseudo-transition*

$$M_S(x,y) = \prod_{i \in I(x,S)} P_\theta(X_i(t+1) = y_i \mid x), \tag{C.4}$$

a product of marginal transitions on $I(x,S) = \{s \in S : x_s + \sum_{j \in \partial s} x_j \geq 1\}$. The aforementioned probabilities are easy to calculate in this case:

$$P_\theta(X_i(t+1) = y_i \mid x) = P_\theta(X_i(t+1) = y_i \mid x_{\{i\} \cup \partial i}).$$

$I(x,S)$ is the set of only the sites x that carry information as in effect, if $x_s + \sum_{t\in\partial s} x_t = 0$ at time t, then $X_{t+1}(s) = 0$.

The *marginal pseudo-likelihood* of temporal observations $\{x_S(0), x_S(1), \ldots, x_S(T)\}$ is thus defined by the product

$$M_{S,T}(\theta) = \prod_{t=0}^{T-1} M_S(x(t), x(t+1)).$$

Conditional on survival of the process, Guyon and Pumo (101) show consistency and asymptotic normality of the estimator associated with this marginal pseudo-likelihood.

Example C.8. Further examples

In definition (C.3), $\log \pi_i(x_i \mid x^i, \alpha)$ can be replaced by other functionals $h_i(x(V_i); \alpha)$ that may be better-adapted and that allow model identification. If the conditional dependency π_i is unbounded and/or not given, we can replace it with an *ad hoc* functional: for example, if $X_i \in \mathbb{R}$ and we are able to calculate $g_i(x(W_i), \alpha) = E_\alpha(X_i \mid x(W_i))$, then

$$h_i(x(V_i), \alpha) = \{x_i - g_i(x(W_i), \alpha)\}^2, \text{ with } V_i = \{i\} \cup W_i$$

leads to a *conditional least squares* (CLS) pseudo-likelihood. A *marginal pseudo-likelihood* for which $h_i(x(V_i), \alpha)$ is the marginal log-density of $X(V_i)$ under α is another example of a contrast.

C.2 Asymptotic properties

Results in the "ergodic" form are due to Dacunha-Castelle and Duflo (54). Other proofs are given by Amemiya (6) and Gourieroux-Montfort (92). The non-ergodic form of the asymptotic properties of minimum contrast estimation is given in ((96), Ch. 3).

C.2.1 Convergence of the estimator

We have the following convergence result (Hardouin (108); Guyon (96)):

Theorem C.1. $\widehat{\theta}_n \xrightarrow{\Pr} \theta$ *under the following conditions:*

(C1) $\alpha \mapsto K(\alpha, \theta)$ *and the contrasts* $\alpha \mapsto U_n(\alpha)$ *are* P_α*-a.s. continuous.*
(C2) (U_n) *satisfies subergodicity condition (C.1).*
(C3) *If* $W_n(\eta)$ *is the modulus of continuity of* $U_n(\cdot)$, $\exists \varepsilon_k \downarrow 0$ *s.t. for each k:*

$$\lim_n P_\theta(W_n(1/k) \geq \varepsilon_k) = 0. \tag{C.5}$$

Corollary C.1. *Let $U_n = \sum_{i=1}^{p} a_{n,i} U_{n,i}$, $a_{n,i} \geq 0$ be a contrast process such that:*

1. *$U_{n,1}$ satisfies (C1-C2-C3).*
2. *Each $U_{n,i}$ satisfies (C3).*
3. *$a = \liminf_n a_{n,1} > 0$.*

Then the minimum contrast estimator converges.

This corollary shows why conditions ensuring convergence of the estimator of a Markov random field for a given coding C implies convergence of the conditional pseudo-likelihood estimator U_n (cf. §5.4.2). In effect, in this case, $U_n = U_{n,C} + \sum_l U_{n,C_l}$ for a given partition $\{C,(C_l)\}$ of S into coding subsets and the conditions of the corollary are indeed satisfied.

For convex contrast processes, we have the following *a.s.* convergence result (Senoussi (196); Guyon (96)):

Proposition C.1. *If Θ is an open convex set in $\mathbb{R}^p$, if contrasts $\theta \mapsto U_n(\theta)$ are convex and if (C.2) holds, then $\widehat{\theta}_n \xrightarrow{a.s.} \theta$.*

Using the Newton-Raphson algorithm to get efficient estimators

Consider an estimator $\widehat{\theta}_n$ of $\theta \in \mathbb{R}^p$, the solution to a system of p equations:

$$F(x(n);\theta) = 0,\ \theta \in \mathbb{R}^p, \tag{C.6}$$

where F takes values in $\mathbb{R}^p$. Minimum contrast estimators fall into this category when $F(\theta) = U_n^{(1)}(\theta)$, the gradient of U_n.

As finding the solution to (C.6) is not always easy, it is useful to use the Newton-Raphson algorithm initialized with a "good" estimator $\widetilde{\theta}_n$ that is easy to obtain. After one step, the algorithm leads to θ_n^*:

$$\theta_n^* = \widetilde{\theta}_n - \mathscr{F}^{-1}(\widetilde{\theta}_n)F(\widetilde{\theta}_n), \tag{C.7}$$

where $\mathscr{F}(\alpha)$ is the $p \times p$ matrix with entries $F_{i,\alpha_j}^{(1)}(\alpha)$ representing the derivative at α_j of component i of F, $i,j = 1,\ldots,p$.

If F is fairly regular and if $\widetilde{\theta}_n$ is consistent at a sufficient rate, Dzhaparidze (75) showed that θ_n^* is asymptotically equivalent to $\widehat{\theta}_n$. More precisely, let $(\upsilon(n))$ be a real-valued sequence going to infinity. We say that an estimator $\overline{\theta}_n$ of θ is $\upsilon(n)$-consistent if

$$\upsilon(n)(\overline{\theta}_n - \theta) = O_P(1).$$

Two estimators $\overline{\theta}_n$ and θ_n^* are said to be asymptotically $v(n)$-equivalent if

$$\lim_n \upsilon(n)(\overline{\theta}_n - \theta_n^*) = o_P(1).$$

Define the following conditions:

(DZ-1) Equation (C.6) has a $\tau(n)$-consistent solution $\widehat{\theta}_n$.
(DZ-2) F is a $\mathscr{C}^2(\mathscr{V}(\theta))$ vector function where $\mathscr{V}(\theta)$ is a neighborhood of θ and there exists a non-stochastic regular matrix $W(\theta)$ such that

$$\lim_n (\mathscr{F}(\theta) - W(\theta)) \overset{P_\theta}{=} 0.$$

(DZ-3) The second derivatives satisfy:

$$\forall \delta > 0, \exists M < \infty \text{ s.t. } \lim_n P_\theta\{\sup\{\left\|F^{(2)}(\alpha)\right\|, \alpha \in \mathscr{V}(\theta)\} < M\} \geq 1 - \delta.$$

Proposition C.2. *(Dzhaparidze (75)) Suppose conditions (DZ) are satisfied. Then, if $\widetilde{\theta}_n$ is a $\widetilde{\tau}(n)$-consistent initial estimator of θ with rate $\widetilde{\tau}(n) = o(\sqrt{\tau(n)})$, the estimator θ_n^* from (C.7) is asymptotically $\tau(n)$-equivalent to $\widehat{\theta}_n$:*

$$\lim_n \tau(n)(\widehat{\theta}_n - \theta_n^*) = o_P(1).$$

C.2.2 Asymptotic normality

Preliminary notation

If h is a real-valued $\mathscr{C}^2$ function in a neighborhood of θ, let $h^{(1)}(\theta)$ denote the gradient of h (vector of first derivatives at θ) and $h^{(2)}(\theta)$ the Hessian matrix of second derivatives at θ. If A and B are symmetric $p \times p$ matrices, we note:

- $\|A - B\| = \sum_{i,j} |A_{ij} - B_{ij}|$.
- $A \geq B$ (resp. $A > B$) if $A - B$ is p.s.d. (resp. p.d.).
- If $A > 0$ has a spectral decomposition $A = PD^tP$ where P is orthogonal and D diagonal, we choose $R = PD^{\frac{1}{2}}$ as the matrix representing the square root of A: $R^tR = {}^tRR = A$.

Hypotheses (N) ensuring asymptotic normality

(N1) There exists a neighborhood V of θ in which U_n is a $\mathscr{C}^2$ function and a real-valued P_θ-integrable random variable h satisfying:

$$\forall \alpha \in V, \left\|U_{n,\alpha^2}(\alpha, x)\right\| \leq h(x).$$

(N2) Matrices $J_n = Var(\sqrt{a_n}U_n^{(1)}(\theta))$ exist, as does a sequence $(a_n) \longrightarrow \infty$ such that:

(N2-1) There exists a p.d. matrix J such that $J_n \geq J$ for large n.

(N2-2) $\sqrt{a_n}J_n^{-1/2}U_n^{(1)}(\theta) \xrightarrow{d} \mathcal{N}_p(0,I_p)$.

(N3) There exists a sequence of non-stochastic matrices (I_n) such that:

(N3-1) There exists a non-stochastic and p.d. matrix I such that for large n, $I_n \geq I$.

(N3-2) $(U_n^{(2)}(\theta) - I_n) \xrightarrow{\Pr} 0$.

Theorem C.2. *Asymptotic normality of $\widehat{\theta}_n$ (108; 96)*
If $\widehat{\theta}_n$ converges and if conditions (N) hold, then:

$$\sqrt{a_n}J_n^{-1/2}I_n(\widehat{\theta}_n - \theta) \xrightarrow{d} \mathcal{N}_p(0,I_p).$$

Corollary C.2. *If U_n is the likelihood contrast, then:*

1. *Under the hypotheses of the previous theorem, we can chose for $I_n = J_n$ the Fisher information matrix:*

$$\sqrt{a_n}I_n^{-1/2}(\widehat{\theta}_n - \theta) \xrightarrow{d} \mathcal{N}_p(0,I_p).$$

2. *Under ergodicity condition $I_n \xrightarrow{\Pr} I(\theta) > 0$,*

$$\sqrt{a_n}(\widehat{\theta}_n - \theta) \xrightarrow{d} \mathcal{N}_p(0,I(\theta)^{-1}).$$

Comments

1. J_n and I_n are pseudo-information matrices that must be positively "lower bounded."
2. Asymptotic normality (N2-2) comes from a CLT for weakly-dependent variables and conditionally centered random fields (cf. Th. B.2).

Example C.9. Additive contrast for mixing random fields on a network (108; 96)

The following conditions (A1–A3) imply (N). They are given relative to a weakly dependent random field defined over a discrete network S in $\mathbb{R}^d$ that is not necessarily regular with additive contrast:

$$U_n(\alpha) = \frac{1}{d_n}\sum_{s\in D_n} g_s(X(V(s)),\alpha).$$

U_n is the sum of local functionals $\{g_s, s \in S\}$, where $\{V(s), s \in S\}$ is a family of bounded neighborhoods of $s \in S$ and d_n the cardinality of D_n. For example, g_s is the (negative of the) log of a conditional or marginal (pseudo) density.

(A1) *For the network $S \subset \mathbb{R}^d$:*
S is infinite, locally finite: $\forall s \in S$ and $\forall r > 0$, $\sharp\{B(s,r) \cap S\} = O(r^d)$.

(A2) *For the random field* X: X is an α-mixing random field with mixing coefficient $\alpha(\cdot) = \alpha_{\infty,\infty}(\cdot)$ (Doukhan (67); Guyon (96), §B.2) satisfying:

(A2-1) $\exists \delta > 0$ s.t. $\sum_{i,j \in D_n} \alpha(d(i,j))^{\frac{\delta}{2+\delta}} = O(d_n)$.
(A2-2) $\sum_{l \geq 0} l^{d-1} \alpha(l) < \infty$.

(A3) *For the functionals* $(g_s, s \in S)$:

(A3-1) (N1) is uniformly satisfied on V by the g_s, $s \in S$.
(A3-2) $\forall s \in S$, $E_\theta(g_s^{(1)}(\theta)) = 0$ and $\sup_{s \in S, \alpha \in \Theta, k=1,2} \left\| g_s^{(k)}(\alpha) \right\|_{2+\delta} < \infty$.
(A3-3) There exists two symmetric p.d. matrices I and J such that for large n: $J_n = Var(\sqrt{d_n} U_n^{(1)}(\theta)) \geq J > 0$ and $I_n = E_\theta(U_n^{(2)}(\theta)) \geq I > 0$.

C.2.2.1 Pseudo-likelihood ratio tests

Let (H_p) be the hypothesis $\theta \in \Theta \subset \mathbb{R}^p$ with dimension p and (H_q), $q < p$ a subhypothesis defined by the functional specification:

$$(H_q) : \alpha = r(\varphi), \varphi \in \Lambda \text{ an open set of } \mathbb{R}^q, \; \theta = r(\phi), \tag{C.8}$$

where $r : \Lambda \longrightarrow \Theta$ is $\mathscr{C}^2(W)$ in a neighborhood W of the true value ϕ of the parameter under (H_q). Suppose that (H_q) has dimension q and that $R = \frac{\partial r}{\partial \alpha}(\phi)$ is of rank q.

There are two ways to deal with the problem of testing the subhypothesis (H_q): the first is to construct a contrast difference test; the second is the Wald test that $\alpha = r(\varphi)$ can be expressed in the constrained form $C(\theta) = 0$.

Test statistic based on difference of contrasts

Note $\overline{U}_n(\varphi) = U_n(r(\varphi))$ the contrast under (H_q) and $\widehat{\varphi}_n$ the associated minimum contrast estimator, $\overline{\theta}_n = r(\widehat{\varphi}_n)$. The contrast difference test uses the statistic:

$$\Delta_n = 2a_n \left[U_n(\overline{\theta}_n) - U_n(\widehat{\theta}_n) \right].$$

Let $\overline{I}_n$, $\overline{J}_n$, $\overline{I}$ and $\overline{J}$ be matrices defined analogously to I_n, J_n, I and J but for $\overline{U}_n$ and:

$$A_n = J_n^{1/2}(I_n^{-1} - R\overline{I}_n^{-1}\,{}^tR)J_n^{1/2}.$$

A_n is a rank $(p-q)$ p.s.d. matrix whose positive eigenvalues we note $\{\lambda_{i,n}, i = 1, \ldots, p-q\}$. Let $(X_n \overset{d}{\sim} Y_n)$ mean that the random variables X_n and Y_n have the same limiting distribution as n goes to infinity.

Theorem C.3. *Asymptotic test for contrast difference (22; 96)*

Suppose that $\widehat{\theta}_n$ *converges and that* (U_n) *(resp.* $(\overline{U}_n)$*) satisfies hypotheses (N) under* (H_p) *(resp. under* (H_q)*). Then, for large n and for independent* χ_1^2*:*

$$\text{under } (H_q) : \Delta_n = 2a_n \left[U_n(\overline{\theta}_n) - U_n(\widehat{\theta}_n) \right] \overset{d}{\sim} \sum_{i=1}^{p-q} \lambda_{i,n} \chi^2_{i,1}.$$

Comments

1. If we can choose $I_n = J_n$, A_n is idempotent with rank $(p-q)$: we therefore obtain, under not necessarily ergodic hypotheses, the χ^2_{p-q} likelihood ratio test. We have $I_n = J_n$ for the following contrasts:
   (a) The likelihood of independent observations.
   (b) The likelihood of not necessarily homogeneous Markov chains.
   (c) The coding contrast of a Markov random field.
   (d) More generally, a model with conditionally independent variables and a subergodicity condition on the conditioning variables.
2. If the model is ergodic, noting I (resp. J, $\overline{I}$) the limit of (I_n) (resp. (J_n), $(\overline{I}_n)$) and $\{\lambda_i, i = 1, \ldots, p-q\}$ the positive eigenvalues of $A = J^{1/2}(I^{-1} - R\overline{I}^{-1}\,{}^tR)J^{1/2}$, we have $\Delta_n \overset{d}{\sim} \sum_{i=1}^{p-q} \lambda_i \chi^2_{i,1}$ under (H_q).

C.2.2.2 Specification tests with constraints

Suppose that (H_q) can be written in the constrained form:

$$(H_q) : \psi = C(\theta) = 0,$$

where $C : \mathbb{R}^p \to \mathbb{R}^{p-q}$ is a $\mathscr{C}^2$ constraint in a neighborhood of θ with rank $(p-q)$ at θ: $\text{rank}(C^{(1)}_\alpha(\theta)) = p - q$. A direct method for testing (H_q) is to use the Wald statistic associated with the constraint: ψ is estimated by $\widehat{\psi}_n = C(\widehat{\theta}_n)$ and the test of (H_q) relies on the statistic

$$\Xi_n = {}^t\widehat{\psi}_n \Sigma_n^{-1} \widehat{\psi}_n \overset{d}{\sim} \chi^2_{p-q} \quad \text{under} \quad (H_q),$$

where $\Sigma_n = C^{(1)}(\widehat{\theta}_n)\widehat{Var}(\widehat{\theta}_n)\,{}^tC^{(1)}(\widehat{\theta}_n)$ is the estimated variance of $\widehat{\psi}_n$ under (H_p).

C.3 Model selection by penalized contrast

Suppose that the parameter space satisfies $\Theta \subseteq \mathbb{R}^M$, where $\mathbb{R}^M$ corresponds to an upper-bounding model, $M < \infty$. A standard choice of family $\mathscr{E}$ of possible models is the family of non-empty subsets of $M = \{1, 2, \ldots, M\}$,

$$\delta = \{\theta = (\theta_i)_{i \in M} \text{ s.t. } \theta_i = 0 \text{ if } i \notin \delta\},$$

or perhaps an increasing sequence of spaces δ. Other choices may be useful for testing random field isotropy hypotheses.

Model selection is the choice of an element $\delta \in \mathscr{E}$ using the data $X(n)$. To do this, if (a_n) is the rate associated with contrast (U_n) (cf. (N2)), we use as decision function the (pseudo-likelihood) contrast *penalized* at rate $c(n)$ by the model dimension $|\delta|$:

$$W_n(\alpha) = U_n(\alpha) + \frac{c(n)}{a_n}\,|\delta(\alpha)|\,.$$

Note:

$$\overline{W}_n(\delta) = \overline{U}_n(\delta) + \frac{c(n)}{a_n}\,|\delta|$$
$$\text{with } \overline{U}_n(\delta) = \operatorname*{argmin}_{\alpha\in\Theta_\delta} U_n(\alpha).$$

According to the Akaike parsimony principle (4) we choose the model:

$$\widehat{\delta}_n = \operatorname*{argmin}_{\delta\in\mathscr{E}} \overline{W}_n(\delta).$$

This choice represents a compromise between goodness of fit (necessitating a relatively "large" model, $\overline{U}_n(\delta)$ decreasing with respect to δ) and a simple, interpretable model.

We say that such a criteria selects the true model if $\widehat{\delta}_n \to \delta_0$: for example, penalized likelihood allows us to select a convex model with i.i.d. observations (196). Similarly, under appropriate conditions, Whittle's contrast for stationary time series selects the true model (199; 107; 34). Selection is also possible without the need for ergodicity hypotheses.

Tools for these types of result rest on a bound on the probability of selecting a 'bad' model and a version of the law of the iterated logarithm for the gradient $U_n^{(1)}$ of the contrast process. Such conditions and results are presented in a general framework in [96, §3.4]. More precisely, Guyon and Yao (102) give, for a large class of models and associated contrasts (regression and least squares, AR and Whittle's contrast, Markov random fields and conditional pseudo-likelihood, infinite variance models) a characterization of the sets of over and under-parametrizations of the models and calculation of their probabilities, leading to conditions ensuring consistency of criteria for each model type. Results on the selection of Markov models via penalized pseudo-likelihood functions are also given in (124) and (52).

C.4 Proof of two results in Chapter 5

C.4.1 Variance of the maximum likelihood estimator for Gaussian regression

Let l be the log-likelihood of a Gaussian regression (§5.3.4), $l^{(1)} = (l_\delta^{(1)}, l_\theta^{(1)})$. After directly calculating that $l_\delta^{(1)} = {}^tZ\Sigma^{-1}(Z\delta - X)$, we use the result $\partial/\partial\theta_i \log(|\Sigma|) = \operatorname{tr}(\Sigma^{-1}\Sigma_i)$ to obtain:

$$(l_\theta^{(1)})_i = 2^{-1}\{\mathrm{tr}(\Sigma^{-1}\Sigma_i) + {}^t(X - Z\delta)\Sigma^i(X - Z\delta)\}.$$

Next, we find both the four blocks of the matrix of second derivatives and their expectations:

(i) $l_{\delta^2}^{(2)} = {}^tZ\Sigma^{-1}Z$ is constant and equal to J_δ.

(ii) The ith column of $l_{(\delta,\theta)}^{(2)}$ is ${}^tZ\Sigma^i(Z\delta - X)$, $E(l_{(\delta,\theta)}^{(2)}) = 0$.

(iii) $(l_{\theta^2}^{(2)})_{ij} = 2^{-1}\{\mathrm{tr}(\Sigma^{-1}\Sigma_{ij} + \Sigma^i\Sigma_j) + {}^t(X - Z\delta)\Sigma^{ij}(X - Z\delta)\}$; but

$$\begin{aligned} E\{{}^t(X - Z\delta)\Sigma^{ij}(X - Z\delta)\} &= \sum_{kl} \Sigma^{ij}(k,l)cov(X_k, X_l) \\ &= \sum_{kl} \Sigma^{ij}(k,l)\Sigma(k,l) = \mathrm{tr}(\Sigma^{ij}\Sigma). \end{aligned}$$

We deduce that $E(l_{\theta^2}^{(2)})_{ij} = 2^{-1}tr(\Sigma^{-1}\Sigma_{ij} + \Sigma^i\Sigma_j + \Sigma^{ij}\Sigma)$. Differentiating at $\theta_i \times \theta_j$ the product $\Sigma^{-1}\Sigma \equiv I$, we find

$$\Sigma^{-1}\Sigma_{ij} + \Sigma^i\Sigma_j + \Sigma^j\Sigma_i + \Sigma^{ij}\Sigma = 0,$$

and therefore: $E(l_{\theta^2}^{(2)})_{ij} = -2^{-1}\mathrm{tr}(\Sigma^j\Sigma_i)$.

□

C.4.2 Consistency of maximum likelihood for stationary Markov random fields

We give here a proof of consistency of the ML estimator for stationary Markov random fields on $\mathbb{Z}^d$ (cf. §5.4.1). The proof involves verifying the general conditions that ensure convergence of minimum contrast estimators (cf. §C.2). To do this, we invoke several properties of Gibbs random fields. The first is that if $\mu = P_\theta \in \mathscr{G}_s(\pi_\theta)$ is stationary, then μ can be expressed as a convex linear combination of extremal elements μ^* of $\mathscr{G}_s(\pi_\theta)$, distributions which are in fact ergodic [85, §7.3 and §14.2]. If we can show the consistency of each component μ^*, we can deduce μ-consistency of the ML estimator as the extremal elements are mutually singular. It therefore suffices to show consistency of one such distribution μ^* (which is ergodic), i.e., show consistency of ML if μ is stationary and ergodic. To this end, noting U_n the contrast equal to the negative of the log-likelihood of X on D_n:

$$U_n(x;\alpha) = \frac{1}{\sharp D_n}\{\log Z_n(x_{\partial D_n};\alpha) - H_n(x;\alpha)\},$$

we will have proved consistency of ML after:

1. Identifying the contrast function: $K(\mu,\alpha) = \lim_n U_n(\alpha)$.
2. Showing its continuity at α and that $\alpha = \theta$ is its unique minimum.
3. Showing that the condition on the modulus of continuity of U_n is satisfied.

In order to study limit behavior of U_n, we make the following preliminary remark: for general potential $\phi = (\phi_A, A \in S)$, define the mean energy at a site by

$$\overline{\phi}_i = \sum_{A:i\in A} \frac{\phi_A}{\sharp A}, \ i \in S.$$

It can be shown that the energy $H_\Lambda(x) = \sum_{A:A\cap\Lambda\neq\emptyset} \phi_A(x)$ of x on Λ satisfies:

$$H_\Lambda(x) = \sum_{i\in\Lambda} \overline{\phi}_i(x) + \varepsilon_\Lambda(x),$$

where

$$\varepsilon_\Lambda(x) = -\sum_{A:A\cap\Lambda\neq\emptyset \text{ and } A\not\subseteq\Lambda} \phi_A(x)\left\{1 - \frac{\sharp(A\cap\Lambda)}{\sharp A}\right\}.$$

For specification (5.23) induced by the $\{\Phi_{A_k}, k = 1,\dots,p\}$, we have:

$$|\varepsilon_\Lambda(x)| \leq \sum_{A:A\cap\Lambda\neq\emptyset \text{ and } A\not\subseteq\Lambda} |\phi_A(x)| \leq p \times (\sharp\partial\Lambda) \times \sup_k \left\|\Phi_{A_k}\right\|_\infty. \qquad \text{(C.9)}$$

In effect, the potentials are bounded and $\sharp\{A : \phi_A \neq 0, A\cap\Lambda \neq \emptyset \text{ and } A \not\subseteq \Lambda\} \leq p \times \sharp\partial\Lambda$. Also, if ϕ is translation-invariant, $\overline{\phi}_i(x) \equiv \overline{\phi}_0(\tau_i(x))$. We infer that:

$$U_n(x;\alpha) = p_n(x;\alpha) - \frac{1}{\sharp D_n}\sum_{i\in D_n} \overline{\Phi_i}(x;\alpha) + \frac{1}{\sharp D_n}\varepsilon_{D_n}(x;\alpha), \qquad \text{(C.10)}$$

where $p_n(x;\alpha) = \frac{1}{\sharp D_n}\log Z_n(x;\alpha)$. The first term of (C.10) converges independently of x to the pressure $p(\alpha)$ of the potential ϕ_α (cf. (85), §15.3). Furthermore, due to the upper bound (C.9), the third term tends to 0 uniformly at x because $\sharp\partial D_n/\sharp D_n \to 0$. As for the second term, it tends to $-E_\mu(\overline{\Phi}_{\{0\}}(\alpha))$ as μ is ergodic and the potential is bounded. We thus obtain:

$$U_n(x;\alpha) - U_n(x,\theta) \longrightarrow K(\mu,\alpha) = p(\alpha) - E_\mu(\overline{\Phi}_{\{0\}}(\alpha)) + h(\mu) \geq 0,$$

where $h(\mu)$ is the specific entropy of μ ((85), §15.4). As representation $\alpha \longmapsto \pi_{\{0\},\alpha}$ is well-defined, $\mathscr{G}(\pi_\theta)\cap\mathscr{G}(\pi_\alpha) = \emptyset$ if $\alpha \neq \theta$; the variational principle ((85), §15.4) therefore gives that $K(\mu,\alpha) > 0$ if $\alpha \neq \theta$.

It remains to show continuity of $\alpha \mapsto K(\mu;\alpha)$ and to verify the condition on the modulus of continuity of U_n. First, we look to bound the p_n term. We have:

$$\begin{aligned} p_n(x;\alpha) - p_n(x;\beta) &= \frac{1}{\sharp D_n}\log\frac{\int_{E^{D_n}} \exp H_n(x;\alpha)\lambda_n(dx_{D_n})}{\int_{E^{D_n}} \exp H_n(x;\beta)\lambda_n(dx_{D_n})} \\ &= \frac{1}{\sharp D_n}\log\int_{E^{D_n}} \exp\{H_n(x;\alpha) - H_n(x;\beta)\}\pi_{D_n}(dx_{D_n}/x;\beta) \\ &\geq \frac{1}{\sharp D_n} E_{\pi_{D_n}(\cdot/x;\beta)}\{H_n(x;\alpha) - H_n(x;\beta)\} \text{ (Jensen)} \\ &= {}^t(\alpha-\beta)\overline{h}_n(x;\beta), \end{aligned}$$

where $\overline{h}_n(x;\beta) = {}^t(\overline{h}_{k,n}(x;\beta), k = 1,\ldots,p)$, $\overline{h}_{k,n}(x;\beta) = \frac{1}{\sharp D_n} E_{\pi_{D_n}(\cdot/x;\beta)}\{h_{k,n}(x)\}$. We deduce that:

$$ {}^t(\alpha-\beta)\overline{h}_n(x;\beta) \le p_n(x;\alpha) - p_n(x;\beta) \le {}^t(\alpha-\beta)\overline{h}_n(x;\alpha). $$

Remarking that if $a \le u \le b$, $|u| \le \max\{|a|,|b|\}$ and that for each $(x;\beta)$, $\overline{h}_{k,n}(x;\beta) \le \|\Phi_k\|_\infty$, we have:

$$ |p_n(x;\alpha) - p_n(x;\beta)| \le \sum_{k=1}^{p} |\alpha_k - \beta_k| \, \|\Phi_k\|_\infty . \tag{C.11} $$

As the same upper bound holds in the limit $p(\alpha) - p(\beta)$, $K(\mu,\alpha)$ is continuous at α. As for the condition on the modulus of continuity of U_n, it results from uniform continuity at (α,x) of U_n, which itself is a consequence of (C.11) and the formulation of the energy as a scalar product:

$$ H_n(x;\alpha) - H_n(x;\beta) = {}^t(\alpha-\beta)h_n(x). \qquad \square $$

Appendix D
Software

We use three software packages to perform calculations illustrating examples in this book: *R*, *OpenBUGS* and *AntsInFields*.

R is a statistical analysis and graphics package created by Ihaka and Gentleman (118). It is both a software package and a language originating from the *S* software created by AT&T Bell Laboratories. More precisely, *R* is an interpreted *object oriented* language.

R is freely available under the terms of the GNU General Public License (GPL) (see `www.r-project.org`). Its development and distribution are managed by the *R Development Core Team*. *R* is available in several forms: the code is principally written in C (though with some Fortran programs) for use with Unix and Linux systems and as precompiled executable versions for Windows, Macintosh and Alpha Unix. Code and executables are available from the internet site `cran.r-project.org` of the *Comprehensive R Archive Network* (*CRAN*).

Various manuals are available alongside *R* in *CRAN*. To get a rapid overview of *R*, we suggest reading "*R* for beginners" by Emmanuel Paradis. Several functions for data analysis are found in the *R* base package but the majority of statistical methods for spatial data analysis in *R* are available in the form of supplementary *packages*.

Such *packages* constitute one of the strong points of *R*: they are collections of functions most often developed by the statisticians who proposed the corresponding methodology. For our spatial data examples, we used the following packages: `geoR`, `RandomFields`, `spatstat` and `spdep`. Other packages are also available and can be browsed by following the 'Spatial' link in the CRAN task views (http://cran.r-project.org/web/views/).

Let us present a short example to give some idea of syntax used in *R*. Consider Example 5.16 examining the spatial distribution of aquatic tulips. We suppose that the data are saved in file `nyssa.txt` whose first two columns give the spatial coordinate and the third the sex (male or female) of the tulip. Reading files with columns of data either in standard ASCII or CSV (comma-separated values) is performed by the function `read.table`. Running this function creates an "object" `nyssa` in a data table format (*data.frame*). This object is made up of a list:

```
nyssa <- read.table("nyssa.txt", header = TRUE)
```

Each object in the list is equivalent to a vector and each element of a vector represents an individual. To discover the names of these objects, the function `names` can be run:

```
names(nyssa)
```

```
[1] "x"      "y"      "genre"
```

and to see the content of the item (for example, *x*), we type `nyssa$x`.

The user can perform operations (arithmetic, logical, comparative) on objects and functions (which are themselves objects):

```
library(spatstat)
X <- ppp(nyssa$x, nyssa$y, marks = nyssa$genre,
         window = owin(xrange = c(-1, 53),
         yrange = c(0, 53)))
plot(X)
```

The first line makes the `spatstat` package available for use. This package, created by Baddeley and Turner (15), contains functions for manipulating and representing data as well as statistical functions for analyzing point data. The second line creates the object `X` of the class `ppp` representing a 2-dimensional spatial distribution in a $[-1,53] \times [0,53]$ window. Lastly, we apply the function `plot` to this object, giving the representation of the configuration shown in Fig. D.1.

To calculate Ripley's *K* function, we use the `Kest` function. The flexibility of *R* means that if we now apply the `plot` function, we get a graphical representation of the *K* function. The final line gives the legend.

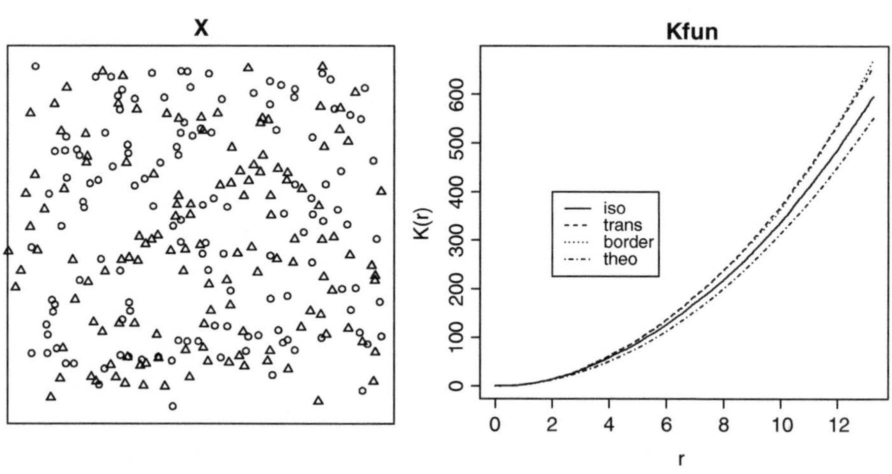

Fig. D.1 (a) Spatial distribution of the species *Nyssa aquatica* based on sex: male (△) or female ◯; (b) theoretic *K* function of a homogeneous Poisson PP and estimates under various boundary effect corrections.

```
Kfun <- Kest(X)
plot(Kfun)
legend(2, 400, legend = c("iso", "trans", "border",
       "theo"))
```

BUGS (*Bayesian inference Using Gibbs Sampling*) and its Windows version *WinBUGS* (146) is a program developed for Bayesian statistical analysis using MCMC simulation methods, notably Gibbs sampling. We have used *OpenBUGS* (212), the open-source version (available at `mathstat.helsinki.fi/openbugs`). It is easy to specify and estimate spatial hierarchical models with *BUGS* using the *GeoBUGS* module. As an example, we can look at specifications of models (5.52), (5.53) and (5.54) from Example 5.23 on lung cancer in Tuscany:

```
model {
  gamma[1:N] ~ car.normal(adj[], weights[], num[],
                          kappa)
  for (i in 1 : N) {
      alpha[i]~dnorm(0.0,tau)
      O[i]  ~ dpois(mu[i])
      log(mu[i]) <- log(E[i]) + beta1  + gamma[i]+
                        alpha[i]
      SMRhat[i] <- mu[i]/E[i]
  }
  beta1 ~ dnorm(0.0, 1.0E-5)
  tau   ~ dgamma(0.5, 0.0005)
  kappa ~ dgamma(0.5, 0.0005)
}
```

R packages `BRugs` and `R2WinBUGS` also provide interfaces linking *R* to *BUGS*.

AntsInFields is a program developed by Felix Friedrich to simulate and estimate Gibbs random fields on networks as well as for image analysis. It is both a good educational resource on simulating and estimating Gibbs random fields and a research tool. Object oriented and modular, it is available at `www.antsinfields.de` under the terms of the GNU Less General Public License (LGPL). *AntsInFields* enables us to simulate using Gibbs sampling and Metropolis-Hastings dynamics, as well as exact simulation and optimization by simulated annealing. It also performs CPL estimation for Ising models, Potts models and Besag auto-models and includes Bayesian methods for image reconstruction.

References

[1] Aarts, E., Korst, J.: Simulated Annealing and Boltzman Machines: Stochastic Approach to Combinatorial and Neural Computing. Wiley, New York (1989)

[2] Abramowitz, M., Stegun, I.A. (eds.): Handbook of Mathematical Functions. Dover, New York (1970)

[3] Adler, R.J.: The Geometry of Random Fields. Wiley, New York (1981)

[4] Akaike, H.: Fitting autoregressive models for prediction. Annals of the Institute of Statistical Mathematics **21**, 243–247 (1969)

[5] Alfò, M., Postiglione, P.: Semiparametric modelling of spatial binary observations. Statistical Modelling **2**, 123–137 (2002)

[6] Amemiya, T.: Advanced Econometric. Basil Blackwell, Oxford (1985)

[7] Anselin, L.: Spatial Econometrics : Methods and Models. Kluwer, Dordrecht (1988)

[8] Arnold, B.C., Castillo, E., Sarabia, J.M.: Conditional Specification of Statistical Models. Springer, New York (1999)

[9] Augustin, N.H., McNicol, J.W., Marriott, C.A.: Using the truncated auto-poisson model for spatially correlated counts of vegetation. Journal of Agricultural, Biological & Environmental Statistics **11**, 1–23 (2006)

[10] Azencott, R. (ed.): Simulated Annealing: Parallelization Techniques. Wiley, New York (1992)

[11] Baddeley, A.J., Gregori, P., Mateu, J., Stoica, R., Stoyan, D. (eds.): Case studies in Spatial Point Processes Modeling. Lecture Notes in Statistics 185. Springer, New York (2006)

[12] Baddeley, A.J., Møller, J.: Nearest-neighbour Markov point processes and random sets. International Statistical Review **57**, 90–121 (1989)

[13] Baddeley, A.J., Møller, J., Waagepetersen, R.P.: Non- and semi-parametric estimation of interaction in inhomogeneous point patterns. Statistica Neerlandica **54**(329-350) (2000)

[14] Baddeley, A.J., Turner, R.: Practical maximum pseudolikelihood for spatial point patterns (with discussion). Australian and New Zealand Journal of Statistics **42**, 283–322 (2000)

[15] Baddeley, A.J., Turner, R.: Spatstat: an R package for analyzing spatial point patterns. Journal of Statistical Software **12**, 1–42 (2005)

[16] Baddeley, A.J., Turner, R., Møller, J., Hazelton, M.: Residual analysis for spatial point processes (with discussion). Journal of the Royal Statistical Society, Series B **67**, 617–666 (2005)

[17] Baddeley, A.J., van Lieshout, M.N.M.: Area-interaction point processes. Annals of the Institute of Statistical Mathematics **46**, 601–619 (1995)

[18] Banerjee, S., Carlin, B.P., Gelfand, A.E.: Hierarchical Modeling and Analysis for Spatial Data. Chapman & Hall/CRC Press, Boca Raton: FL (2004)

[19] Barker, A.A.: Monte C arlo calculations of the radial distribution functions for a proton-electron plasma. Australian Journal of Physics **18**, 119–133 (1965)

[20] Bartlett, M.S.: Physical nearest-neighbour models and non-linear time series (I). Journal of Applied Probability **8**, 222–232 (1971)

[21] Bartlett, M.S.: Physical nearest-neighbour models and non-linear time series (II). Journal of Applied Probability **9**, 76–86 (1972)

[22] Bayomog, S., Guyon, X., Hardouin, C., Yao, J.: Test de différence de contraste et somme pondérée de Chi 2. Canadian Journal of Statistics **24**, 115–130 (1996)

[23] Benveniste, A., Métivier, M., Priouret, P.: Adaptive Algorithms and Stochastic Approximations. Springer, New York (1990)

[24] Besag, J.: On the correlation structure of some two dimensional stationary processes. Biometrika **59**, 43–48 (1972)

[25] Besag, J.: Spatial interaction and the statistical analysis of lattice systems. Journal of the Royal Statistical Society, Series B **36**, 192–236 (1974)

[26] Besag, J.: Efficiency of pseudo likelihood estimation for simple Gaussian fields. Biometrika **64**, 616–618 (1977)

[27] Besag, J., P., M.P.A.: On the estimation and testing of spatial interaction for Gaussian lattice processes. Biometrika **62**, 555–562 (1975)
[28] Besag, J., York, J., Mollié, A.: Bayesian image restoration, with two applications in spatial statistics (with discussion. Annals of the Institute of Statistical Mathematics **43**, 1–59 (1991)
[29] Bochner, N.: Lectures on Fourier Integrals. Princeton University Press, Princeton: NJ (1959)
[30] Bolthausen, E.: On the central limit theorem for stationary mixing random fields. Annals of Probability **10**, 1047–1050 (1982])
[31] Bouthemy, P., Hardouin, C., Piriou, G., Yao, J.: Mixed-state auto-models and motion texture modeling. Journal of Mathematical Imaging and Vision **25**, 387–402 (2006)
[32] Breiman, L.: Probability. SIAM Classics in Applied Mathematics 7 (1992)
[33] Brillinger, D.R.: Estimation of the second-order intensities of a bivariate stationary point process. Journal of the Royal Statistical Sociery, Series B **38**, 60–66 (1976)
[34] Brockwell, P.J., Davis, R.A.: Time Series Analysis: Theory and Methods. Springer, New York (1992)
[35] Brook, D.: On the distinction between the conditional probability and joint probability approaches in the specification of nearest neighbour systems. Biometrika **51**, 481–483 (1964)
[36] Brown, P.E., Kåresen, K.F., Roberts, G.O., Tonellato, S.: Blur-generated non-separable space-time models. Journal of the Royal Statistical Society, Series B **62**, 847–860 (2000)
[37] Cairoli, R., Walsh, J.B.: Stochastic integral in the plane. Acta Mathematica **134**, 111–183 (1975)
[38] Casson, E., Coles, S.G.: Spatial regression models for extremes. Extremes **1**, 449–468 (1999)
[39] Catelan, D., Biggeri, A., Dreassi, E., Lagazio, C.: Space-cohort Bayesian models in ecological studies. Statistical Modelling **6**, 159–173 (2006)
[40] Catoni, O.: Rough large deviation estimates for simulated annealing : application to exponential schedules. Annals of Probability **20**, 1109–1146 (1992)
[41] Chadoeuf, J., Nandris, D., Geiger, J., Nicole, M., Pierrat, J.C.: Modélisation spatio-temporelle d'une épidémie par un processus de Gibbs : estimation et tests. Biometrics **48**, 1165–1175 (1992)
[42] Chalmond, B.: Eléments de modélisation pour l'analyse d'image. Springer, Paris (2000)
[43] Chilès, J.P., Delfiner, P.: Geostatistics. Wiley, New York (1999)
[44] Christensen, O.F., Roberts, G.O., Sköld, M.: Robust Markov chain Monte Carlo methods for spatial generalised linear mixed models. Journal of Computational and Graphical Statistics **15**, 1–17 (2006)
[45] Cliff, A.D., Ord, J.K.: Spatial Processes: Models and Applications. Pion, London (1981)
[46] Comets, F.: On consistency of a class of estimators for exponential families of Markov random fields on a lattice. Annals of Statistics **20**, 455–468 (1992)
[47] Comets, F., Janzura, M.: A central limit theorem for conditionally centered random fields with an application to Markov fields. Journal of Applied Probability **35**, 608–621 (1998)
[48] Cressie, N.A.C.: Statistics for Spatial Data, 2nd edn. Wiley, New York (1993)
[49] Cressie, N.A.C., Hawkins, D.M.: Robust estimation of the variogram, I. Journal of the International Association of Mathematical Geology **12**, 115–125 (1980)
[50] Cressie, N.A.C., Huang, H.C.: Classes of nonseparable, spatio-temporal stationary covariance functions. Journal of the American Statistical Association **94**, 1330–1340 (1999)
[51] Cross, G.R., Jain, A.K.: Markov field texture models. IEEE Transactions on Pattern Analysis and Machine Intelligence **5**, 155–169 (1983)
[52] Csiszar, I., Talata, Z.: Consistent estimation of the basic neighborhood of Markov random fields. Annals of Statistics **34**, 123–145 (2006)
[53] Cuzick, J., Edwards, R.: Spatial clustering for inhomogeneous populations. Journal of the Royal Statistical Society, Series B **52**, 73–104 (1990)
[54] Dacunha-Castelle, D., Duflo, M.: Probabilités et Statistiques, Tome 2: Problèmes à temps mobile. Masson, Paris (1993)
[55] Dacunha-Castelle, D., Duflo, M.: Probabilités et Statistiques. Tome 1: Problèmes à temps fixe. Masson, Paris (1994)

[56] Daley, D., Vere-Jones, D.: An Introduction to The Theory of Point Processes, Vol. I, Elementary Theory and Methods, 2nd edn. Springer, New York (2003)
[57] Dalhaus, R., Künsch, H.R.: Edge effect and efficient parameter estimation for stationary random fields. Biometrika **74**, 877–882 (1987)
[58] De Iaco, S., Myers, D.E., Posa, T.: Nonseparable space-time covariance models: some parametric families. Mathematical Geology **34**, 23–42 (2002)
[59] Devroye, L.: Non Uniform Random Variable Generation. Springer, New York (1986)
[60] Diaconis, P., Freedman, D.: Iterated random gunctions. SIAM Review **41**, 45–76 (1999)
[61] Diaconis, P., Graham, R., Morrison, J.: Asymptotic analysis of a random walk on an hypercube with many dimensions. Random Structure Algorithms **1**, 51–72 (1990)
[62] Diggle, P.J.: Statistical Analysis of Spatial Point Patterns. Oxford University Press, Oxford (2003)
[63] Diggle, P.J., Ribeiro, P.J.: Model-based Geostatistics. Springer, New York (2007)
[64] Diggle, P.J., Tawn, J.A., Moyeed, R.A.: Model-based geostatistics (with discussion). Applied Statistics **47**, 299–350 (1998)
[65] Dobrushin, R.L.: Central limit theorems for non stationary Markov chains I, II. Theory of Probability and its Applications **1**, 65–80, 329–383 (1956)
[66] Dobrushin, R.L.: The description of a random field by means of conditional probabilities and condition of its regularity. Theory of Probability and its Applications **13**, 197–224 (1968)
[67] Doukhan, P.: Mixing: Properties and Examples. Lecture Notes in Statistics 85. Springer, Berlin (1994)
[68] Droesbeke, J.J., Fine, J., Saporta, G. (eds.): Méthodes bayésiennes en statistique. Technip, Paris (2002)
[69] Droesbeke, J.J., Lejeune, M., Saporta, G. (eds.): Analyse statistique des données spatiales. Technip, Paris (2006)
[70] Dubois, G., Malczewski, J., De Cort, M.: Spatial Interpolation Comparison 1997. Journal of Geographic Information and Decision Analysis **2** (1998)
[71] Duflo, M.: Algorithmes stochastiques. Mathématiques et Applications. Springer, Paris (1996)
[72] Duflo, M.: Random Iterative Models. Springer, New York (1997)
[73] Durrett, R.: Ten lectures on particle systems. In: P. Bernard (ed.) École d'Été de St. Flour XXIII, Lecture Notes in Mathematics 1608, pp. 97–201. Springer-Verlag, New York (1995)
[74] Durrett, R., Levin, S.A.: Stochastic spatial models : a user's guide to ecological applications. Philosophical Transactions of the Royal Society of London, series B **343**, 329–350 (1994)
[75] Dzhaparidze, K.O.: On simplified estimators of unknown parameters with good asymptotic properties. Theory of Probability and its Applications **19**, 347–358 (1974)
[76] E., P.P., Deutsch, S.J.: Identification and interpretation of first order space-time arma models. Technometrics **22**, 397–408 (1980)
[77] Eriksson, M., Siska, P.P.: Understanding anisotropy computations. Mathematical Geology **32**, 683–700 (2000)
[78] Ferrandiz, J., Lopez, A., Llopis, A., Morales, M., Tejerizo, M.L.: Spatial interaction between neighbouring counties : cancer mortality data in Valencia (Spain). Biometrics **51**, 665–678 (1995)
[79] Fuentes, M.: Approximate likelihood for large irregularly spaced spatial data. Journal of the American Statistical Association **102**, 321–331 (2007)
[80] Gelman, A., Rubin, D.B.: Inference from iterative simulation using multiple sequences. Statistical Science **7**, 457–511 (1992)
[81] Geman, D.: Random fields and inverse problem in imaging. In: P.L. Hennequin (ed.) École d' Eté de Probabilités de Saint-Flour XVIII, Lecture Notes in Mathematics 1427, pp. 113–193. Springer, New York (1990)
[82] Geman, D., Geman, S.: Stochastic relaxation, Gibbs distributions and the bayesian restoration of images. IEEE Transactions on Pattern Analysis and Machine Intelligence **6**, 721–741 (1984)

[83] Geman, S., Graffigne, C.: Markov random fields models and their applications to computer vision. In: A.M. Gleason (ed.) Proceedings of the International Congress of Mathematicians 1986, pp. 1496–1517. American Mathematical Society, Providence: RI (1987)
[84] Georgii, H.O.: Canonical and grand canonical Gibbs states for continuum systems. Communications of Mathematical Physics **48**, 31–51 (1976)
[85] Georgii, H.O.: Gibbs measure and phase transitions. De Gruyter, Berlin (1988)
[86] Geyer, C.J.: On the convergence of Monte Carlo maximum likelihood calculations. Journal of the Royal Statistical Society, Series B **56**, 261–274 (1994)
[87] Geyer, C.J.: Likelihood inference for spatial point processes. In: O.E. Barndorff-Nielsen, W.S. Kendall, M.N.M. Van Lieshout (eds.) Stochastic geometry: likelihood and computation, pp. 79–140. Chapman & Hall/CRC, Florida (1999)
[88] Geyer, C.J., Møller, J.: Simulation procedures and likelihood inference for spatial point processes. Scandinavian Journal of Statistics **21**, 359–373 (1994)
[89] Gilks, W.R., Richardson, S., Spiegelhalter, D.J. (eds.): Markov Chain Monte Carlo in Practice. Chapman & Hall, London (1996)
[90] Gneiting, T.: Nonseparable, stationary covariance functions for space-time data. Journal of the American Statistical Association **97**, 590–600 (2002)
[91] Gneiting, T., Genton, M.G., Guttorp, P.: Geostatistical space-time models, stationarity, separability and full symmetry. In: B. Finkenstadt, L. Held, V. Isham (eds.) Statistical Methods for Spatio-Temporal Systems, pp. 151–175. Chapman & Hall/CRC, Boca Raton: FL (2007)
[92] Gourieroux, C., Monfort, A.: Statistiques et modèles économétriques, Tomes 1 et 2. Economica, Paris (1992)
[93] Green, P.J., Richardon, S.: Hidden Markov models and disease mapping. Journal of the American Statistical Association **97**, 1055–1070 (2002)
[94] Greig, D.M., Porteous, B.T., Seheult, A.H.: Exact maximum a posteriori estimation for binary images. Journal of the Royal Statistical Society, Series B **51**, 271–279 (1989)
[95] Guan, Y., Sherman, M.: On least squares fitting for stationary spatial point processes. Journal of the Royal Statistical Society, Series B **69**, 31–49 (2007)
[96] Guyon, X.: Random Fields on a Network: Modeling, Statistics and Applications. Springer, New York (1995)
[97] Guyon, X., Hardouin, C.: The Chi-2 difference of coding test for testing Markov random field hypothesis. In: P. Barone, A. Frigessi, M. Piccioni (eds.) Stochastic Models, Statistical Methods and Algorithms in Image Analysis, Lecture Notes in Statistics 74, pp. 165–176. Spinger, Berlin (1992)
[98] Guyon, X., Hardouin, C.: Markov chain Markov field dynamics: models and statistics. Statistics **36**, 339–363 (2002)
[99] Guyon, X., Künsch, H.R.: Asymptotic comparison of estimator of the Ising model. In: P. Barone, A. Frigessi, M. Piccioni (eds.) Stochastic Models, Statistical Methods and Algorithms in Image Analysis, Lecture Notes in Statistics 74, pp. 177–198. Spinger, Berlin (1992)
[100] Guyon, X., Pumo, B.: Estimation spatio-temporelle d'un modèle de système de particule. Comptes rendus de l'Académie des sciences Paris **I-340**, 619–622 (2005)
[101] Guyon, X., Pumo, B.: Space-time estimation of a particle system model. Statistics **41**, 395–407 (2007)
[102] Guyon, X., Yao, J.F.: On the underfitting and overfitting sets of models chosen by order selection criteria. Journal of Multivariate Analysis **70**, 221–249 (1999)
[103] Häggström, O.: Finite Markov Chains and Algorithmic Applications. Cambridge University Press, Cambridge (2002)
[104] Häggström, O., van Lieshout, M.N.M., Møller, J.: Characterisation results and Markov chain Monte Carlo algorithms including exact simulation for some spatial point processes. Bernoulli **5**, 641–658 (1999)
[105] Haining, R.: Spatial Data Analysis in the Social and Environmental Sciences. Cambridge University Press, Cambridge (1990)

[106] Hajek, B.: Cooling schedules for optimal annealing. Mathematics of Operations Research **13**, 311–329 (1999)

[107] Hannan, E.J.: The estimation of the order of an ARMA process. Annals of Statistics **8**, 1071–1081 (1980)

[108] Hardouin, C.: Quelques résultats nouveaux en statistique des processus: contraste fort, régressions à rélog-périodogramme. Ph.D. thesis, Université Paris VII, Paris (1992)

[109] Hardouin, C., Yao, J.: Multi-parameter auto-models and their application. Biometrika (2008). To appear

[110] Hastings, W.: Monte Carlo sampling methods using Markov chains and their applications. Biometrika **57**, 97–109 (1970)

[111] Heinrich, L.: Minimum contrast estimates for parameters of spatial ergodic point processes. In: Transactions of the 11th Prague Conference on Random Processes, Information Theory and Statistical Decision Functions, pp. 479–492. Academic Publishing House, Prague (1992)

[112] Higdon, D.: Space and space-time modeling using process convolutions. In: C. Anderson, V. Barnett, P.C. Chatwin, A. El-Shaarawi (eds.) Quantitative Methods for Current Environmental Issues, pp. 37–56. Springer-Verlag, London (2002)

[113] Higdon, D.M., Swall, J., Kern, J.: Non-stationary spatial modeling. In: J.M. Bernado, J.O. Berger, A.P. Dawid, A.F.M. Smith (eds.) Bayesian Statistics 6, pp. 761–768. Oxford University Press, Oxford (1999)

[114] Hoeting, A., Davis, A., Merton, A., Thompson, S.: Model selection for geostastistical models. Ecological Applications **16**, 87–98 (2006)

[115] Huang, F., Ogata, Y.: Improvements of the maximum pseudo-likelihood estimators in various spatial statistical models. Journal of Computational and Graphical Statistics **8**, 510–530 (1999)

[116] Ibragimov, I.A., Linnik, Y.V.: Independent and Stationary Sequences of Random Variables. Wolters-Noordhoff Publishing, Groningen (1971)

[117] Ibragimov, I.A., Rozanov, Y.A.: Processus aléatoires gaussiens. MIR, Moscou (1974)

[118] Ihaka, R., Gentleman, R.: R: a language for data analysis and graphics. Journal of Computational and Graphical Statistics **5**, 299–314 (1996)

[119] Illig, A.: Une modélisation de données spatio-temporelles par AR spatiaux. Journal de la société francaise de statistique **147**, 47–64 (2006)

[120] Isaacson, D.L., Madsen, R.Q.: Markov Chains: Theory and Application. Wiley, New York (1976)

[121] Jensen, J., Møller, J.: Pseudolikelihood for exponential family of spatial point processes. Annals of Applied Probability **3**, 445–461 (1991)

[122] Jensen, J.L.: Asymptotic normality of estimates in spatial point processes. Scandinavian Journal of Statistics **20**, 97–109 (1993)

[123] Jensen, J.L., Künsch, H.R.: On asymptotic normality of pseudo-likelihood estimates for pairwise interaction processes. Annals of the Institute of Statistical Mathematics **46**, 475–486 (1994)

[124] Ji, C., Seymour, L.: A consistent model selection procedure for Markov random fields based on penalized pseudo-likelihood. Annals of Applied Probability **6**, 423–443 (1996)

[125] Jolivet, E.: Central limit theorem and convergence of empirical processes for stationary point processes. In: P. Bastfai, J. Tomko (eds.) Point Processes and Queuing Problems, pp. 117–161. North-Holland, Amsterdam (1978)

[126] Jones, R., Zhang, Y.: Models for continuous stationary space-time processes. In: T.G. Gregoire, D.R. Brillinger, P.J. Diggle, E. Russek-Cohen, W.G. Warren, R.D. Wolfinger (eds.) Modelling Longitudinal and Spatially Correlated Data, Lecture Notes in Statistics 122, pp. 289–298. Springer, New York (1997)

[127] Kaiser, M.S., Cressie, N.A.C.: Modeling Poisson variables with positive spatial dependence. Statistics and Probability Letters **35**, 423–432 (1997)

[128] Keilson, J.: Markov chain models: Rarity and Exponentiality. Springer, New York (1979)

[129] Kemeny, G., Snell, J.L.: Finite Markov Chains. Van Nostrand, Princeton: NJ (1960)

[130] Kendall, W.S., Møller, J.: Perfect simulation using dominating processes on ordered state spaces, with application to locally stable point processes. Advances in Applied Probability **32**, 844–865 (2000)

[131] Klein, D.: Dobrushin uniqueness techniques and the decay of correlation in continuum statistical mechanics. Communications in Mathematical Physics **86**, 227–246 (1982)

[132] Koehler, J.B., Owen, A.B.: Computer experiments. In: S. Ghosh, C.R. Rao (eds.) Handbook of Statistics,Vol 13, pp. 261–308. North-Holland, New York (1996)

[133] Kolovos, A., Christakos, G., Hristopulos, D.T., Serre, M.L.: Methods for generating non-separable spatiotemporal covariance models with potential environmental applications. Advances in Water Resources **27**, 815–830 (2004)

[134] Krige, D.: A statistical approach to some basic mine valuation problems on the Witwatersrand. Journal of the Chemical, Metallurgical and Mining Society of South Africa **52**, 119–139 (1951)

[135] Kutoyants, Y.A.: Statistical Inference for Spatial Poisson Processes. Springer, New York (1998)

[136] Kyriakidis, P.C., Journel, A.G.: Geostatistical space-time models: a review. Mathematical Geology **31**, 651–684 (1999)

[137] Lahiri, S.N.: CLT for weighted sums of a spatial process under a class of stochastic and fixed designs. Sankhya A **65**, 356–388 (2003)

[138] Lahiri, S.N., Lee, Y., C., C.N.A.: On asymptotic distribution and asymptotic efficiency of least squares estimators of spatial variogram parameters. Journal Statistical Planning and Inference **103**, 65–85 (2002)

[139] Lantuèjoul, C.: Geostatistical Simulation. Springer, Berlin (2002)

[140] Laslett, M.: Kriging and splines: and empirical comparison of their predictive performance in some applications. Journal of the Americal Statistical Association **89**, 391–409 (1994)

[141] Lawson, A.B.: Statistical Methods in Spatial Epidemiology. Wiley, New York (2001)

[142] Le, N.D., Zidek, J.V.: Statistical Analysis of Environmental Space-Time Processes. Springer, New York (2006)

[143] Lee, H.K., Higdon, D.M., Calder, C.A., Holloman, C.H.: Efficient models for correlated data via convolutions of intrinsic processes. Statistical Modelling **5**, 53–74 (2005)

[144] Lee, Y.D., Lahiri, S.N.: Least square variogram fitting by spatial subsampling. Journal of the Royal Statistical Society, Series B **64**, 837–854 (2002)

[145] Loève, M.: Probability Theory II. Springer, New York (1978)

[146] Lunn, D.J., Thomas, A., Best, N., Spiegelhalter, D.J.: WinBUGS - a Bayesian modelling framework: concepts, structure, and extensibility. Statistics and Computing **10**, 325–337 (2000)

[147] Ma, C.: Families of spatio-temporal stationary covariance models. Journal of Statistical Planning and Inference **116**, 489–501 (2003)

[148] Mardia, K.V., Goodall, C., Redfern, E.J., Alonso, F.J.: The Kriged Kalman filter (with discussion). Test **7**, 217–252 (1998)

[149] Mardia, K.V., Marshall, J.: Maximum likelihood estimation of models for residual covariance in spatial regression. Biometrika **71**, 289–295 (1984)

[150] Marroquin, J., Mitter, S., Poggio, T.: Probabilistic solution of ill posed problem in computational vision. Journal of the American Statistical Association **82**, 76–89 (1987)

[151] Mase, S.: Marked Gibbs processes and asymptotic normality of maximum pseudo-likelihood estimators. Mathematische Nachrichten **209**, 151–169 (1999)

[152] Matheron, G.: Traité de géostatistique appliquée, Tome 1. Mémoires du BRGM, n. 14. Technip, Paris (1962)

[153] Matheron, G.: The intrinsic random function and their applications. Advances in Applied Probability **5**, 439–468 (1973)

[154] Matérn, B.: Spatial Variation: Stochastic Models and their Applications to Some Problems in Forest Surveys and Other Sampling Investigations, 2nd edn. Springer, Heidelberg (1986)

[155] McCullagh, P., Nelder, J.A.: Generalized Linear Models. Chapman & Hall, London (1989)

[156] Mercer, W.B., Hall, A.D.: The experimental error of field trials. The experimental error of field trials **4**, 107–132 (1973)
[157] Meyn, S.P., Tweedie, R.L.: Markov Chains and Stochastic Stability. Springer, New York (1993)
[158] Mitchell, T., Morris, M., Ylvisaker, D.: Existence of smoothed process on an interval. Stochastic Processes and their Applications **35**, 109–119 (1990)
[159] Møller, J., Syversveen, A.R., Waagepetersen, R.P.: Log-gaussian Cox processes. Scandinavian Journal of Statistics **25**, 451–82 (1998)
[160] Møller, J., Waagepetersen, R.P.: Statistical Inference and Simulation for Spatial Point Processes. Chapman & Hall/CRC, Boca Raton: FL (2004)
[161] Møller, J., Waagepetersen, R.P.: Modern statistics for spatial point processes. Scandinavian Journal of Statistics **34**, 643–684 (2007)
[162] Moyeed, R.A., Baddeley, A.J.: Stochastic approximation of the MLE for a spatial point pattern. Scandinavian Journal of Statistics **18**, 39–50 (1991)
[163] Neyman, J., Scott, E.L.: Statistical approach to problems of cosmology. Journal of the Royal Statistical Society, Series B **20**, 1–43 (1958)
[164] Nguyen, X.X., Zessin, H.: Ergodic theorems for spatial processes. Probability Theory and Related Fields **48**, 133–158 (1979)
[165] Nguyen, X.X., Zessin, H.: Integral and differential characterization of the Gibbs process. Mathematische Nachrichten **88**, 105–115 (1979)
[166] Onsager, L.: Crystal statistics I : A two dimensional model with order-disorder transition. Physical Review **65**, 117–149 (1944)
[167] Ord, J.K.: Estimation methods for models of spatial interaction. Journal of the American Statistical Association **70**, 120–126 (1975)
[168] Papangelou, F.: The conditional intensity of general point processes and application to line processes. Zeitschrift für Wahscheinlichkeitstheorie und verwandte Gebiete **28**, 207–227 (1974)
[169] Penttinen, A.: Modelling interaction in spatial point patterns: parameter estimation by the maximum likelihood method. Jyväskylä Studies in Computer Science, Economics and Statistics **7** (1984)
[170] Perrin, O., Meiring, W.: Identifiability for non-stationary spatial structure. Journal of Applied Probability **36**, 1244–1250 (1999)
[171] Perrin, O., Senoussi, R.: Reducing non-stationary random fields to stationary and isotropy using space deformation. Statistics and Probability Letters **48**, 23–32 (2000)
[172] Peskun, P.: Optimum Monte Carlo sampling using Markov chains. Biometrika **60**, 607–612 (1973)
[173] Peyrard, N., Calonnec, A., Bonnot, F., Chadoeuf, J.: Explorer un jeu de données sur grille par test de permutation. Revue de Statistique Appliquée **LIII**, 59–78 (2005)
[174] Pfeifer, P.E., Deutsch, S.J.: A three-stage iterative procedure for space-time modeling. Technometrics **22**, 93–117 (1980)
[175] Ploner, A.: The use of the variogram cloud in geostatistical modelling. Environmentrics **10**, 413–437 (1999)
[176] Preston, C.: Random Fields. Lecture Notes in Mathematics 534. Springer, Berlin (1976)
[177] Propp, J.G., Wilson, D.B.: Exact sampling with coupled Markov chains and applications to statistical mechanics. Random Structures and Algorithms **9**, 223–252 (1996)
[178] R Development Core Team: R: A Language and Environment for Statistical Computing. R Foundation for Statistical Computing, Vienna, Austria (2007). URL `http://www.R-project.org`
[179] Rathbun, S.L., Cressie, N.A.C.: Asymptotic properties of estimators for the parameters of spatial inohomogeneous Poisson point processes. Advances in Applied Probability **26**, 122–154 (1994)
[180] Revuz, D.: Probabilités. Herman, Paris (1997)
[181] Ribeiro, P., Diggle, P.J.: geoR: a package for geostatistical analysis. R-NEWS **1**, 14–18 (2001)

[182] Richarson, S., Guihenneuc, C., Lasserre, V.: Spatial linear models with autocorrelated error structure. The Statistician **41**, 539–557 (1992)

[183] Ripley, B.D.: The second-order analysis of stationary point processes. Journal of Applied Probability **13**, 255–266 (1976)

[184] Ripley, B.D.: Statistical Inference for Spatial Processes. Cambridge University Press, Cambridge (1988)

[185] Ripley, B.D.: Spatial Statistics. Wiley, New York (1991)

[186] Ripley, B.D., Kelly, F.P.: Markov point processes. Journal of the London Mathematical Society **15**, 188–192 (1977)

[187] Robert, C.P.: L'analyse statistique bayésienne. Economica, Paris (1992)

[188] Robert, C.P., Casella, G.: Monte-Carlo Statistical Methods. Springer, New York (1999)

[189] Rue, H., Held, L.: Gaussian Markov Random Fields: Theory and Applications. Chapman & Hall/CRC, Boca Raton: FL (2005)

[190] Ruelle, D.: Statistical Mechanics. Benjamin, New York (1969)

[191] Saloff-Coste, L.: Lectures on finite Markov chains. In: P. Bernard (ed.) Lectures on Probability Theory and Statistics. Ecole d'été de Probabilité de St. Flour XXVI, Lecture Notes in Mathematics 1665, pp. 301–408. Springer (1997)

[192] Sampson, P., Guttorp, P.: Nonparametric estimation of nonstationary spatial covariance structure. Journal of the American Statistical Association **87**, 108–119 (1992)

[193] Santer, T.J., Williams, B.J., Notz, W.I.: The Design and Analysis of Computers Experiments. Springer, New York (2003)

[194] Schabenberger, O., Gotway, C.A.: Statistical Methods for Spatial Data Analysis. Chapman & Hall/CRC, Boca Raton: FL (2004)

[195] Schlather, M.: Introduction to positive definite functions and to unconditional simulation of random fields. Tech. Rep. ST 99-10, Lancaster University, Lancaster (1999)

[196] Senoussi, R.: Statistique asymptotique presque sûre des modèles statistiques convexes. Annales de l'Institut Henri Poincaré **26**, 19–44 (1990)

[197] Serra, J.: Image Analysis and Mathematical Morphology. Academic Press, New York (1982)

[198] Shea, M.M., Dixon, P.M., R., S.R.: Size differences, sex ratio, and spatial distribution of male and female water tupelo, nyssa aquatica (nyssaceae). American Journal of Botany **80**, 26–30 (1993)

[199] Shibata, R.: Selection of the order of an autoregessive model by Akaike's information criterion. Biometrika **63**, 117–126 (1976)

[200] Stein, M.L.: Interpolation of Spatial Data: Some Theory for Kriging. Springer, New York (1999)

[201] Stein, M.L.: Statistical methods for regular monitoring data. Journal of the Royal Statistical Society, Series B **67**, 667–687 (2005)

[202] Storvik, G., Frigessi, A., Hirst, D.: Stationary space-time gaussian fields and their time autoregressive representation. Stochastic Modelling **2**, 139–161 (2002)

[203] Stoyan, D., Grabarnik, P.: Second-order characteristics for stochastic structures connected with Gibbs point processes. Mathematische Nachrichten **151**, 95–100 (1991)

[204] Stoyan, D., Kendall, W.S., Mecke, J. (eds.): Stochastic Geometry and its Applications, 2nd edn. Wiley, New York (1995)

[205] Strathford, J.A., Robinson, W.D.: Distribution of neotropical migratory bird species across an urbanizing landscape. Urban Ecosystems **8**, 59–77 (2005)

[206] Strauss, D.J.: A model for clustering. Biometrika **62**, 467–475 (1975)

[207] Strauss, D.J.: Clustering on colored lattice. Journal of Applied Probability **14**, 135–143 (1977)

[208] Stroud, J.R., Müller, P., Sansó, B.: Dynamic models for spatio-temporal data. Journal of the Royal Statistical Society, Series B **63**, 673–689 (2001)

[209] Sturtz, S., Ligges, U., Gelman, A.: R2WinBUGS: a package for running WinBUGS from R. Journal of Statistical Software **2**, 1–16 (2005)

[210] Sweeting, T.J.: Uniform asymptotic normality of the maximum likelihood estimator. Annals of Statistics **8**, 1375–1381 (1980)

[211] Tempelman, A.A.: Ergodic theorems for general dynamical systems. Transactions of the Moscow Mathematical Society **26**, 94–132 (1972)

[212] Thomas, A., O' Hara, B., Ligges, U., Sturtz, S.: Making BUGS open. R News **6**, 12–17 (2006)

[213] Thomas, M.: A generalisation of Poisson's binomial limit for use in ecology. Biometrika **36**, 18–25 (1949)

[214] Tierney, L.: Markov chains for exploring posterior distributions (with discussion). Annals of Statistics **22**, 1701–1762 (1994)

[215] Tierney, L.: A note on Metropolis-Hastings kernels for general state space. Annals of Applied Probability **3**, 1–9 (1998)

[216] Tuckey, J.W.: Spectral Analysis Time Series. Wiley, New York (1967)

[217] Van Lieshout, M.N.M.: Markov Point Processes and their Applications. Imperial College Press, London (2000)

[218] Van Lieshout, M.N.M., Baddeley, A.J.: Indices of dependence between types in multivariate point patterns. Scandinavian Journal of Statistics **26**, 511–532 (1999)

[219] Ver Hoef, J., Barry, R.P.: Constructing and fitting models for cokriging and multivariable spatial prediction. Journal of Statistical Planning and Inference **69**, 275–294 (1998)

[220] Waagepetersen, R.P.: An estimating function approach to inference for inhomogeneous Neyman-Scott processes. Biometrics **63**, 252–258 (2007)

[221] Wackernagel, H.: Multivariate Geostatistics: A n Introduction with Applications, 3rd edn. Springer, New York (2003)

[222] Whittle, P.: On stationary processes in the plane. Biometrika **41**, 434–449 (1954)

[223] Wikle, C.K., Cressie, N.A.C.: A dimension-reduced approach to space-time Kalman filtering. Biometrika **86**, 815–829 (1999)

[224] Winkler, G.: Image Analysis, Random Fields and Markov Chain Monte Carlo Methods, 2nd edn. Springer (2003)

[225] Wolpert, R.L., Icktadt, K.: Poisson/Gamma random fields models for spatial statistics. Biometrika **85**, 251–267 (1998)

[226] Wu, H., Huffer, F.W.: Modelling the distribution of plant species using the autologistic regression model. Environmental and Ecological Statistics **4**, 49–64 (1997)

[227] Yaglom, A.M.: Correlation Theory of Stationary and Related Random Functions. Volume I: Basic Results. Springer, New York (1987)

[228] Yao, J.F.: On constrained simulation and optimisation by Metropolis chains. Statistics and Probability Letters **46**, 187–193 (2000)

[229] Ycart, B.: Modèles et algorithmes markoviens. Mathématiques et Applications. Springer, Paris (2002)

[230] Younes, L.: Estimation and annealing for Gibbsian fields. Annales de l'Institut Henri Poincaré (B). Probabilités et Statistiques **2**, 269–294 (1988)

[231] Zhang, H., Zimmerman, D.L.: Towards reconciling two asymptotic frameworks in spatial statistics. Biometrika **92**, 921–936 (2005)

[232] Zimmerman, D., Zimmerman, M.: A comparison of spatial semivariogram estimators and corresponding ordinary kriging predictors. Technometric **33**, 77–91 (1991)

Index

MIX
Papier aus verantwortungsvollen Quellen
Paper from responsible sources
FSC® C105338

If you have any concerns about our products,
you can contact us on
ProductSafety@springernature.com

In case Publisher is established outside the EU,
the EU authorized representative is:
Springer Nature Customer Service Center GmbH
Europaplatz 3, 69115 Heidelberg, Germany

Printed by Libri Plureos GmbH
in Hamburg, Germany